ALLE ZEIT WACH 1842

Syntheses of Fluoroorganic Compounds

Edited by
I. L. Knunyants and G. G. Yakobson

Springer-Verlag
Berlin Heidelberg New York Tokyo 1985

Professor I. L. Knunyants
Professor G. G. Yakobson †

Institute of Organic Chemistry,
SU-630090 Novosibirsk

ISBN-13: 978-3-642-70209-9 e-ISBN-13: 978-3-642-70207-5
DOI:10.1007/978-3-642-70207-5

Library of Congress Cataloging in Publication Data. Main entry under title:
Syntheses of fluoroorganic compounds. Bibliography: p. Includes index.
1. Organofluoron compounds. 2. Chemistry, Organic—Synthesis.
I. Knunyants, Ivan Ludvigovich. II. Yakobson, G. G. (Georgi Gustavovich)
QD412.FIS95 1985 547'.02 84-26875

Softcover reprint of the hardcover 1st edition 1985

2131/3020-543210

Preface

Recently there has been a tremendous growth in the chemistry of fluoroorganic compounds, which find wide application in various fields of technique and are used to solve some basic theoretical problems. The chemical properties of these compounds are rather specific, as well as the methods for their preparation. We realized that no good handbooks on the preparation of fluoroorganic compounds were available, whereas the methods of preparation are scattered in numerous scientific papers and patents. Even such a well-known publication as *Organic Syntheses* contains just a few methods of preparation of fluoroorganic compounds. As a consequence, not only for newcomers but even for specialists, searching for simple and convenient methods for the preparation of the required compounds is tedious.

To alleviate the problem, we undertook to prepare this book, which presents detailed preparation methods of more than 300 fluoroorganic compounds. Emphasis has been laid on syntheses of polyfluorinated compounds, since they are of major interest with respect to preparation and utility.

Preparation of fluoroorganic compounds is based both on the classical methods of organic chemistry and on the specific ones, which is due to the difficulty or even impossibility of the direct introduction of fluorine into a definite position of organic molecule. Most organic compounds of fluorine are prepared by a succession of conversions of a relatively small number of starting compounds.

We attempted, on the one hand, to exemplify the synthetic methods used in the chemistry of fluoroorganic compounds and, on the other hand, to describe the preparation of compounds that may be used as starting materials in further syntheses. The preparation methods reported here may be reproduced in any chemical laboratory. Accordingly, we have intentionally excluded such methods as fluorination with elemental fluorine, electrochemical fluorination and some others, which require special equipment. Moreover, many such compounds as may be prepared by these methods are commercially available at a reasonable price.

This book is based on the two-volume Russian edition of *Syntheses of Fluoroorganic Compounds* ("Sintezy Ftororganicheskih Soedinenii", Khimia Publishers, Moscow), which covers more than 500 preparation methods of fluoroorganic compounds. The present edition includes the preparation methods of compounds that seemed to be the most interesting ones; these are complemented with some new preparations worked out in recent years at the laboratories of the authors.

The book is divided into three sections: fluoroaliphatic compounds, fluoroaromatic compounds, and aromatic compounds with fluorinated side chains. The preparations are arranged according to two principles. On the one hand, it seemed appropriate to give first the preparation methods of fluorinated hydrocarbons, then of compounds containing other halogens and, finally, of compounds with various functional groups. On the other hand, it appeared reasonable to give together compounds obtained by similar procedures. Thus the preparation methods of some phenyl-containing compounds are included in the first section of the book.

Every procedure is preceded by an enumeration of methods for the preparation of the given compound. Only the principal ones, although not exhaustively, are included.

Whenever identical procedures affording compounds of similar structures are feasible, this is indicated after the procedure, together with yields and constants for these compounds. All the constants are given as they appear in References.

Since naming of polyfluoroorganic compounds does not strictly follow the IUPAC nomenclature, the compounds included in the book have widely accepted names, generally those used in the cited papers.

Most preparations have been checked at the laboratories of A. V. Fokin, I. L. Knunyants, L. M. Yagupolskii and G. G. Yakobson and modified, where necessary.

Syntheses of fluoroaliphatic compounds were compiled for this publication by Dr. S. R. Sterlin, of fluoroaromatic compounds by Dr. G. G. Furin, and of aromatic compounds with fluorinated side chain by Prof. L. M. Yagupolskii and Dr. Yu. A. Fialkov.

We would like to thank Mrs. L. G. Smolina for translating the original into English and Dr. V. V. Bardin for checking references and typing the manuscript in English. We also wish to make use of this opportunity to express our deep gratitude to Mrs M. N. Paštushenko, editor of Khimia Publishers, whose patience and goodwill greatly promoted publication of *Syntheses of Fluoroorganic Compounds* in Russian.

I. L. Knunyants,
G. G. Yakobson.

Contents

Safety Measures 1

Fluoroaliphatic Compounds 3
G. G. Belenki, A. V. Fokin, D. S. Rondarev, R. M. Ryazanova, S. V. Sokolov, S. R. Sterlin, A. N. Voronkov, Yu. V. Zeifman

References 103

Fluoroaromatic Compounds 109
G. G. Furin, G. F. Grebenshchikova, A. Yu. Lvova, V. M. Vlasov, G. G. Yakobson

References 225

Aromatic Compounds with Fluorinated Sid Chains 233
Yu. A. Fialkov, L. M. Yagupolski

References 286

Subject Index 291

Formula Index 297

Contents

Safety Measures

Some syntheses described in the book involve toxic and dangerous compounds. The safety measures for handling them are generally included in the corresponding procedures. However, such compounds as hydrogen fluoride, sulphur tetrafluoride, tetrafluoroethylene and perfluoroisobutene are used in many syntheses, therefore it seemed reasonable to consider the safety measures pertaining to these chemicals, and first-aid treatment of injuries, in a separate chapter to avoid repetitions. These measures are to be perused before starting work with the above-mentioned compounds.

Hydrogen Fluoride

Hydrogen fluoride is a colourless liquid fuming in air Mp −83 °C, bp 19.5 °C, critical temperature 230 °C, critical pressure 66.2 atm, density 1.00 (0 °C).
Anhydrous HF may be handled in a steel, polyethylene or polytetrafluoroethylene vessel. Even diluted HF in contact with the skin produces painful and slowly healing wounds. Its vapours are irritant to the respiratory passage and mucous membranes. It is therefore imperative to protect the hands with *undamaged* rubber gloves (preferably made of synthetic rubber), to shield the face with a protective shield of transparent plastic, the body with a rubber or plastic apron, and the feet with rubber boots. In small laboratory experiments at least the face shield and rubber gloves must be used. It is advisable to test the gloves for leakages from time to time. Even an insignificant leak in the glove may cause serious trouble. Work with anhydrous hydrogen fluoride must be carried out under efficient hoods or in strongly ventilated rooms since its vapours attack the lungs and the mucous membranes, especially those of the eyes.

Anhydrous hydrogen fluoride reacts violently, almost explosively, with water. Disposal of spent HF must therefore be carried out with caution and with protective devices against splattering. Smaller amounts are disposed of by evaporating in the draft of a hood, or by pouring into a sink and flushing rapidly with a strong stream of water, larger amounts by treatment with a lime slurry.

When the skin is stained with anhydrous hydrogen fluoride, the spot must be immediately washed with large amounts of water. Immersion of the burnt spot in ice-cold saturated solution of magnesium sulfate or in ice-cold aqueous ethanol for about an hour is recommended to retard the passage of the fluoride ion through the tissues. In this respect, ice-cold aqueous or alcoholic solutions of high-molecular-weight quaternary ammonium compounds (Hyamine, Phemorol, Zephirane, etc.) are even better because of their bactericidal effects (2 g of Hyamine 1622 in 1 liter of distilled water is recommended).

Finally a paste is applied, prepared by mixing 20% of magnesium sulphate, 6% of magnesium oxide, 18% of glycerol, 55% of water, and 1% of procaine hydrochloride, or by mixing 3 parts of magnesium oxide, 4 parts of mineral oil, and 11 parts of white vaseline. Even a simple slurry of magnesium oxide in glycerol slightly diluted with a small amount of water (to reduce the viscosity) is satisfactory. More extensive burns should be treated by subcutaneous injections of calcium gluconate solution.

Eyes affected by hydrogen fluoride should be washed long enough with luke warm water, then with a 2% solution of sodium bicarbonate, and finally exposed for a quarter of an hour to the action of 2–3 drops of a 0.5% solution of pantocaine. Ointments are not recommended [1].

Sulphur Tetrafluoride

Sulphur tetrafluoride is a colourless gas with a heavy smell. Np —124 °C, bp —40.4 °C, critical temperature 90 °C, density 1.92 (—73 °C), vapour pressure 12 atm at 25 °C. Sulphur tetrafluoride is easily hydrolysed, forming hydrogen fluoride and thionyl fluoride. It is unreactive with sulphur, mercury and rubber and does not affect thoroughly dried glassware. Sulphur tetrafluoride should be stored in steel cylinders with steel valves. Copper and brass valves in contact with SF_4 get destroyed.

The toxic effect of sulphur tetrafluoride is similar to that of difluorophosgene and about twice as strong as that of phosgene. It is enhanced by HF, released when SF_4 has been inhaled or has been in contact with skin, therefore skin should be protected, work should be done under a good hood and a gas mask should be worn.

The first-aid treatment of poisoning is the same as of difluorophosgene poisoning (see. p. 60). ARTIFICIAL RESPIRATION IS FORBIDDEN! The injured person should be hospitalized without delay.

Tetrafluoroethylene

Tetrafluoroethylene is a colourless gas. Mp —142.5 °C, bp —76.3 °C, critical temperature 33.3 °C, density 1.533 (—80 °C). Tetrafluoroethylene is explosive under pressure, therefore syntheses should be conducted only in autoclave provided with anti-explosion diaphragms or in armed boxes.

The hazard may be minimized by heating the charged autoclave during 16 h, at 20°–25 °C, shaking it the while, with subsequent heating to the required temperature.

The experience of one of the authors of the book who worked many years in the laboratory and a pilot plant shows that in flow systems tetrafluoroethylene may be used without any additional safety measures. But it should be kept in mind that the pyrolysis of that compound may yield, among other products, toxic perfluoroisobutene in small amounts.

Perfluoroisobutene

For the physical properties of perfluoroisobutene see p. 9. Perfluoroisobutene is highly toxic. Its toxic activity is comparable to that of phosgene. Work with this compound should be conducted only under a efficient hood. When poisoned by perfluoroisobutene, one should inhale ethanol vapours till easy respiration is restored.

References

1. Hudlicky, M.: Chemistry of Fluorine Compounds. New York etc. John Wiley & Sons, 1976, p. 13–14

Fluoroaliphatic Compounds

G. G. Belenki, A. V. Fokin, D. S. Rondarev, R. M. Ryazanova, S. V. Sokolov, S. R. Sterlin, A. N. Voronkov, Yu. V. Zeifman

2-Iodoheptafluoropropane

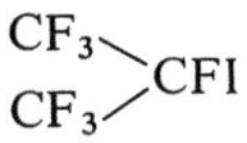

C_3F_7I MW 296

2-Iodoheptafluoropropane is a low-boiling volatile colourless liquid; when exposed to light, assumes a rosy colour. Bp 40 °C, n_D^{20} 1.3237 [1].

Method of Preparation

2-Iodoheptafluoropropane is obtained in the reaction of perfluoropropylene with iodine and potassium iodine in acetonitrile (63% yield) [2], or by the reaction of perfluoropropylene with iodine and iodine pentafluoride (77% yield) [1].

Procedure [1]

$$CF_3CF{=}CF_2 + I_2 + IF_5 \xrightarrow{SbF_5} (CF_3)_2CFI$$

In a 250-ml steel autoclave are placed 15 g (0.065 mol) of iodine pentafluoride, 30 g (0.24 mol) of finely powdered iodine and 2 or 3 ml of SbF_5. The autoclave is cooled with liquid nitrogen, evacuated and 50 g (0.3 mol) of perfluoropropylene is condensed into it; then it is heated in a tilting furnace at 150 °C during 10 h, whereupon it is cooled to room temperature. The contents are poured onto ice, the organic layer is separated and washed with aqueous solution of $NaHCO_3$, with two portions of 25% aqueous KOH and with water, then dried over P_2O_5 and distilled.

The yield of perfluoroisopropyl iodide is 70 g (80%); bp 40 °C, n_D^{20} 1.3237.

1-Fluorobutane

$$CH_3{-}CH_2{-}CH_2{-}CH_2F$$

C_4H_9F MW 76

1-Fluorobutane is a low-boiling liquid. Bp 31.5 °C/731 mm Hg; n_D^{20} 1.3410 [3].

Method of Preparation

1-Fluorobutane is obtained by the reaction of 1-bromobutane with HgF_2 [4], or by the reaction of n-butyl alcohol with (2-chloro-1,1,2-trifluoroethyl)diethylamine (75% yield) [3].

Procedure [3]

$$CF_2{=}CFCl \xrightarrow{(C_2H_5)_2NH} CHFCl{-}CF_2{-}N(C_2H_5)_2 \xrightarrow{n\text{-}C_4H_9OH}$$

$$CH_3(CH_2)_2CH_2F$$

(2-Chloro-1,1,2-trifluoroethyl)diethylamine. In a Drechsel bottle cooled to −5° to −10 °C is placed 40 g (0.56 mol) of diethylamine, then 70 g (0.6 mol) of thoroughly dried chlorotrifluoroethylene is bubbled through it during 10 h. The reaction is distilled in vacuum. The yield of (chlorotrifluoroethyl)diethyl amine is 70 g (66%); bp 32°–33 °C/6 mm Hg.

1-Fluorobutane. In a three-necked flask provided with a stirrer, a dropping funnel, and a Wurtz column connected to a downward Liebig condenser, is placed 37.8 g (0.2 mol) of (chlorotrifluoroethyl)diethylamine, to which is added dropwise 14.8 g (0.2 mol) of n-butyl alcohol. The reaction proceeds with evolution of heat; 1-fluorobutane formed in the reaction distils off from the flask. The product is washed with ice-water, dried over $CaCl_2$ and distilled. The yield of 1-fluorobutane is 10.1 g (75%); bp 31.5 °C/731 mm Hg; n_D^{20} 1.3410.

1-Fluorohexane

$$CH_3{-}CH_2{-}CH_2{-}CH_2{-}CH_2{-}CH_2F$$

$C_6H_{13}F$ MW 104

1-Fluorohexane is a colourless liquid. Bp 91.1 °C/755 mm Hg [4], 91°–93.5 °C [5]; n_D^{20} 1.3748; d_4^{20} 0.8002 [4].

Method of Preparation

1-Fluorohexane is a colourless liquid. Bp 91.1 °C/755 mm Hg [4], 91°–93.5 °C [5]; fluoride in the absence of solvents (20% yield) [6], or in a diethylene glycol — ethylene glycol mixture (54% yield) [5].

Procedure [5]

$$CH_3(CH_2)_4CH_2Cl \xrightarrow{KF} CH_3(CH_2)_3CH_2CH_2F$$

In a 1-l three-necked flask provided with a stirrer, a dropping funnel, and a fractionating column (300 mm long) carrying a reflux condensing head, are placed 116 g (2 mol)

of finely divided dry KF (Note), 200 g of ethylene glycol and 50 g of diethylene glycol. The mixture is heated to 180 °C and 120.6 g (1 mol) of n-hexyl chloride is added to it dropwise at the above temperature, with vigorous stirring during 8 h. 1-Fluorohexane, formed in the reaction, distils off through the column (the temperature of vapour in the column should not rise above 93°–94 °C). Then the reaction mixture is cooled to 110 °C, whereupon a slow flow of air is let into the flask to remove the remaining 1-fluorohexane. The crude product (79 g), containing n-hexene and unchanged n-hexyl chloride, is distilled. The yield of 1-fluorohexane is 56.3 g (54%); bp 91°–93.5 °C.
Note. KF is calcinated in a cast-iron pan, then it is thoroughly powdered and dried in vacuum at 200 °C (for the last procedure a glass apparatus may be used).

1,1,1-Trifluoro-2,2-bis(trifluoromethyl)hexane

$$(CF_3)_3C{-}CH_2{-}CH_2{-}CH_2{-}CH_3$$

$C_8H_9F_9$ MW 276

1,1,1-Trifluoro-2,2-bis(trifluoromethyl)hexane is a volatile colourless liquid, soluble in common organic solvents. Bp 107°–109 °C; n_D^{20} 1.3190 [7].

Method of Preparation

1,1,1-Trifluoro-2,2-bis(trifluoromethyl)hexane is obtained by the reaction of perfluoroisobutylene with butyl bromide in the presence of CsF (90%) [7].

Procedure

$$(CF_3)_2C{=}CF_2 + BrCH_2(CH_2)_2CH_3 \xrightarrow{CsF} (CF_3)_3C(CH_2)_3CH_3$$

Caution! Perfluoroisobutylene is extremely toxic. It should be handled under an efficient hood (see p. 2).

Freshly calcinated, powdered CsF (15.2 g, 0.1 mol) and 40 ml of absolute diglyme are placed in a 150-ml four-necked flask provided with a stirrer, a dropping funnel, a gas inlet tube, and a dry-ice acetone reflux condenser; then 20 g (0.1 mol) of perfluoroisobutylene is added with vigorous stirring. The reaction mixture is heated at 40 °C for 2 h and to the resulting solution is gradually added, with continuous stirring, 13.7 g (0.1 mol) of dry n-butyl bromide. The mixture is stirred at 50°–60 °C for 2 h, then distilled. The yield of 1,1,1-trifluoro-2,2-bis(trifluoromethyl)hexane is 24.3 g (90%); bp 107°–109 °C; n_D^{20} 1.3190.

Likewise, perfluoroisobutylene and methyl iodide give 1,1,1-trifluoro-2,2-bis(trifluoromethyl)propane (mp 20 °C; 86% yield) [7], benzyl chloride yields 1,1,-trifluoro-2,2-bis(trifluoromethyl)-4-phenylpropane (bp 159°–161 °C; n_D^{20} 1.3901; 74.5% yield) [7], and allyl iodide affords 5,5,5-trifluoro-4,4-bis(trifluoromethyl)-pentene-1 (bp 75° to 76 °C; 80% yield) [7].

7,7-Difluoronorcarane

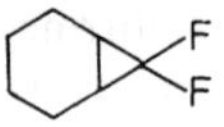

$C_7H_{10}F_2$ MW 132

7,7-Difluoronorcarane is a colourless liquid with a faint odour. Bp 122 °C [8, 9].

Method of Preparation

7,7-Difluoronorcarane is prepared by the reaction of cyclohexene with bis(trifluoromethyl)mercury in the presence of NaI (35% yield) [8], with trimethyltrifluoromethyltin in the presence of NaI (73% yield) [9], with trifluorobromomethane and n-butyllithium [10]; by decomposition of sodium difluorochloroacetate in the presence of cyclohexene (60%–65% yield) [11]; or by heating a mixture of perfluorocyclopropane and cyclohexene at 160°–170 °C (86%–99% yield) [12].

Procedure [8]

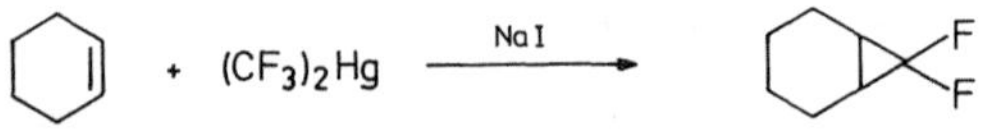

Cyclohexene (7.4 g, 0.09 mol), freshly calcinated NaI (9 g, 0.06 mol) and anhydrous tetrahydrofuran (12 ml) are placed in a 100-ml four-necked flask fitted with a stirrer, a reflux condenser and a dropping funnel; then a solution of 10.2 g (0.03 mol) of bis-(trifluoromethyl)mercury (Note) in 5 ml of tetrahydrofuran is added with stirring and heating of the mixture to 80°–82 °C over a period of 1 h. Stirring is continued at the same temperature during 7 h, whereupon further 3 g (0.02 mol) of NaI is added and the mixture is stirred at 80°–82 °C till bis(trifluoromethyl)mercury signal in [^{19}F] NMR spectrum has disappeared (4–5 h). After the reaction has been completed, the volatile products are distilled off from the reaction mixture into a dry-ice acetone trap under vacuum and the contents of the trap are distilled on a fractionating column. 7,7-Difluoronorcarane (2.8 g) (35% yield) distils off at 122 °C.

Likewise, isopropenylacetate gives 1,1-difluoro-2-acetoxy-3-methylcyclopropane (bp 124 °C; 38% yield) [8].

Note. Preparation of bis(trifluoromethyl)mercury and the safety measures are found on p. 97.

Iodotrifluoroethylene

$CF_2{=}CFI$

C_2F_3I MW 208

Iodotrifluoroethylene is a highly volatile liquid. Bp 28 °C/627 mm Hg [13], 30 °C/760 mm Hg [14]; n_D^{20} 1.3974; d_4^{20} 2.076 [14].

Method of Preparation

Iodotrifluoroethylene is prepared by dehydrochlorination of 1-iodo-2-chloro-1,2,2-trifluoroethane with an aqueous alkaline solution (90% yield) [14].

Procedure

$$CF_2{=}CFH \xrightarrow{ICl} ClCF_2{-}CFHI \xrightarrow{KOH} CF_2{=}CFI$$

1-Iodo-2-chloro-1,2,2-trifluoroethane. A 100-ml steel autoclave is charged with 60 g (0.37 mol) of ICl (for the preparation see p. 64), then 20 ml (0.31 mol) of liquid trifluoroethylene is added with cooling to —90 °C. The autoclave is closed and heated at 70°–80 °C for 6 h, then cooled to —90 °C and opened. Unchanged gas (0.5 l) is collected in a gasometer; the reaction product, a dark violet liquid, is washed several times with water, with a solution of sodium thiosulphate to complete decolorization, then again with water, dried over $MgSO_4$ and distilled to give 6 g of 1,2-dichloro-1,2,2-trifluoroethane, bp 28°–30 °C, and 49 g (64%) of 1-iodo-2-chloro-1,2,2-trifluoroethane; bp 82°–84 °C; n_D^{20} 1.4312; d_4^{20} 2.1790.

Iodotrifluoroethylene. In a three-necked flask fitted with a stirrer, a dropping funnel, and a dephlegmator, are placed 60 ml of 50% aqueous solution of KOH. The solution is heated to 70°–80 °C and 39 g (0.16 mol) of 1-iodo-2-chloro-1,2,2-trifluoroethane is dropped in with vigorous stirring, at such a rate that the temperature of the distillate should not rise above 40°–45 °C. In the course of the reaction 7–10 g of dry KOH is added to the reaction mixture. The distillate is collected in a trap cooled to —50 °C, washed with water, then dried over $MgSO_4$ and redistilled. The yield of iodotrifluoroethylene is 30.4 g (90%); bp 30 °C.

symm-Dichlorodifluoroethylene

CFCl=CFCl

$C_2Cl_2F_2$ MW 133

symm-Dichlorodifluoroethylene is a colourless liquid. Bp 21°–22 °C [15]; n_D^0 1.3777 to 1.3798 [16].

Method of Preparation

symm-Dichlorodifluoroethylene is prepared by dehalogenation of tetrachlorodifluoroethane with zinc in ethanol (90%–95% yield) [15].

Procedure

$$CFCl_2{-}CFCl_2 \xrightarrow{\text{Zn; ethanol or dioxan}} CFCl{=}CFCl$$

In a three-necked flask (Note 1) provided with a stirrer, a dropping funnel and a reflux condenser (Note 2) attached to a downward Liebig condenser, are placed

1.5 kg of zinc dust, 1 l of anhydrous ethanol or dioxan and 5–10 ml of conc. HCl. The mixture is boiled during 30 min with stirring, then the solution of 1 kg of 1,2-difluorotetrachloroethane in 500 ml of anhydrous ethanol (dioxan) is added at such a rate that the reaction mixture should continuously boil. The condensate is collected in a water trap cooled to 0°–5 °C (Note 3). The organic layer is separated from water, dried over $CaCl_2$ and distilled. The yield of *symm*-dichlorodifluoroethylene is 550 g (86%); bp 21°–22 °C.

Notes.

1. The reaction may be conducted in a two-necked flask bearing a column packed with granulated zinc. In this case, the solution of Freon 112 in absolute ethanol is added dropwise to an equal volume of absolute ethanol. The reaction is conducted at boiling point of ethanol.
2. The reflux condenser should be cooled with warm water.
3. The receiver should be attached to a trap cooled to —78 °C.

1,1,3,3,3-Pentafluoropropylene

$CF_3{-}CH{=}CF_2$

C_3HF_5 MW 132

1,1,3,3,3-Pentafluoropropylene is a gas. Bp —21 °C; mp —153 °C [17].

Method of Preparation

1,1,3,3,3-Pentafluoropropylene is prepared by dehydrohalogenation of 3-iodo-1,1,1,3,3-pentafluoropropane (85% yield) obtained by the reaction of iodotrifluoromethane with vinylidene fluoride [17]. Pentafluoropropylene is also obtained by the reaction of 3-bromo-1,1,3,3-tetrafluoropropylene with hydrogen fluoride (60% yield) [18] and by decomposition of α-hydroperfluoroisobutyrate (70% yield) [19].

Procedure [19]

$$(CF_3)_2CHCOOK \xrightarrow{CH_3COOC_2H_5} CF_3{-}CH{=}CF_2$$

Potassium α-hydroperfluoroisobutyrate. A mixture of 100 g (0.5 mol) of α-hydroperfluoroisobutyric acid and 60 g (0.6 mol) of $KHCO_3$ is thoroughly powdered in a porcelain mortar. The resulting salt is dried over P_2O_5 in a vacuum-desiccator. The yield of the salt is 123 g (quantitative).

1,1,3,3,3-Pentafluoropropylene. Absolute ethylacetate (500 ml) and potassium α-hydroperfluoroisobutyrate (150 g, 0.6 mol) are placed in a two-necked flask provided with an effective reflux condenser. The mixture is refluxed during 5 to 6 h. The liberated gas is washed with concentrated H_2SO_4 and condensed in a trap cooled to —78 °C. The condensate is distilled on a low-temperature distillation column and a fraction boiling at −21° to −19 °C is collected. The yield of 1,1,3,3,3-pentafluoropropylene is 50 g (70%).

Perfluoroisobutylene

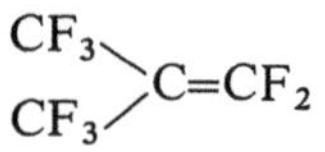

C_4F_8 MW 200

Perfluoroisobutylene is a highly toxic colourless gas (~10 times as toxic as phosgene [20]) with a specific odour, partially soluble in ether, acetone, acetonitrile, insoluble in water. Bp 6.5 °C [21], 7 °C [22]; d_4^0 1.5922 [21].

Method of Preparation

Perfluoroisobutylene is prepared by the pyrolysis of tetrafluoroethylene (44%–47% yield) [23], Teflon (34%–60% yield) [23, 24], hexafluoropropylene (80% yield) [23] or perfluorocyclobutane (45%–70% yield) [21, 23, 25]; or by dechlorination of 1,2-dichloro-2-trifluoromethyl-1,1,3,3,3-pentafluoropropane by heating it with zinc dust in dioxan (88% yield) [22].

Procedure

1. Pyrolysis of Perfluorocyclobutane [23]

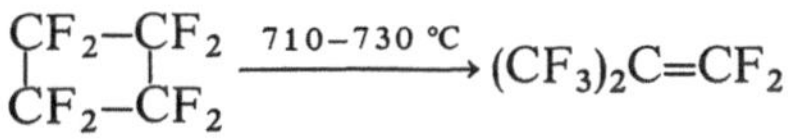

Caution! Perfluoroisobutylene is toxic. It should be handled only under an efficient hood (see p. 2).

The reaction is carried out in the apparatus that consists of a series-connected drying column packed with P_2O_5 on glass wool, a flow meter (concentrated H_2SO_4 as a manometer liquid), a stainless steel tube (27 mm i.d.), placed in an electrical furnace with 600 mm zone of heating, an intermediate bottle for collection of liquid products and a trap cooled to −78 °C. The temperature is measured with a thermocouple placed in the centre of the heating zone between the outer surface of the pyrolysis tube and the inner surface of the furnace.

Perfluorocyclobutane is passed from a cylinder through the drying column and a flow meter at the rate of 5 g/min (the contact time ~25 s) into the pyrolysis tube, which has been heated to 710°–730 °C. After 460 g (2.3 mol) of perfluorocyclobutane has been passed, 420 g of the product is collected in a cooled trap. Distillation of the product on a glass-packed column with a low-temperature distillation head gives 180 g (45%) of crude perfluoroisobutylene, bp 5°–9 °C. In order to remove perfluorobutenes, the product, placed together with excess dry bromine into a sealed glass tube, is irradiated with a mercury lamp during 6 to 8 h and redistilled (Note).

2. Dechlorination of 1,2-dichloro-2-trifluoromethyl-1,1,3,3,3-pentafluoropropane (a modification of the procedure described in [22])

$$(CF_3)_2CCl{-}CF_2Cl \xrightarrow{Zn} (CF_3)_2C{=}CF_2$$

Apparatus for dechlorination consists of a 1-l two-necked flask provided with a dropping funnel, series- connected column (500 mm long, 35–40 mm i.d.) covered with a heat insulator, a reflux condenser, a gas-washing bottle with concentrated H_2SO_4, and a trap cooled to −78 °C.

The column is packed with granulated zinc, which is activated with a small amount (1.5–2 ml) of concentrated HCl; 300–350 ml of acetone is placed into a flask and heated to reflux. 1,2-Dichloro-2-trifluoromethyl-1,1,3,3,3-pentafluoropropane (30 g) is then added and after the gaseous products have started to evolve, further 240 g of the dichloride is gradually added during 2.5–3 h to the refluxed mixture. The product, which has condensed in the trap, is distilled (the still residue contains ~30 g of the starting dichloride) and purified by bromination as described in the foregoing preparation to give 136 g of perfluoroisobutylene (85% conversion, 80% yield); bp 6–8 °C.

Note. In order to remove excess bromine, the reaction product is frozen to −78 °C, the liquid layer is decanted and distilled over sodium thiosulphate.

1,3,3,3-Tetrafluoro-1-chloro-2-trifluoromethylpropylene-1

$(CF_3)_2C{=}CFCl$

C_4ClF_7 MW 216.5

1,3,3,3-Tetrafluoro-1-chloro-2-trifluoromethylpropylene-1 is a low-boiling colourless liquid, partially soluble in polar solvents (dimethylformamide, diglyme, tetrahydrofuran, acetonitrile), insoluble in water. Bp 40°–42 °C; n_D^{20} 1.3005 [26].

Method of Preparation

1,3,3,3-Tetrafluoro-1-chloro-2-trifluoromethylpropylene-1 is prepared by the reaction of perfluoroisobutylene with phosphorus oxychloride in the presence of triethylbenzylammonium chloride (34% yield) [26], with benzoyl chloride in the presence of pyridine (49% yield) [26], or with anhydrous $AlCl_3$ (16% yield) [26].

Procedure

$$(CF_3)_2C{=}CF_2 + C_6H_5COCl \xrightarrow[C_5H_5N]{} (CF_3)_2C{=}CFCl$$

Caution! Perfluoroisobutylene is extremely toxic. It should be handled only under an efficient hood (see p. 2).

Caution! 1,3,3,3-Tetrafluoro-1-chloro-2-trifluoromethylpropylene, 3,3,3-trifluoro-1,1-dichloro-2-trifluoromethylpropylene, and 1,3,3,3-tetrafluoro-1-bromo-2-trifluoromethylpropylene are highly toxic. Work with these compounds requires using the same safety measures as work with perfluoroisobutylene.

In a 200-ml rocking stainless steel autoclave are placed 35 ml of benzoyl chloride and 1.5 ml of absolute pyridine. The autoclave is closed, cooled with liquid nitrogen and evacuated. Perfluoroisobutylene (34 g, 0.17 mol) is then condensed into the autoclave and heated at 160°–170 °C over a period 25 h. The autoclave is opened, the reaction mixture is transferred into a distillation flask and distilled. The yield of 1.3.3.3-tetrafluoro-1-chloro-2-trifluoromethylpropylene-1 is 18.0 g (49%); bp 40°–42 °C; n_D^{20} 1.3005.

Apart from 1,3,3,3-tetrafluoro-1-chloro-2-trifluoromethylpropylene-1, 8.7 g (21%) of 3,3,3-trifluoro-1,1-dichloro-2-trifluoromethylpropylene is formed (bp 71°–74 °C).

Likewise, the reaction of perfluoroisobutylene and benzoyl bromide in the presence of pyridine (in an autoclave at 170°–180 °C, during 18 h) affords 1,3,3,3-tetrafluoro-1-bromo-2-trifluoromethylpropylene (bp 56°–58 °C, n_D^{20} 1.3260).

4,4,4-Trifluoro-3,3-bis(trifluoromethyl)butene-1

[(Perfluoro-tert-butyl)ethylene]

$$(CF_3)_3C{-}CH{=}CH_2$$

$C_6H_3F_9$ MW 246

(Perfluoro-*tert*-butyl)ethylene is a volatile colourless liquid, soluble in common organic solvents. Bp 53°–54 °C [27, 28].

Method of Preparation

(Perfluoro-*tert*-butyl)ethylene is prepared by dehydroiodination of 1,1,1-trifluoro-4-iodo-2,2-bis(trifluoromethyl)butane (70% yield) [28], dehydrobromination of 1,1,1-trifluoro-4-bromo-2,2-bis(trifluoromethyl)butane (60% yield) [27]; by pyrolytic decomposition of *β*-(perfluoro-*tert*-butyl)ethylacetate (75% yield) [27]; or by the reaction of perfluoroisobutylene with ethylene oxide in the presence of CsF (77% yield) [29].

Procedure

1. Interaction of Perfluoroisobutylene with Ethylene Oxide in the Presence of CsF [29]

$$(CF_3)_2C{=}CF_2 + H_2C\overset{O}{-}CH_2 \xrightarrow{CsF} (CF_3)_3CH{=}CH_2 + (CF_3)_3CH +$$

$$+ (CF_3)_2C \cdots \underset{F}{\overline{C}} \cdots O\,Cs^+ \rightarrow (CF_3)_2CHCOOH$$

Caution! Perfluoroisobutylene is extremely toxic. It should be handled only under an efficient hood (see p. 2).

Caution! Ethylene oxide is toxic. It should be handled only under hood.

In a 100-ml three-necked flask provided with a stirrer, a gas inlet tube and a dry-ice acetone reflux condenser, are placed 9 g (0.06 mol) of freshly calcinated powdered CsF, 50 ml of absolute diglyme; then 30 g (0.15 mol) of perfluoroisobutylene is gradually added with stirring. After the gas has been introduced, the inlet tube is replaced with a dropping funnel and a solution of 2.2 g (0.05 mol) of ethylene oxide in 10 ml of diglyme is added dropwise, with vigorous stirring. The reaction mixture is allowed to stand overnight, whereupon it is heated at 40°–50 °C for 3 h. The reflux condenser is replaced with a downward Liebig condenser and a low-boiling fraction, a mixture of monohydroperfluoroisobutane (44%) and (perfluoro-*tert*-butyl)ethylene (56%) (as shown by GLC), is distilled off into a trap cooled to —70 °C, with heating to 40°–50 °C, under vacuum (10 mm Hg). Distillation of the mixture affords 9.5 g (77%) of (perfluoro-*tert*-butyl)ethylene (bp 53°–54 °C).

After the low-boiling fraction has been distilled off, the residue is washed with diluted HCl, the oily layer is separated and distilled over concentrated H_2SO_4. The yield of α-hydroperfluoroisobutyric acid is 7 g (70%); bp 126°–134 °C.

2. Pyrolytic Decomposition of β-(Perfluoro-tert-butyl)ethylacetate [27]

$$(CF_3)_2C{=}CF_2 + BrCH_2CH_2OCOCH_3 \xrightarrow{CsF} (CF_3)_3CCH_2CH_2OCOCH_3\,;$$

$$(CF_3)_3CCH_2CH_2OCOCH_3 \xrightarrow[540\ ^\circ C]{-CH_3COOH} (CF_3)_3CCH{=}CH_2$$

2-(Perfluoro-tert-butyl)ethylacetate. 2-Bromoethylacetate (94 g, 0.56 mol) is added to a mixture of 200 g (1 mol) of perfluoroisobutylene, 152 g (1 mol) of CsF and 300 ml of dry diglyme; the temperature of the mixture rises to 65 °C. The mixture is stirred for 2 h, the reaction mixture is poured into water, the oily layer is dissolved in ether, washed with water and dried over $MgSO_4$. The ether is distilled off and the residue is distilled. The yield of 2-(perfluoro-*tert*-butyl)ethylacetate is 124 g (72%), bp 140° to 142 °C; n_D^{20} 1.3630.

(Perfluoro-tert-butyl)ethylene. 2-(Perfluoro-*tert*-butyl)ethylacetate (30.5 g) is passed during 40 min in a slow flow of argon (6 ml/min), through a quartz tube (650 mm long, 15 mm i.d.) filled with crushed porcelain and heated to 540 °C. Pyrolysate (30.0 g) is washed with water, dried over $MgSO_4$ and distilled. A fraction (18 g) boiling at 50°–60 °C is collected. Redistillation gives pure (perfluoro-tert-butyl)ethylene, bp 55°–58 °C, n_D^{20} 1.3940. The residue is the starting acetate. The yield of (perfluoro-*tert*-butyl)ethylene is 14.2 g (73.2%) conversion 78.5%.

Perfluorobutadiene

$$CF_2{=}CF{-}CF{=}CF_2$$

C_4F_6 MW 162

Perfluorobutadiene is a colourless gas. Bp 5.8 °C [30].

Method of Preparation

Perfluorobutadiene is prepared by dehalogenation of 3,4-dichlorohexafluorobutene-1 [31], formed (in a low yield) in the pyrolysis of chlorotrifluoroethylene at 500 °C [31]. Perfluorobutadiene is also obtained by decarboxylation of disodium salt of octafluoroadipic acid (30% yield), or by dehalogenation of 1,2,3,4-tetrachlorohexafluorobutane (98% yield) [30, 33]. 1,2,3,4-Tetrachlorohexafluorobutane is obtained either by the reaction of 1-iodo-1,2-dichloro-1,2,2-trifluoroethane with mercury, or by UV irradiation [30], or in the reaction of 1-iodo-1,2-dichloro-1,2,2-trifluoroethane with zinc in acetic oxide [33].

Procedure

$$CF_2Cl{-}CFICl \xrightarrow[(CH_3CO)_2O]{Zn;} CF_2Cl{-}CFCl{-}CFCl{-}CF_2Cl \rightarrow$$

$$\xrightarrow[C_2H_5OH]{Zn;} CF_2{=}CF{-}CF{=}CF_2$$

1,2,3,4-Tetrachlorohexafluorobutane [33]. In a three-necked flask fitted with a stirrer, a thermometer, and a reflux condenser, are placed 80 ml of acetic oxide, 80 ml of methylene chloride, 32.7 g (0.5 mol) of granulated zinc and 140 g (0.5 mol) of 1-iodo-1,2-dichloro-1,2,2-trifluoroethane. The reaction starts spontaneously; the reaction temperature is controlled by intermittent cooling of the flask with an ice-water mixture. The reaction mixture is stirred for 3 h, the unaltered zinc is then filtered off and to the filtrate is added dropwise 150 ml of diluted H_2SO_4, with stirring and cooling of the flask to 0 °C. The lower organic layer of the filtrate is separated, neutralized (Note), washed with water, dried and distilled. The yield of 1,2,3,4-tetrachlorohexafluorobutane is 38.1 g (51%); bp 133°–135 °C, n_D^{14} 1.3832.

Perfluorobutadiene [30]. In a three-necked flask equipped with a stirrer, a dropping funnel, and an efficient reflux condenser, are placed 100 g of zinc dust and 100 ml of absolute ethanol. To this mixture is added dropwise with vigorous stirring, during 5 h, a mixture of 30.4 g (0.1 mol) of 1,2,3,4-tetrachlorohexafluorobutane and 10 ml of absolute ethanol. The liberated gas is collected in a trap (—78 °C). The condensed product is washed by passing the gas through an aqueous 5% NaOH solution and distilled. The yield of perfluorobutadiene is 15.8 g (98%); bp 5.8 °C.

Note. The reaction mixture should be thoroughly neutralized, since it is difficult to separate 1,2,3,4-tetrachlorohexafluorobutane from acetic acid by ordinary distillation.

Perfluoro(2,3-dimethylbutadiene-1,3)

2,3-Bis(trifluoromethyl)perfluorobutadiene-1,3

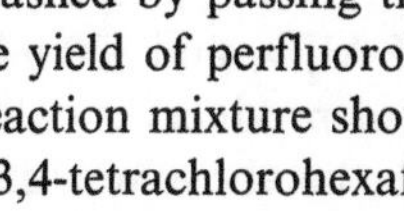

C_6F_{10} MW 262

Perfluoro(2,3-dimethylbutadiene-1,3) is a volatile colourless liquid. Mp —90 °C [34], bp 49°–51 °C; n_D^{20} 1.2887; d_4^{20} 1.56 [35].

Method of Preparation

Perfluoro(2,3-dimethylbutadiene-1,3) is prepared by the pyrolysis of perfluoroisopropenylsilver (56% yield), formed in the reaction of silver fluoride with tetrafluoroallene [36]; by copyrolysis of perfluorobutyne-2 and trifluoromethyldifluorophosphine (8% yield) [34]; or by dehydrofluorination of 2,3-dihydroperfluoro(2,3-dimethylbutane) (39% yield) [35].

The following procedure is a modification of the one described in [35].

Procedure

$$(CF_3)_2CHCOOH \xrightarrow{\text{electrolysis}} (CF_3)_2CHCH(CF_3)_2 \xrightarrow[Al_2O_3;\ H_2O]{NaOH;} CF_2{=}C(CF_3){-}C(CF_3){=}CF_2$$

2,3-Dihydroperfluoro(2,3-dimethylbutane) [35]. A solution of 40 g (0.2 mol) of α-hydroperfluoroisobutyric acid and 1 g of its potassium salt in 90 ml of water is electrolysed in electrolyser with two platinum anodes and a platinum cathode placed between them. Electrolysis is conducted at 6 A; the voltage is gradually increased from 9 to 30 V. The gaseous products of the reaction are condensed in a trap (—78 °C). After the electrolysis has been completed (the duration of the process is about 4 h), the product condensed in the trap is combined with the organic layer, formed at the bottom of electrolyser, and distilled. The yield of 2,3-dihydroperfluoro(2,3-dimethylbutane) is 20.8 g (67%); bp 61°–63 °C, n_D^{20} 1.2740, d_4^{20} 1.665.

Perfluoro(2,3-dimethylbutadiene-1,3) [35]. In a 2-l three-necked flask provided with a stirrer, a dropping funnel and a reflux condenser attached to a trap (—78 °C), is placed 604 g (2 mol) of 2,3-dihydroperfluoro(2,3-dimethylbutane); then a suspension prepared from 330 g (8.25 mol) of NaOH, 51 g (0.5 mol) of Al_2O_3 and 600 ml of water, is gradually added with stirring. After the exothermic reaction has been completed, the reaction mixture is stirred for an additional 48 h and distilled; the products are collected in a dry-ice receiver.

The resulting mixture (about 450 g), which consists, as shown by GLC, of perfluoro-(2,3-dimethylbutadiene-1,3) (60%) and 3-hydroperfluoro(2,3-dimethylbutene-1) (40%), is treated again with 165 g of NaOH, 25.5 g of Al_2O_3 and 300 ml of water as described above. After 48 g, 220–340 g of a mixture of products is collected as described above, which is again treated during 48 h with the above mixture of NaOH, Al_2O_3 and water, and distilled. The yield of perfluoro(2,3-dimethylbutadiene-1,3) is 180–210 g (35%–40%); bp 49°–51 °C.

1,4-Divinylperfluorobutane

$$CH_2{=}CH{-}CF_2{-}CF_2{-}CF_2{-}CF_2{-}CH{=}CH_2$$

$C_8H_6F_8$ MW 254

1,4-Divinylperfluorobutane is a colourless liquid. Bp 123 °C; n_D^{20} 1.3356, d_4^{20} 1.361 [37].

Method of Preparation

1,4-Divinylperfluorobutane is prepared by treatment of 3,3,4,4,5,5,6,6-octafluoro-1,8-diiodooctane with an alcoholic solution of an alkali (48% yield) [37].

Procedure

$$CF_2{=}CF_2 \xrightarrow{I_2} ICF_2{-}CF_2I \xrightarrow{250-300\,°C} I(CF_2CF_2)_2I \xrightarrow{CH_2{=}CH_2}$$

$$\rightarrow ICH_2CH_2(CF_2CF_2)_2CH_2CH_2I \xrightarrow{KOH} CH_2{=}CH(CF_2CF_2)_2CH{=}CH_2$$

Tetrafluoro-1,2-diiodoethane [37, 38]. Iodine (127 g, 0.5 mol) is placed in a 250-ml stainless steel autoclave. The autoclave is closed, flushed with argon, cooled with liquid nitrogen, evacuated, and 50 g (0.5 mol) of tetrafluoroethylene is condensed into it. Then the autoclave is heated to 160 °C over a period of 5 to 6 h The product is washed with $Na_2S_2O_3$ solution, with water, then dried over $MgSO_4$ and distilled to afford 147.7 g (83%) of tetrafluoro-1,2-diiodoethane; bp 63 °C, n_D^{20} 1.4880, d_4^{20} 2.7064.

Octafluoro-1,4-diiodobutane [37]. In a 200-ml steel autoclave is placed 250 g (0.7 mol) of tetrafluoro-1,2-diiodoethane; the autoclave is flushed with nitrogen, closed, and heated at 250°–300 °C during 7 to 8 h. The product is washed with $Na_2S_2O_3$ solution, then with water, dried over $MgSO_4$ and distilled. The yield of octafluoro-1,4-diiodobutane is 50 g (35%); bp 80 °C/82 mm Hg, n_D^{20} 1.4300, d_4^{20} 2.4870.

3,3,4,4,5,5,6,6-Octafluoro-1,8-diiodooctane [37]. In a 100-ml steel autoclave is placed 57 g (0.125 mol) of octafluoro-1,4-diiodobutane. The autoclave is cooled with liquid nitrogen, evacuated, then 9 l (0.4 mol) of ethylene is condensed into it and the autoclave is heated, with shaking, at 200°–240 °C during 5 to 6 h. The product is crystallized from heptane. The yield is 49 g (77%); mp 93 °C.

1,4-Divinylperfluorobutane [37]. To a solution of 51 g (0.1 mol) of 3,3,4,4,5,5,6,6-octafluoro-1,8-diiodooctane in 50 ml of ethanol is added, with stirring during 1 h, 123 ml of 10% alcoholic solution of KOH. The reaction mixture is boiled for 2 h, then diluted with water to 500 ml volume and extracted with ether. The ether extract is washed with $CaCl_2$ solution, dried over $MgSO_4$ and distilled. The yield of 1,4-divinylperfluorobutane is 12 g (48%); bp 123 °C, n_D^{20} 1.3356, d_4^{20} 1.361.

Perfluorooctadiene-1,7

$$CF_2{=}CF{-}CF_2{-}CF_2{-}CF_2{-}CF_2{-}CF{=}CF_2$$

C_8F_{14} MW 362

Perfluorooctadiene-1,7 is a colourless liquid. Bp 106 °C, n_D^{20} 1.3020, d_4^{20} 1.695 [39].

Method of Preparation

Perfluorooctadiene-1,7 is prepared by dechlorination of tetradecafluoro-1,2,7,8-tetrachlorooctane with zinc in acetic acid (72% yield) [39].

Procedure

$$CF_2{=}CFCl \xrightarrow{ICl} CF_2Cl{-}CFICl \xrightarrow[(C_6H_5COO)_2]{CF_2{=}CF_2;} CF_2Cl{-}CFCl{-}CF_2{-}CF_2I \xrightarrow{Zn}$$

$$\rightarrow CF_2Cl{-}CFCl{-}(CF_2CF_2)_2{-}CFCl{-}CF_2Cl$$

$$\xrightarrow{Zn} CF_2{=}CF{-}(CF_2CF_2)_2{-}CF{=}CF_2$$

1-Iodo-1,2-dichloro-1,2,2-trifluoroethane [40]. Chlorotrifluoroethylene (450 g, 3.56 mol) is condensed into a three-necked round-bottom flask fitted with a stirrer, a dropping funnel, and a reflux condenser cooled with dry-ice acetone mixture. The contents of the flask is cooled to —78 °C and a solution of 500 g (3 mol) of ICI in 750 ml of methylene chloride is added at that temperature, with stirring during 45 min. The temperature of the reaction mixture is slowly raised to —8 to —50 °C (Note 1). Cooling of the reflux condenser is stopped and unchanged olefin is condensed in a trap cooled to —78 °C (Note 2). The reaction mixture is washed with water, aqueous solution of $NaHSO_3$, then once more with water and dried over $CaCl_2$. The solvent is distilled off to give 430 g of the product containing 98% of 1-iodo-1,2-dichloro-1,2,2-trifluoroethane; bp 101 °C (45 °C/100 mm Hg); n_D^{25} 1.4493; yield 45%.

Heptafluoro-1,2-dichloro-4-iodobutane. In a 250-ml shaking steel autoclave are placed 200 g (0.72 mol) of 1-iodo-1,2-dichloro-1,2,2-trifluoroethane and 2 g of benzoyl peroxide. The autoclave is flushed with nitrogen, cooled with liquid nitrogen and evacuated, then 14 l (0.7 mol) of tetrafluoroethylene is condensed into it and the autoclave is heated at 130 °C for 4 h. The reaction product is distilled; a fraction boiling at 62°–70 °C/60–70 mm Hg is collected. Redistillation gives 72 g of heptafluoro-1,2-dichloro-4-iodobutane (57% based on the amount of converted 1-iode-1,2-dichloro-1,2,2-trifluoroethane); bp 66 °C/67 mm Hg, n_D^{20} 1.4046, d_4^{20} 2.135.

Tetradecafluoro-1,2,7,8-tetrachlorooctane. To a mixture of 10 g (0.15 mol) of granulated zinc, 31 g (0.3 mol) of acetic oxide and 50 ml of methylene chloride is added, with stirring, 56.9 g (0.15 mol) of heptafluoro-1,2-dichloro-4-iodobutane, at such a rate as to maintain the temperature of the reaction mixture at about 40 °C. Stirring is continued at 40 °C for 6 h, then the reaction mixture is cooled, 100 ml of water is added, the lower layer is separated, washed with water, dried over $MgSO_4$ and distilled. The yield of tetradecafluoro-1,2,7,8-tetrachlorooctane is 29.9 g (80%); bp 67°–68 °C/3 mm Hg, n_D^{20} 1.3604, d_4^{20} 1.838.

Perfluorooctadiene-1,7. To a mixture of 180 ml of acetic acid, 60 ml of acetic oxide and 40 g (0.6 mol) of granulated zinc is gradually added, with stirring at 118 °C, 50.4 g (0.1 mol) of tetradecafluoro-1,2,7,8-tetrachlorooctane. The mixture is boiled for 12 h; the boiling point of the mixture gradually decreases. As soon as it has reached 103°–107 °C, the mixture is distilled, the distillate is washed with water, with $NaHCO_3$ solution, then again with water, dried over $MgSO_4$ and distilled. The yield of perfluorooctadiene-1,7 is 26.2 g (72.5%); bp 106 °C, n_D^{20} 1.3020, d_4^{20} 1.695.

Notes

1. When the reaction is conducted at a higher temperature, the content of 1-iodo-2,2-dichloro-1,1,2-trifluoroethane (the by-product) increases.
2. Unchanged olefin may be recycled.

Tetrakis(trifluoromethyl)allene

$(CF_3)_2C{=}C{=}C(CF_3)_2$

C_7F_{12} MW 312

Tetrakis(trifluoromethyl)allene is a colourless liquid. Bp 41 °C/743 mm Hg; n_D^{20} 1.275 [41].

Method of Preparation

Tetrakis(trifluoromethyl)allene is prepared by the reaction of bis(trifluoromethyl)-ketene with triethylphosphite (74% yield); by the pyrolysis of bis(trifluoromethyl)-ketene dimer (80% yield); or by the reaction of that dimer with triethylphosphite (65% yield) [41].

Procedure

$$(CF_3)_2C{=}C{=}O \xrightarrow{P(OC_2H_5)_3} (CF_3)_2C{=}C{=}C(CF_3)_2$$

Bis(trifluoromethyl)ketene (80.7 g, 0.45 mol) is placed in a three-necked flask provided with a stirrer, a dropping funnel, and a dry-ice reflux condenser, the upper end of which is connected with a trap cooled to −78 °C, a gas-washing bottle filled with concentrated H_2SO_4, and a gasometer (Note 1). The flask is cooled to 0 °C and 32 g of triethylphosphite is added dropwise, with stirring, during 1.5 h (Note 2). The mixture is then allowed to stand and in 30 min a gas starts to evolve. The mixture is stirred at 0 °C for further 3 h, then the temperature is raised to that of the room and stirred for 1 h till the gas has ceased to evolve. The reaction mixture is allowed to stand overnight, whereupon the lower colourless layer is separated and distilled. The yield of tetrakis(trifluoromethyl)allene is 50.7 g (74%); bp 41 °C/743 mm Hg, n_D^{20} 1.275 (Note 3).

Notes

1. In the gasometer, 4.5 l of carbon monoxide is collected.
2. Before use, triethylphosphite is stored over sodium.
3. Vacuum distillation of the upper layer of the reaction mixture gives further 3 g of the allene.

3,3,3-Trifluoropropyne

$CF_3{-}C{\equiv}CH$

C_3HF_3 MW 94

Trifluoropropyne is a colourless gas with the odour resembling that of acetylene. Bp —48 ° to —47 °C/705 mm Hg [42].

Method of Preparation

Trifluoropropyne is prepared by dehydrohalogenation of 1-iodo-3,3,3-trifluoropropylene-1 (70% yield) [42], obtained by the reaction of iodotrifluoromethane with acetylene at 220 °C and under increased pressure [42]. Trifluoropropyne is also prepared by the reaction of 1,1,2-trichloro-3,3,3-trifluoropropylene-1 with zinc in dimethylacetamide (78% yield) [43]; or by the reaction of acetylenecarboxylic acid with sulphur tetrafluoride (60% yield) [44].

Procedure [44]

$$HC{\equiv}C{-}COOH \xrightarrow{SF_4} HC{\equiv}C{-}CF_3$$

Sulphur tetrafluoride [45]. Finely ground sulphur (87 g, 2.71 mol) is placed in a 1-l steel autoclave. The autoclave is flushed with nitrogen, cooled to —78 °C and 479 g (2.16 mol) of IF_5 is added at that temperature. The mixture is heated, with stirring, at 100 °C for 5 h, then at 200 °C for 5 h, and at 300 °C for 2 h, whereupon it is cooled to room temperature and the low-boiling products are condensed in a metallic trap cooled to —78 °C (Notes 1 and 2). The condensed product is distilled to give 230 g of the product containing 93% of SF_4 (Note 3).

Sulphur tetrafluoride [46]. Finely powdered anhydrous NaF (840 g, 20 mol) and 1.5 l of absolute acetonitrile are placed in a thoroughly dried 3-l four-necked flask fitted with a stirrer, a thermometer, a dropping funnel, and an efficient reflux condenser cooled to —20° to —30 °C with dry-ice chloroform mixture; the flask is flushed with nitrogen. Then freshly distilled SCl_2 (1040 g, 10 mol) is added to the stirred mixture during 30–40 min. The mixture is heated to 50 °C during 1 h, then at 65°–70 °C for 1–1.5 h (Note 4). The liberated gases are purified (Note 5) and condensed, with cooling to —70 °C, into a steel cylinder provided with a siphon tube. The yield of SF_4 is 260–320 g (70%–80%). The product contains 10% to 15% of thionyl fluoride and HF.

Trifluoropropyne [44]. In a steel autoclave is placed 18.8 g (0.27 mol) of acetylenecarboxylic acid. The autoclave is cooled in liquid nitrogen and evacuated. Then 85 g (0.78 mol) of SF_4 is condensed into the autoclave and the mixture is heated at 120 °C for 3 h. The autoclave is then cooled to room temperature, the low-boiling products are passed through a wash bottle containing a buffer solution (pH 8.5) (Note 6) and condensed in a trap (—78 °C). The yield of trifluoropropyne is 15 g (60%).

Notes

1. The autoclave is connected with the trap by means of copper or polyethylene tubes.
2. A steel or quartz bottle filled with NaF pellets is placed between the autoclave and the trap to remove HF from SF_4.
3. Sulphur tetrafluoride thus obtained contains thionyl fluoride as an impurity.
4. In order to stop the reaction in an emergency, it is necessary to quickly cool the reaction mixture to the temperature below 40 °C.
5. A system for purification of SF_4 consists of three traps connected in series by means of a thick rubber tubing. In the first trap cooled at —25° to —30 °C, sulphur chloride carried off by gases from the reactor is collected. The second and the third traps are filled with 100 g of rod sulphur (the size of granules 3 to 5 mm) and 100 g of dry, activated coal, respectively; they serve to purify SF_4 from sulphur chlorides and chlorine. The trap with activated coal is cooled to —30 °C.
6. In order to prepare the buffer solution with pH 8.5, 450 g of $Na_2H_2PO_4 \cdot H_2O$ and 220 g of KOH are dissolved in 4 l of water.

Perfluorobutyne-2

$$CF_3{-}C{\equiv}C{-}CF_3$$

C_4F_6 MW 162

Perfluorobutyne-2 is a colourless gas. Bp —24.5 °C [47].

Method of Preparation

Perfluorobutyne-2 is prepared by dechlorination of hexafluoro-2,3-dichlorobutene-2 with zinc dust in ethanol (60% yield) (Note 1) [47], or in acetic oxide (63% yield) [48]; or by thermal isomerization of perfluorocyclobutane [49].

Procedure

$$CCl_2{=}CCl{-}CCl{=}CCl_2 \xrightarrow{SbF_3Cl_2} CF_3CCl{=}CClCF_3$$

$$\xrightarrow[(CH_3CO)_2O]{Zn;} CF_3C{\equiv}CCF_3$$

Hexafluoro-2,3-dichlorobutene-2 [50]. In a 2-l steel autoclave provided with a manometer (up to 20 atm) and a needle valve, is placed 1225 g (6.8 mol) of SbF_3; the autoclave is closed, placed into a furnace mounted on a mechanical shaker, and connected by a flexible copper tube with a chlorine cylinder, whereupon chlorine is let in with heating of the autoclave to 135 °C and constant shaking, till a 480 g (6.8 mol) gain in weight has been reached. The chlorine inlet tube is then disconnected, the excess chlorine is let out through the valve, the autoclave is opened, and further 305 g (1.7 mol) of SbF_3 together with 1044 g (4 mol) of perchlorobutadiene is placed in it. Then the autoclave is closed, heated during 30 min in a boiling water-bath, than placed once more into a furnace and heated, with constant shaking, at 155 °C for further 2 h; the pressure in the autoclave has reached about 6 atm.

The autoclave is cooled, opened, the cap is replaced with a steel tube 0.5 m long (a dephlegmator) connected with a dowmward steel (or copper) condenser, to which are connected in series two 2 l Bunsen flasks, half-filled with water (Note 2), and a trap cooled to —78 °C. The autoclave is heated to 200 °C on an open flame; all the organic products are completely distilled off.

The organic layers contained in the two flasks and the contents of the trap are combined and steam-distilled, the product is dried over $CaCl_2$ and distilled on a fractionating column. The yield of hexafluoro-2,3-dichlorobutene-2 is 680 g (73%); bp 66°–68 °C/750 mm Hg (Note 3).

Perfluorobutyne-2 [48]. Acetic anhydride (80 ml) and zinc dust (60 g, 0.9 mol) are placed in a three-necked flask equipped with a stirrer, a dropping funnel, and an efficient reflux condenser connected with a trap cooled to —78 °C. The mixture is heated, with vigorous stirring, till acetic anhydride starts to reflux. When the mixture begins to reflux vigorously, a solution of 84 g (0.36 mol) of hexafluoro-2,3-dichlorobutene-2 in 50 ml of acetic anhydride is added dropwise to the reaction mixture during 4 h; the liberated gas is collected in the trap. Then 30 g (0.45 mol) of zinc dust is added and the mixture is refluxed with stirring during an additional 3 h. The product collected in the trap is distilled on a fractionating column with a low-temperature distilling head. The yield of hexafluorobutyne-2 is 34 g (63%); bp —24.5 °C.

Notes

1. Dechlorination of hexafluoro-2,3-dichlorobutene-2 in ethanol may result in the formation of considerable amounts of the reduction products.
2. The end of the adapter should be immersed in water; the same for the tube connecting the flasks.
3. Hexafluoro-2,3-dichlorobutene-2 may be prepared from perchlorobutadiene and SbF_5. In a flask provided with a dephlegmator and a downward Liebig condenser (the apparatus is all-copper), is placed SbF_5, then perchlorobutadiene is added dropwise at 60°–65 °C (the reaction is mildly exothermic); the organic products are collected under water. The yield of hexafluoro-2,3-dichlorobutene-2 is about 40%.

Perfluoro(4,4-dimethylpentyne-2)

Hexafluoro-bis(trifluoromethyl)pentyne-2

$$(CF_3)_3C{-}C{\equiv}C{-}CF_3$$

C_7F_{12} MW 312

Perfluoro(4,4-dimethylpentyne-2) is an extremely volatile colourless liquid, soluble in reons, slightly soluble in common organic solvents. Bp 47°–49 °C [51].

Method of Preparation

Perfluoro(4,4-dimethylpentyne-2) is prepared by the reaction of perfluoroisobutylene with trifluoro-1-chloropropyne in the presence of CsF (90% yield) [51].

Procedure

$$CF_3CH_2CH_2Cl \xrightarrow{Cl_2} CF_3CH_2CCl_3 \xrightarrow{KOH} CF_3C{\equiv}CCl \xrightarrow[CsF]{(CF_3)_2C=CF_2}$$

$$\rightarrow (CF_3)_3C{-}C{\equiv}C{-}CF_3$$

1,1,1-Trifluoro-3,3,3-trichloropropane. In a quartz flask provided with a gas inlet tube and a reflux condenser cooled with ice-water, is placed 1,1,1-trifluoro-3-chloropropane and 0.1% $FeCl_3$, then chlorine is bubbled through the mixture under irradiation, while heating the flask with a gas burner till 1,1,1-trifluoro-3-chloropropane starts to boil. Further boiling is kept up by the heat of the reaction. The rate of the process depends on the capacity of the reflux condenser. After the density of the reaction mixture has become 1.55–1.56 g/ml (Note 1), chlorination is stopped and the mixture is distilled to afford 1,1,1-trifluoro-3,3,3-trichloropropane; bp 95°–96 °C, n_D^{20} 1.3910, d_4^{20} 1.550.

Trifluoro-1-chloropropyne. 1,1,1-Trifluoro-3,3,3-trichloropropane (70–80 g) is passed during 2 h in a flow of dry nitrogen through the melted 85% KOH (120 g) at 240°–250 °C (Note 2). The products are collected in two traps arranged in series and cooled to 0° and —78 °C respectively. After passing is completed, the melt is flushed with nitrogen during 5 min. The first trap is heated to 40 °C and the product is recondensed into another trap. The product contained in the second trap is evaporated through P_2O_5 layer on asbestos or glass wool into the still of the Podbielniak column and distilled. The yield of trifluoro-1-chloropropyne is 30–35 g (85%); bp 2.5°–3 °C.

The still residue and the product from the first trap (from which water has been removed), are combined, dried over $CaCl_2$, and distilled to give 15 g of 1,1,1-trifluoro-3,3-dichloropropylene; bp 54°–55 °C (Note 3).

Perfluoro(4,4-dimethylpentyne-2) [51]. In a three-necked flask fitted with a stirrer, a gas inlet tube, and a dry-ice reflux condenser, are placed 7.5 g (0.05 mol) of freshly calcinated CsF powder and 40 ml of absolute diglyme, then 10 g (0.05 mol) of perfluoroisobutylene is added with stirring (Note 4). The mixture is stirred during 20 min, then 6 g (0.048 mol) of trifluoro-1-chloropropyne is added and stirring is continued at 40 °C for 1 h. The reflux condenser is replaced with a downward Liebig condenser and the mixture is distilled. The yield of perfluoro(4,4-dimethylpentyne-2) is 13.3 g (90%); bp 47°–49 °C.

Heating of perfluoroisobutylene, chlorophenylacetylene and CsF in diglyme (in an autoclave, at 100 °C, during 12 h) affords 4,4,4-trifluoro-3,3-bis(trifluoromethyl)-1-phenylbutyne-1; bp 66°–67 °C/16 mm Hg [51].

Notes

1. A float is placed into the reaction mixture; when the required density has been achieved, the float goes up.

2. The reaction is conducted in a stainless steel (or copper) reactor.
3. 1,1,1-Trifluoro-3,3-dichloropropylene may be obtained by treating 1,1,1-trifluoro-3,3,3-trichloropropane with 30% solution of KOH or NaOH in methanol at 0°–2 °C; yield 90%.
4. Perfluoroisobutylene is extremely toxic. Work with this compound should be carried out under an efficient hood.

Monofluoroethyl Alcohol

$CH_2F{-}CH_2OH$

C_2H_5FO MW 64

Monofluoroethyl alcohol is a transparent colourless liquid with a faint, pleasant odour. Bp 102°–104 °C [52], 103.3 °C [54]; n_D^{20} 1.364; d_4^{20} 1.102 [52].

Method of Preparation

Monofluoroethyl alcohol is prepared by the reaction of ethylene chlorohydrine with KF in a high-boiling solvent at atmospheric pressure [5, 53], or in an autoclave in the absence of a solvent (40% yield) [54]; or by the reaction of ethylene oxide with HF in ether (40% yield) [52].

Procedure [52]

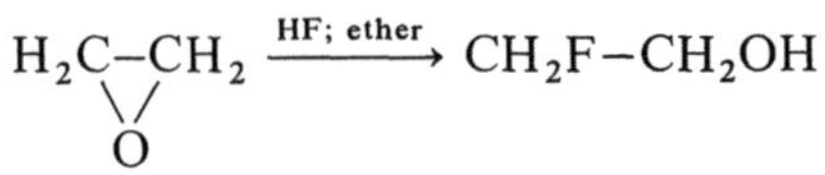

Ether (500 ml) is placed in a 1 l steel autoclave and cooled to −78 °C. Anhydrous HF (20 g, 1 mol) and ethylene oxide (20 g, 0.45 mol) are carefully introduced into the autoclave at the above temperature (Note 1). The autoclave is closed and heated during 6 h in a boiling water-bath, then cooled; the reaction mixture is neutralized with sodium carbonate. The neutral ethereal solution is transferred into a distillation flask fitted with a long dephlegmator. The ether is distilled off, the residue is dried over anhydrous KF and distilled. Synthesis is repeated several times; 190 g of ethylene oxide gives 122.5 of monofluoroethanol, bp 98°–105 °C. Redistillation affords the product with bp 102°–104 °C; yield 102 g (40%) (Notes 2,3).

Notes

1. Ethylene oxide should be dry.
2. Monofluoroethanol, monofluoroacetic acid and its derivatives are extremely toxic: LD_{50} of monofluoroethanol is 10 mg/kg (for mice), LD_{50} of monofluoroacetic acid 0.6 mg/kg. These compounds should be handled only under an efficient hood. Hands should be protected with rubber gloves. First-aid treatment of poisoning: emetic or stomach-pump, monoacetine solution (100 ml in 500 ml of

water), warmth, in case of convulsions intravenous or intramuscular introduction of barbiturates. Monoacetine (glycerol monoacetate) (0.1–0.5 ml/kg) may be introduced intramuscularly (twice with a 30 min interval). Monoacetine may be substituted by acetamide solution in physiological solution (the same dose) [80].

3. Apart from monofluoroethanol, the reaction gives 37 g of glycol 2-fluoroethyl ether (bp 172°–174 °C).

Trifluoroethanol

$CF_3 - CH_2OH$

$C_2H_3F_3O$ MW 100

Trifluoroethanol is a colourless liquid. Mp —43.5 °C [81], bp 73.9°–74.1 °C at 750 mm Hg; d_4^{20} 1.3842 [55].

Method of Preparation

Trifluoroethanol is prepared by the reaction of trifluoroacetate or trifluoroacetyl chloride with $LiAlH_4$ in anhydrous ether (50%–70% yield) [55, 56]; by hydration of trifluoroacetamide with hydrogen on a platinum catalyst (77% yield) [57]; by oxidation of vinylidene fluoride with chromic anhydride, potassium permanganate or peracetic acid in anhydrous hydrogen fluoride (20%–30% yield) [58–60]; or by treatment of fluoral with isopropylmagnesium bromide (70% yield) [261].

Procedure [55]

$$CF_3COOC_4H_9 \xrightarrow{LiAlH_4;\ ether} CF_3{-}CH_2OH$$

In a 1-l three-necked flask provided with a stirrer, a dropping funnel and a reflux condenser, is placed a solution of 17.8 g (0.47 mol) of $LiAlH_4$ in 500 ml of anhydrous ether, then 116 g (0.68 mol) of butyl trifluoroacetate is added dropwise, with vigorous stirring during 2 h. The reaction mixture heats to reflux. After all the butyl trifluoroacetate has been added, the reaction mixture is refluxed for 15 min. Excess $LiAlH_4$ is cautiously decomposed with water and the contents of the flask are poured into a mixture of ice and diluted H_2SO_4. The ether layer is separated and aqueous layer extracted with ether (2 × 100 ml). The combined ether extract is dried over $MgSO_4$ and the ether is distilled off (Note). The residue is distilled over P_2O_5. The yield of trifluoroethyl alcohol is 52 g (76%); bp 73.9°–74.1 °C at 750 mm Hg; d_4^{20} 1.3842.

Note. The ether should better be distilled off on a column, since trifluoroethyl alcohol is rather volatile.

2-Hydroperfluoroisopropyl Alcohol

$$\begin{matrix} CF_3 \\ CF_3 \end{matrix} \rangle CHOH$$

$C_3H_2F_6O$ MW 168

2-Hydroperfluoroisopropyl alcohol is a highly volatile colourless liquid. Bp 59 °C; $n_D^{20} < 1.3$ [61].

Method of Preparation

2-Hydroperfluoroisopropyl alcohol is prepared by the reduction of hexafluoroacetone with sodium borohydride [62] or $LiAlH_4$ (60% yield) [61]; by the reduction of hexafluoroacetone with hydrogen on a platinum catalyst (90% yield) [61]; or by the reaction of hexafluoroacetone with certain amines [63].

Procedure [61]

$$(CF_3)_2C{=}O \xrightarrow[THF]{LiAlH_4;} (CF_3)_2CHOH$$

Hexafluoroacetone (15 ml), which has been condensed at −78 °C, is slowly passed through the solution of 4 g of $LiAlH_4$ in 100 ml of anhydrous tetrahydrofuran. The rate of flow of hexafluoroacetone is regulated so that the temperature of the reaction mixture should not rise above 50 °C. Then the reaction mixture is decomposed with wet tetrahydrofuran and 10% HCl. The organic layer is separated, washed twice with water, dried over silica gel, and distilled to afford a 1:1 complex of 2-hydroperfluoroisopropyl alcohol with tetrahydrofuran (colourless oil, bp 99°–100 °C, n_D^{25} 1.3233). The yield of the complex is 22 g (61%). To obtain pure 2-hydroperfluoroisopropyl alcohol, the complex is distilled twice over 20% oleum under reduced pressure. This procedure gives pure 2-hydroperfluoroisopropyl alcohol, bp 59 °C, $n_D^{20} < 1.3$.

Perfluoro-*tert*-butyl Alcohol

$(CF_3)_3C{-}OH$

C_4HF_9O MW 236

Perfluoro-*tert*-butyl alcohol is a highly volatile colourless liquid with a specific odour, partially soluble in water, soluble in polar organic solvents. Bp 45 °C; n_D^{20} 1.27, d_4^{20} 1.6930 [64].

Method of Preparation

Perfluoro-*tert*-butyl alcohol is prepared by the reaction of hexafluoroacetone with CsF in aqueous diglyme (25%–30% yield) [65]; by treatment of perfluoroisobutylene oxide with HF in the presence of SbF_5 (quantitative yield) [66]; by treatment of hexafluoro-2-nitroso-2-trifluoromethylpropane with nitrogen tetroxide (78% yield) [64]; or by oxidation of hexafluoro-2-nitroso-2-trifluoromethylpropane with oxygen in the presence of nitrogen tetroxide (78% yield) [67].

Procedure

$$(CF_3)_2C{=}CF_2 \xrightarrow[KF]{NOCl;} (CF_3)_3CNO \xrightarrow[2.\ H_2SO_4]{1.\ O_2;\ N_2O_4} (CF_3)_3C{-}OH$$

Hexafluoro-2-nitroso-2-trifluoromethylpropane (nitrosononafluoro-tert-butane) [68]. (Note 1) In a 1-l four-necked flask provided with a stirrer, two gas inlet tubes and a dry-ice acetone reflux condenser, are placed 500 ml of anhydrous dimethylformamide and 80 g (1.3 mol) of freshly calcinated KF powder. To the mixture cooled to —12° to —15 °C are simultaneously added, with vigorous stirring, 200 g (1 mol) of perfluoroisobutylene (Note 2) and 70 g (1.1 mol) of nitrosyl chloride (Notes 3, 4). After all the perfluoroisobutylene and nitrosyl chloride have been added, the reaction mixture is allowed to warm to 10°–15 °C and stirred at that temperature during 10 to 15 min (Note 5). The flask is then again cooled to —5° ÷ —8 °C and the reaction mixture is poured onto 1.5 kg of ice (Note 6). The blue oily layer is separated and distilled over concentrated H_2SO_4 on a fractionating column with a low-temperature distilling head. The yield of hexafluoro-2-nitroso-2-trifluoromethylpropane is 195 g (80%); bp 24 °C.

Perfluoro-tert-butyl alcohol. A mixture of 14.5 g (0.06 mol) of hexafluoro-2-nitroso-2-trifluoromethylpropane and 1.5 g (0.025 mol) of nitrogen oxides is placed in a gas-washing bottle connected to a quartz tube (500 to 600 mm long, 8 mm i.d.) heated to 160°–190 °C, which is attached to a receiver cooled to —78 °C. Oxygen is bubbled through the mixture at the rate of 70–90 ml/min (Note 7). After the reaction is complete, the products condensed in the receiver are distilled twice over two volumes of concentrated H_2SO_4 (Note 8); the distillate is collected in the ice-water-cooled receiver. This gives 13.8 g of the product, bp 42°–47 °C, containing 80% of perfluoro-*tert*-butyl alcohol, as shown by GLC and [^{19}F]NMR, yield 78%.

Notes

1. The procedure suggested here is a modification of the one described in [68].
2. Perfluoroisobutylene is extremely toxic. Work with this compound should be carried out under an efficient hood (see p. 2).
3. Air moisture should be kept out of the reaction flask.
4. It is preferable to regulate the rate of addition of perfluoroisobutylene and nitrosyl chloride, so that the reagents are added in about stoichiometric ratio.
5. At the higher temperature, hexafluoro-2-nitroso-2-trifluoromethylpropane freezes and clogs the reflux condenser.
6. Ice is previously crushed and placed in a separatory funnel. The reaction mixture should be poured onto ice slowly; otherwise it heats, which results in a loss of hexafluoro-2-nitroso-2-trifluoromethylpropane.

7. The mixture of hexafluoro-2-nitroso-2-trifluoromethylpropane and nitrogen oxides may be introduced into a flow of oxygen from the dropping funnel with a pressure-equalizing tube.
8. The resulting distillate, containing nitrogen oxides, is first shaken with concentrated H_2SO_4 to decolorization, then distilled.

Perfluoropropenol-2

$$CF_3-\underset{\underset{OH}{|}}{C}=CF_2$$

C_3HF_5O MW 148

Perfluoropropenol-2 is a highly volatile colourless liquid with a specific strong odour, soluble in ethers, acetonitrile, methylene chloride, partially soluble in water (Note). Bp 55 °C [69, 70].

Method of Preparation

Perfluoropropenol-2 is prepared by thermal decomposition of diisopropyl(perfluoropropenyl-2) phosphate (by heating it to 200°–210 °C/70–80 mm Hg; yield 71%) [70]; or by treatment of dialkyl (perfluoropropenyl-2) phosphates with concentrated H_2SO_4 (65% yield) [70].

Procedure

$$CF_3COCF_2Cl \xrightarrow{[(CH_3)_2CHO]_2P(O)H} CF_2=\underset{\underset{CF_3}{|}}{C}-O-P(O)[OCH(CH_3)_2]_2 \rightarrow$$

$$\xrightarrow[-2\,CH_3CH=CH_2]{170\,°C} CF_2=\underset{\underset{CF_3}{|}}{C}-O-P(O)(OH)_2 \xrightarrow{200-210\,°C/70\,mm\,Hg}$$

$$CF_3-\underset{\underset{OH}{|}}{C}=CF_2$$

Diisopropyl (perfluoropropenyl-2) phosphate [70]. In a 50-ml three-necked flask fitted with a stirrer, a gas inlet tube, and a dry-ice acetone reflux condenser, is placed 10 g (0.06 mol) of diisopropyl phosphite; then 15 g (0.08 mol) of chloropentafluoroacetone is gradually added with cooling of the mixture to 0 °C. The reaction mixture is heated at 80 °C till hydrogen chloride has ceased to evolve, then distilled. The yield of diisopropyl(perfluoropropenyl-2) phosphate is 9 g (47%); bp 70°–72 °C/4 mm Hg.

Likewise, from dimethyl phosphite and chloropentafluoroacetone, dimethyl(perfluoropropenyl-2) phosphate is obtained (bp 68°–72 °C at 10 mm Hg; 53% yield) [70]; from diisopropyl phosphite and tetrafluoro-1,3-dichloroacetone: diisopropyl-(tetrafluoro-3-chloropropenyl-2) phosphate (bp 86°–90 °C/4 mm Hg; 43% yield)

[70]; from diisopropyl phosphite and 1,1,3,3-tetrafluoro-3-chloroacetone: diisopropyl(3-hydroperfluoropropenyl-2) phosphate (bp 80 °C/2 mm Hg; 25% yield) [70]; from diisopropyl phosphite and 1,1,3-trichloro-1,3,3-trifluoroacetone: diisopropyl(trifluoro-1,3-dichloropropenyl-2) phosphate (bp 87°–92 °C/1 mm Hg; 52% yield) [70]; from diisopropyl phosphite and 1,3,3-trifluoro-1,3-dichloroacetone: diisopropyl(1,3,3-trifluoro-3-chloropropenyl-2) phosphate (bp 108 °C/10 mm Hg, 35% yield) [70].

Perfluoropropenol-2 [70]

A) Diisopropyl(perfluoropropenyl-2) phosphate (15 g, 0.048 mol) is placed in a 50-ml flask connected to a trap cooled to —78 °C, and heated at 170 °C during 3 min [3 g (77%) is collected in a trap]. Then the reaction flask is connected with another trap, attached to a water-jet pump and also cooled to —78 °C, and heated under vacuum (70–80 mm Hg) at 200°–210 °C. The volatile products condensed in the trap are distilled. The yield of perfluoropropenol-2 is 4.2 g (71%); bp 54°–55 °C.

Likewise, from diisopropyl(tetrafluoro-3-chloropropenyl-2) phosphate, tetrafluoro-3-chloropropenol-2 is obtained (bp 80 °C; 64% yield) [70].

B) A mixture of 5.8 g (0.0186 mol) of diisopropyl(perfluoropronenyl-2) phosphate and 15 ml of concentrated H_2SO_4 is heated and a fraction boiling at 45° to 60 °C is collected in a receiver cooled with ice-water. The distillate is redistilled. The yield of perfluoropropenol-2 is 2 g (72%); bp 54°–55 °C.

Likewise, dimethyl(perfluoropropenyl-2) phosphate gives perfluoropropenol-2 in a 65% yield.

Note. Perfluoropropenol-2 dissolved in water is slowly isomerized to pentafluoroacetone.

Perfluoropinacon

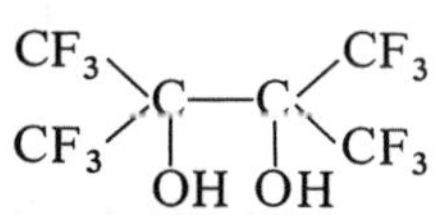

$C_6H_2F_{12}O_2$ MW 334

Perfluoropinacon is a highly volatile, highly toxic liquid with a strong odour, partially soluble in water, soluble in alcohols. Bp 129 °C [61], 59°–60 °C/60 mm Hg [71]; mp 18°–20 °C [71], 26 °C [61]; n_D^{25} 1.3102; pK_a 3.95 (in water) [61].

Method of Preparation

Perfluoropinacon is prepared by hydrodimerization of hexafluoracetone, under irradiation in the presence of isopropanol (63% yield) [61], or by acidolysis of 2,2,2-triethoxy-4,4,5,5-tetrakis(trifluoromethyl)-1,3,2-dioxaphospholane (65%–72% yield) [71].

Procedure

1. Acidolysis of 2,2,2-Triethoxy-4,4,5,5-tetrakis(trifluoromethyl)-1,3,2-dioxaphospholane [71]

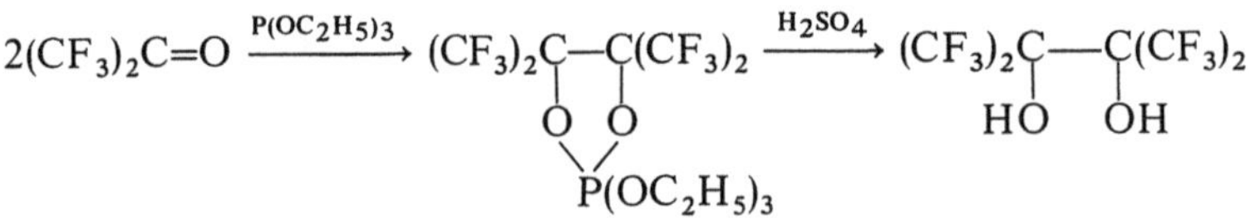

Triethylphosphite (17 g, 0.1 mol) (Note 1) is placed in a 100-ml two-necked flask equipped with a gas inlet tube and dry-ice reflux condenser (Note 2). Then hexafluoroacetone (32 g, 20 ml, at −78 °C) (0.2 mol) is bubbled in, with intermittent shaking and cooling of the flask with ice-water. The reaction mixture is allowed to stand for 2–3 h at room temperature, then distilled. The yield of dioxaphospholane is 40.2 g (96%); bp 61°–62 °C/0.5 mm Hg (Note 3).

Dioxaphospholane is added dropwise, with stirring, to 50 ml of concentrated H_2SO_4. The mixture is heated under downward condenser in an oil-bath; the temperature of the bath is gradually raised to 190 °C. As a result, 22–24 g of crude perfluoropinacon distils off (Note 4), which is redistilled over P_2O_5. The yield of perfluoropinacon is 18–20 g (65%–72%); bp 58°–60 °C/60 mm Hg.

2. Hydrodimerization of Hexafluoroacetone [61]

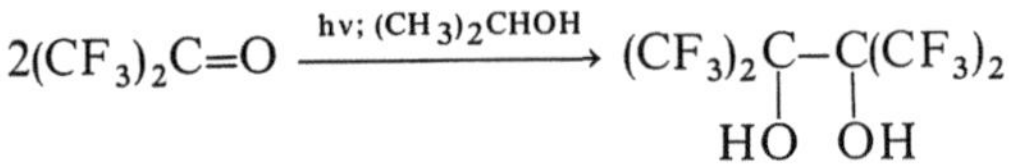

In a quartz flask is placed 50 ml of isopropanol, then 88 g (50 ml at −78 °C; 0.53 mol) of hexafluoroacetone is bubbled through it, with cooling. The mixture is then irradiated with a mercury lamp at room temperature over a period of 4 days. Distillation gives a fraction boiling at 130°–160 °C, which is mixed with an equal volume of concentrated H_2SO_4 and redistilled. The yield of perfluoropinacon is 55 g (63%); bp 129 °C.

Notes

1. Triethylphosphite should be previously kept over sodium and distilled.
2. Air moisture should be kept out of the reactor.
3. Distillation at atmospheric pressure leads to decomposition of the product.
4. Perfluoropinacon should be handled with caution, as it is rather toxic.

Difluoronitroethanol

$O_2NCF_2{-}CH_2OH$

$C_2H_3F_2NO_3$ MW 127

Difluoronitroethanol is a colourless liquid, soluble in organic solvents, partially soluble in water; toxic. Bp 64°–65 °C/25 mm Hg [72, 73], 65 °C/40 mm Hg [74],

47 °C/12 mm Hg [75], 64 °C/20 mm Hg [75], 59°–61 °C/22 mm Hg [76]; n_D^{20} 1.3780, d_4^{20} 1.4792 [72].

Method of Preparation

Difluoronitroethanol is prepared by condensation of difluoronitromethane with paraform in the presence of K_2CO_3 (yield up to 88%) [72–74]; by nitration of vinylidene fluoride with nitrogen tetroxide in the solution of reon 112 in an autoclave (12% yield) [75]; or by the reduction of difluoronitroacetate with sodium borohydride in an aqueous solution (70% yield) [76, 77].

Procedure [72]

$$O_2NCHF_2 + CH_2O \rightarrow O_2NCF_2CH_2OH$$

A) In a 200-ml three-necked flask fitted with a reflux condenser, a dropping funnel and a thermometer, are placed 39 g of 40% formalin, 1 g (0.007 mol) of K_2CO_3 and 48 g (0.5 mol) of difluoronitromethane. The reaction mixture is heated at 90 °C for 2 h and cooled, then extracted with ether. The ether extract is dried over $MgSO_4$ and distilled. The yield of difluoronitroethanol is 44.5 g (71.5%); bp 65°–66 °C/40 mm Hg(Note). B) In a glass ampoule is placed 13 g (0.134 mol) of difluoronitromethane, 4 g (0.134 mol) of paraform, 0.15 g (0.011 mol) of K_2CO_3 and two or three drops of water. The ampoule is sealed and heated at 100 °C during 1 h, whereupon it is cooled, opened, and the reaction mixture is distilled. The yield of difluoronitroethanol is 15 g (88%); bp 65 °C/40 mm Hg (Note).

Likewise, condensation of difluoronitromethane with the carbonyl compounds gives 1,1-difluoro-1-nitropropanol-2 (bp 60 °C/18 mm Hg; n_D^{20} 1.3825, d_4^{20} 1.3900; yield 65%), 1,1-difluoro-1-nitro-2-methylpropanol-2 (bp 55 °C/35 mm Hg; n_D^{20} 1.3915, d_4^{20} 1.2950; yield 44%) [72, 74], 1,1-difluoro-1-nitropentanol-2 (bp 49 °C/3 mm Hg; n_D^{20} 1.4005, d_4^{20} 1.238; yield 40%) [78].

Note. Difluoronitroethanol can cause temporary insensitiveness of the skin. During work with this compound hands should be protected with rubber gloves; face, with a screen of transparent plastic. The skin stained with difluoronitroethanol should be washed with running water, then with 10% Na_2CO_3 solution, and finally with water and soap. The glassware should be degassed with a warm alkaline solution.

2,2-Difluoro-2-nitroethyl-2'-hydroxyethyl Ether

$$O_2NCF_2{-}CH_2{-}O{-}CH_2{-}CH_2OH$$

$C_4H_7F_2NO_4$ MW 271

2,2-Difluoro-2-nitroethyl-2'-hydroxyethyl ether is a colourless liquid. Bp 72°–74 °C/1 mm Hg; n_D^{20} 1.4008, d_4^{20} 1.3798 [79].

Method of Preparation

2,2-Difluoro-2-nitroethyl-2'-hydroxyethyl ether is prepared by the reaction of difluoronitroethanol with ethylene oxide at 0 °C in the presence of an aqueous alkali (32% yield) [79].

Procedure

$$O_2NCF_2CH_2OH + H_2C\underset{O}{\diagdown\!\diagup}CH_2 \xrightarrow{NaOH} O_2NCF_2CH_2OCH_2CH_2OH$$

In a 250-ml three-necked flask provided with a stirrer, a dropping funnel, and a reflux condenser, is placed a solution of 6.5 g (0.162 mol) of NaOH in 100 ml of water; than a cooled mixture of 20 g (0.157 mol) of difluoronitroethanol and 8 g (0.151 mol) of ethylene oxide is added dropwise with stirring and cooling of the flask to 0°–5 °C. The reaction mixture is stirred at 0 °C during 16 h, then extracted with methylene chloride. The extract is dried over $MgSO_4$ and distilled under vacuum. The yield of 2,2-difluoro-2-nitroethyl-2'-hydroxyethyl ether is 8.7 g (32%); bp 72°–74 °C/1 mm Hg.

Bis-3,3,3-trifluoropropyl Ether

$$CF_3{-}CH_2{-}CH_2{-}O{-}CH_2{-}CH_2{-}CF_3$$

$C_6H_8F_6O$ MW 210

Bis-3,3,3-trifluoropropyl ether is a colourless liquid. Bp 110°–112 °C/750 mm Hg; n_D^{20} 1.3175; d_4^{20} 1.2969 [82].

Method of Preparation

Bis-3,3,3-trifluoropropyl ether is prepared by the reaction of vinylidene fluoride with formaldehyde in HF (45%–47% yield) [82, 83].

Procedure [82]

$$CF_2{=}CH_2 \xrightarrow[HF]{CH_2O} (CF_3CH_2CH_2)_2O$$

In a steel or polyethylene reactor fitted with a stirrer, a dry-ice reflux condenser, and a gas inlet tube, is placed 100 g of anhydrous hydrogen fluoride (Note 1). The reactor is cooled to —78 °C and 15 g (0.5 mol) of paraform is added at that temperature, whereupon vinylidene fluoride (12 l, 0.5 mol) is passed in it at 0°–5 °C, with vigorous stirring during 4 h. The major part of unchanged hydrogen fluoride is evaporated in air (Note 2) and the remaining part is cautiously decomposed with a mixture of ice and an aqueous solution of ammonia (Note 3). The resulting aqueous-organic mixture is distilled till the distillate does not separate into layers any longer. The organic layer is separated, washed with water, dried over $MgSO_4$ and distilled. The yield of bis-3,3,3-trifluoropropyl ether is 26.5 g (47%); bp 110°–112 °C/750 mm Hg; n_D^{20} 1.3175.

Notes

1. For the safety measures used in handling hydrogen fluoride see p. 1.
2. Evaporation of HF should preferably be conducted in a wide-necked polyethylene beaker.
3. At the end of decomposition of the reaction mixture the medium should be alkaline.

2-Chloroethyl-1′,1′-dihydroperfluorobutyl Ether

$$CF_3—CF_2—CF_2—CH_2—O—CH_2—CH_2Cl$$

$C_6H_6ClF_7O$ MW 266.5

2-Chloroethyl-1′,1′-dihydroperfluorobutyl ether is a colourless liquid, soluble in common organic solvents. Bp 63°–64 °C/52 mm Hg [84].

Method of Preparation

2-Chloroethyl-1′,1′-dihydroperluorobutyl ether is prepared by chlorination of 2-ethoxyethyl-1′,1′-dihydroperfluorobutyl ether with thionyl chloride in pyridine (76% yield) [84].

Procedure

$$C_3F_7CH_2OH + \underset{\text{(ethylene oxide)}}{H_2C{-}CH_2} \xrightarrow{KOH} C_3F_7CH_2OCH_2CH_2OH \xrightarrow{SOCl_2}$$

$$\rightarrow C_3F_7CH_2OCH_2CH_2Cl$$

2-Hydroxyethyl-1′,1′-dihydroperfluorobutyl ether. In a 250-ml steel autoclave are placed 150 g (0.75 mol) of 1,1-dihydroperfluorobutanol and 2.5 g (0.04 mol) of KOH. The autoclave is closed, cooled with liquid nitrogen and evacuated, then 33 g (0.75 mol) ethylene oxide is condensed into it (Note). The autoclave is heated at 70 °C for 4 h, then cooled to room temperature, opened, and the reaction mixture is distilled. The yield of 2-hydroxyethyl-1′,1′-dihydroperfluorobutyl ether is 128 g (70%); bp 91°–92 °C/54 mm Hg; n_D^{25} 1.3310.

2-Chloroethyl-1′,1′-dihydroperfluorobutyl ether. In a 100 ml three-necked flask provided with a stirrer, a reflux condenser, and a dropping funnel, are placed 15 g (0.06 mol) of 2-hydroxyethyl-1′,1′-dihydroperfluorobutyl ether, 7.3 g (0.09 mol) of pyridine and 8 ml of anhydrous ether. The flask is cooled to 0 °C and 11 g (0.09 mol) of $SOCl_2$ is gradually added with stirring. The reaction mixture is then heated during 1 h in a water-bath at 60 °C, then cooled to room temperature and washed with diluted HCl, with water, with 5% solution of $NaHCO_3$, and again with water. The organic layer is separated, dried over $MgSO_4$, and the ether is distilled off. The residue is distilled in vacuum to give 12.2 g (76%) of 2-chloroethyl-1′,1′-dihydroperfluorobutylether; bp 63°–64 °C/52 mm Hg.

Note. Ethylene oxide is rather toxic. Work with it should be carried out under on efficient hood.

Ethylperfluorovinyl Ether

$$CF_2{=}CF{-}O{-}C_2H_5$$

$C_4H_5F_3O$ MW 126

Ethylperfluorovinyl ether is a low-boiling colourless liquid with a pleasant faint odour. Bp 38.5 °C [85]; 36°–39 °C [86]; n_D^{20} 1.3150; d_4^{20} 1.322 [86].

Method of Preparation

Ethylperfluorovinyl ether is prepared by the reaction of tetrafluoroethylene with sodium ethoxide in dioxan, in an autoclave at 20 °C [85], or at 50°–60 °C [86].

Procedure [86]

$$CF_2{=}CF_2 \xrightarrow[\text{dioxan}]{C_2H_5ONa;} CF_2{=}CF{-}O{-}C_2H_5$$

In a 250-ml stainless steel autoclave are placed 80 ml of anhydrous dioxan. 14 g (0.2 mol) of sodium ethoxide (Note 1) and 2 g of hydroquinone. The autoclave is cooled with liquid nitrogen, evacuated and 8.5 l (0.3 mol) of tetrafluoroethylene (Notes 2, 3) is condensed into it via a gas-washing bottle with concentrated H_2SO_4. The autoclave is heated during 20 h in a tilting furnace at 50°–60 °C, then cooled to −78 °C and excess is let out into a gasometer. The product is poured out of the autoclave into a distillation flask and distilled on a 25 cm column; a fraction boiling at 35°–60 °C is collected. The yield of ethylperfluorovinyl ether after redistillation is 7 g (65%, 26% conversion); bp 36°–39 °C; n_D^{20} 1.3150.

Notes

1. Anhydrous dioxan and sodium ethoxide should be free from ethanol; otherwise the product of addition of ethanol across the double bond of tetrafluoroethylene is formed in considerable amount.
2. Tetrafluoroethylene is liable to polymerize and detonate under pressure. In order to prevent an explosion, a small amount of hydroquinone is loaded into the autoclave. After the autoclave, cooled with liquid nitrogen, has been charged with tetrafluoroethylene, it is evacuated once more to remove traces of oxygen.
3. The cooled autoclave is immediately transferred to an armoured box and heated (see 2).

2,2-Difluoro-2-nitroethylallyl Ether

$$O_2NCF_2{-}CH_2{-}O{-}CH_2CH{=}CH_2$$

$C_5H_7F_2NO_3$ MW 167

2,2-Difluoro-2-nitroethylallyl ether is a colourless liquid, soluble in organic solvents; bp 54°–55 °C/20 mm Hg; n_D^{20} 1.3889, d_4^{20} 1.2173 [87].

Method of Preparation

2,2-Difluoro-2-nitroethylallyl ether is prepared by the reaction of difluoronitroethanol with allyl bromide in an aqueous alkaline solution (75.5%) [87].

Procedure

$$O_2NCF_2CH_2OH + BrCH_2CH{=}CH_2 \xrightarrow{NaOH} O_2NCF_2CH_2OCH_2CH{=}CH_2$$

In a 100-ml three-necked flask provided with a stirrer, a thermometer, and a reflux condenser, are placed a solution of 2.05 g (0.051 mol) of NaOH in 40 ml of water, 6.4 g (0.051 mol) of difluoronitroethanol and 6.2 g (0.051 mol) of allyl bromide. The reaction mixture is stirred at 25 °C for 3 h, then extracted with methylene chloride, the extract is dried over $CaCl_2$ and distilled. The yield of 2,2-difluoro-2-nitroethylallyl ether is 6.5 g (76%); bp 54°–55 °C/20 mm Hg.

Likewise, difluoronitroethanol and propargyl bromide give 2,2-difluoro-2-nitroethylpropargyl ether (bp 84°–85 °C/70 mm Hg; yield 27%) [87].

Bis(2,2-difluoro-2-triethyl)Formal

$$O_2NCF_2{-}CH_2{-}O{-}CH_2{-}O{-}CH_2{-}CF_2NO_2$$

$C_5H_6F_4N_2O_6$ MW 266

Bis(2,2-difluoro-2-nitroethyl)formal is a colourless liquid, soluble in common organic solvents. Bp 67°–68 °C/0.6 mm Hg [76].

Method of Preparation

Bis(2,2-difluoro-2-nitroethyl)formal is prepared by condensation of difluoronitroethanol with formaldehyde in the presence of sulphuric acid (60% yield) [76].

Procedure

$$O_2NCF_2CH_2OH + CH_2O \xrightarrow{H_2SO_4} CH_2(OCH_2CF_2NO_2)_2$$

In a 50-ml three-necked flask equipped with an efficient stirrer, a reflux condenser, and a dropping funnel, is placed a solution of 1.59 g (0.05 mol) of paraform in 3.5 ml of

90% H_2SO_4, then 8.98 g (0.07 mol) of difluoronitroethanol is added with vigorous stirring during 2 min. The reaction mixture is stirred for 1 h, poured into ice-water, extracted with methylene chloride, the extract dried over $MgSO_4$ and distilled. The yield of bis(2,2-difluoro-2-nitroethyl)formal is 5.5 g (60%); bp 67°–68 °C at 0.6 mm Hg.

Perfluoropropylene Oxide

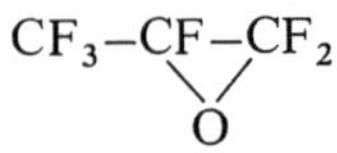

C_3F_6O MW 166

Perfluoropropylene oxide is a colourless gas. Bp —28 °C [88], —29 °C to —30 °C [89].

Method of Preparation

Perfluoropropylene oxide is prepared by the reaction of perfluoropropylene with hydrogen peroxide in an alcoholic solution of an alkali (30% yield) [88, 89]; by oxidation of perfluoropropylene with oxygen under UV irradiation, in a small yield [90], or with potassium permanganate in anhydrous HF (29% yield) [60]; or by passing a mixture of perfluoropropylene and fluorine oxide through a heated copper tube with a copper gauze covered with AgF_2 [91].

Procedure [60]

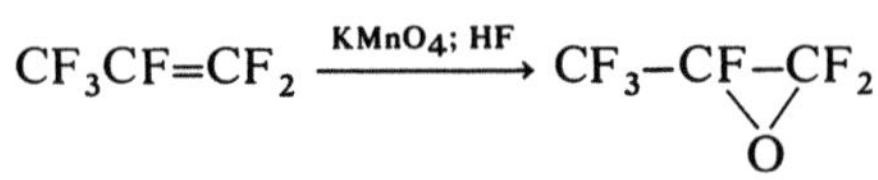

In a copper flask fitted with a metallic stirrer, a gas inlet tube, and a dry-ice reflux condenser, is placed 150 g (7.5 mol) of anhydrous HF (Note). The flask is cooled to —78 °C, then 45 g (0.3 mol) of $KMnO_4$ is added at that temperature. Perfluoropropylene (37 g, 0.25 mol) is condensed into the mixture with stirring and cooling it to —70 °C, whereupon the temperature is slowly raised to that of the room. The liberated gas is passed through a gas-washing bottle with H_2SO_4, condensed in a trap cooled to —78 °C and distilled. The yield of perfluoropropylene oxide is 12 g (29%); bp —29 °C to —30 °C.

Note. For the safety measures to be taken during work with anhydrous hydrogen fluoride see p. 1.

Perfluoroisobutylene Oxide

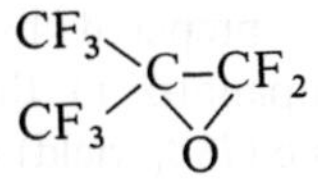

C_4F_8O MW 216

Perfluoroisobutylene oxide is a colourless gas. Bp 6°–7 °C [89].

Method of Preparation

Perfluoroisobutylene oxide is prepared by oxidation of perfluoroisobutylene with hydrogen peroxide in an aqueous alkaline solution (32% yield) [89], or in aqueous acetone in a phosphate buffer (45%–60% yield) [92, 93].

Procedure [89]

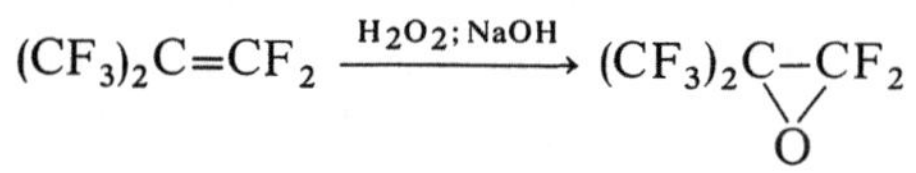

Caution! Perfluoroisobutylene and its oxide are highly toxic gases. Work with these compounds should be carried out under an efficient hood (see p. 2).

In a 250-ml four funnel flask provide with a stirrer, a thermometer, a dropping funnel, and a dry-ice reflux condenser, are placed 100 ml of water and 28 ml of 30% H_2O_2. The flask is cooled to —15 °C, then 30 g (0.15 mol) of perfluoroisobutylene is added with stirring, whereupon a solution of 14 g of NaOH in 20 ml of water is slowly added dropwise. The reaction mixture is stirred at —5 °C for 4 h. Then the temperature is gradually raised to that of the room. The liberated gas is condensed in a trap cooled to —78 °C. The product (14 g) contains 80% of perfluoroisobutylene oxide and 15% of unchanged perfluoroisobutylene; it is passed through a gas-washing bottle with ethanol to remove unchanged perfluoroisobutylene. The gas which has passed through the gas-washing bottle, is condensed and distilled. The yield of perfluoroisobutylene oxide is 10 g (30%, based on the starting amount of perfluoroisobutylene); bp 6°–7 °C.

1,3,3-Trifluoro-1,3-dichloro-1,2-epoxy-2-methylpropane

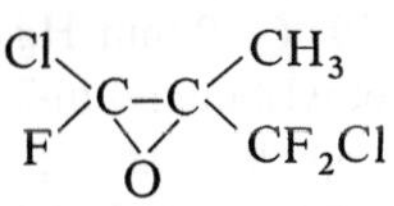

$C_4H_3Cl_2F_3O$ MW 195

1,3,3-Trifluoro-1,3-dichloro-1,2-epoxy-2-methylpropane is a colourless liquid with a faint odour, soluble in common organic solvents. Bp 89°–90 °C [94].

Method of Preparation

1,3,3-Trifluoro-1,3-dichloro-1,2-epoxy-2-methylpropane is prepared by dehydrochlorination of 1,1,3-trifluoro-1,3,3-trichloro-2-methylpropanol-2 in the presence of an aqueous alkali with heating for a short period of time (78% yield) [94].

Procedure

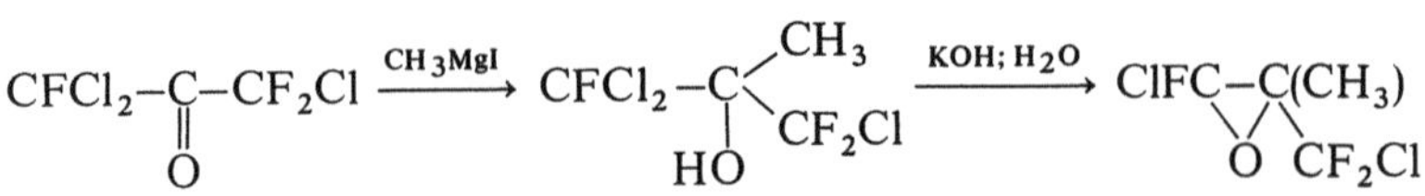

1,1,3-Trifluoro-1,3,3-trichloro-2-methylpropanol-2 [95]. In a 500-ml four-necked flask provided with a thermometer, a stirrer, a reflux condenser, and a dropping funnel, is placed a solution of methylmagnesium iodide obtained from 30 g (0.21 mol) of methyl iodide and 7 g (0.29 mol) of magnesium in 250 ml of ether. The flask is cooled to 5°–10 °C, then 30 g (0.14 mol) of 1,1,3-trifluoro-1,3,3-trichloroacetone is dropped in. The reaction mixture is stirred at room temperature during 1 h, then decomposed with 10% HCl. The ether layer is separated and the aqueous one extracted with ether. The ethereal solutions are combined, dried over $MgSO_4$, the ether is distilled off and the residue is vacuum-distilled. The yield of 1,1,3-trifluoro-1,3,3-trichloro-2-methylpropanol-2 is 26 g (79%); bp 70 °C/40 mm Hg.

A similar treatment of 1,1,3-trifluoro-1,3,3-trichloroacetone with phenylmagnesium bromide gives 1,1,3-trifluoro-1,3,3-trichloro-2-phenylpropanol-2 (bp 86°–88 °C/3 mm Hg; 68% yield) [94], with trifluorovinyllithium: tetrafluoro-1,1-dichloro-2-(difluorochloromethyl)buten-3-ol-2 (bp 85 °C/75 mm Hg, 67% yield) [96], with vinylmagnesium bromide: 1,1-difluoro-1-chloro-2-(fluorodichloromethyl)buten-3-ol-2 (bp 75 °C/40 mm Hg, 54% yield) [96], with pentafluorophenyllithium: 1,1,3-trichloroperfluoro(2-phenyl-propanol-2) (bp 112 °C/12 mm Hg, 78% yield) [97].

1,3,3-Trifluoro-1,3-dichloro-1,2-epoxy-2-methylpropane. A solution of 5 g (0.088 mol) of KOH in 30 ml of water is slowly added, with stirring, to 10.7 g (0.055 mol) of 1,1,3-trifluoro-1,3,3-trichloro-2-methylpropanol-2. The reaction mixture is heated at 70°–80 °C during 3 min; the mixture separates into layers. The lower layer is separated, dried over $MgSO_4$ and distilled. The yield of 1,3,3-trifluoro-1,3-dichloro-1,2-epoxy-2-methylpropane is 7 g (78%); bp 89°–90 °C.

Likewise, from 1,1,3-trifluoro-1,3,3-trichloro-2-phenylpropanol-2, trifluoro-1,3-dichloro-1,2-epoxy-2-phenylpropane is obtained (bp 79 °C/12 mm Hg, 66% yield) [94]; from tetrafluoro-1,1-dichloro-2-(difluorochloromethyl)buten-3-ol-2: tetrafluoro-1-chloro1,2-epoxy-2-(difluorochloromethyl)butene-3 (bp 90°–92 °C; 50% yield) [96]; from 1,1-difluoro-1-chloro-2-(fluorodichloromethyl)-buten-3-ol-2: 1-fluoro-1-chloro-1,2-epoxy-2-(difluorochloromethyl)butene-3 (a mixture of *cis-* and *trans-* isomers; bp 106°–108 °C, 70% yield) [96]; from 1,1,3-trichloroperfluoro(2-phenyl-propanol-2): 1,3-dichloro-1,2-epoxyperfluoro(2-phenylpropane) (bp 86 °C/20 mm Hg, 77% yield) [97].

Trifluoroacetaldehyde

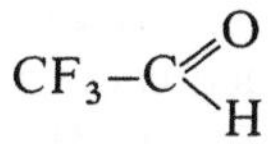

C_2HF_3O MW 98

Trifluoroacetaldehyde is a gas. Bp —19.6° to —18 °C/746 mm Hg [98].

Method of Preparation

Trifluoroacetaldehyde is prepared by oxidative gas phase nitration of 1,1,1-trifluoropropane (20%–24% yield) [99], by the reduction of trifluoroacetonitrile (46% yield) [98] or of trifluoroacetic acid with $LiAlH_4$ (yield depends on the order of addition of the reagents; best results are obtained when $LiAlH_4$ is added to the acid) [100].

Procedure [100]

$$CF_3COOH \xrightarrow{LiAlH_4} CF_3CHO$$

In a three-necked flask provided with a stirrer, a dropping funnel, and a reflux condenser, is placed a solution of 350 g (3 mol) of trifluoroacetic acid in 2 l of anhydrous ether. The flask is flushed with dry nitrogen, cooled to —12 °C, and a suspension of 67 g of $LiAlH_4$ in 800 ml of anhydrous ether is added at that temperature in portions, with vigorous stirring over a period of 1.5 h. The reaction mixture is stirred at —10° to —5 °C for 2.5 h, whereupon 530 ml of water is dropped into the flask with stirring and cooling, and a solution of 244 ml of concentrated H_2SO_4 in 1 l of water. The ether layer is separated and the aqueous one extracted with ether. The combined extracts are dried over $MgSO_4$, the ether is partially distilled off and the residue is distilled on a column; a fraction boiling at 99°–106 °C is collected (Note). The product is heated, with stirring, with a equal volume of concentrated H_2SO_4, in an oil-bath at 120 °C. The liberated gas is condensed in a trap cooled to —78 °C and distilled. The yield of trifluoroacetaldehyde is 156 g (50%).

Note. The fraction boiling at 80° to 90 °C contains trifluoroethanol.

1,1,1-Trifluoroacetone

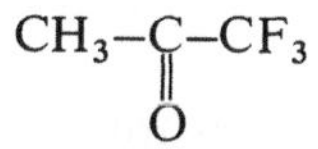

$C_3H_3F_3O$ MW 112

1,1,1-Trifluoroacetone is a highly volatile colourless liquid with a specific odour, soluble in water and common organic solvents. Bp 21 °C [101], 21.9 °C [102], 21.5° to 22.5 °C [103]; d_4^0 1.282 [102].

Method of Preparation

1,1,1-Trifluoroacetone is prepared by acid hydrolysis of 3,3,3-trifluoroacetoacetic ester (75%–90% yield) [101, 102, 104] (see p. 69); or by the reaction of trifluoroacetic acid with methylmagnesium iodide in dibutyl ether (56%–65% yield) [103].

Procedure [103]

$$CF_2COOH + CH_3MgI \xrightarrow{(C_4H_9)_2O} CF_3COCH_3$$

In a four-necked flask equipped with a stirrer, a thermometer, a reflux condenser, and a dropping funnel, is prepared 49.8 g (0.3 mol) of methylmagnesium iodide (from 7.2 g of magnesium and 42.6 g of methyl iodide) in 75 ml of anhydrous dibutyl ether. The resulting solution is cooled to 10 °C in an ice-bath, then 11.4 g (0.1 mol) of trifluoroacetic acid in 25 ml of dibutyl ether is added dropwise with vigorous stirring (Note 1). The reaction mixture is stirred at 10°–15 °C during 1.5 h (Note 2). The water condenser is then replaced with a dry-ice reflux condenser. The reaction mixture is cooled to 0 °C and, first 25 ml of water, then 75 ml of 7 N HCl are added to it dropwise, with stirring. The aqueous layer is separated and extracted three times with dibutyl ether. The combined ethereal solutions are dried over $MgSO_4$ at −10° to 0 °C and distilled on a fractionating column with a glass packing. The yield of 1,1,1-trifluoroacetone is 6.3 g (56%); bp 21.5°–22.5 °C.

Notes

1. Before use, trifluoroacetic acid is distilled over concentrated sulphuric acid.
2. The yield of 1,1,1-trifluoroacetone may be increased by stirring the reaction mixture at 10°–15 °C during 19 h till it is hydrolysed.

Pentafluoroacetone

$$\begin{array}{c} CF_3-C-CF_2H \\ \| \\ O \end{array}$$

C_3HF_5O MW 148

Pentafluoroacetone is a colourless gas. Bp 14°–15 °C [105].

Method of Preparation

Pentafluoroacetone is prepared by the pyrolysis of sodium bis(trifluoromethyl)glycolate (90% yield) [105]; by the reaction of chloropentafluoroacetone with triethylphosphite (the Perkow reaction) with its subsequent hydrolysis with sulphuric acid (85% yield) [106]; by the ketone decomposition of pentafluoroacetoacetic ester (35% yield) [107]; by the Grignard reaction, from trifluoromethylmagnesium iodide and difluoroacetyl chloride (59% yield) [108].

Procedure [105]

$$(CF_3)_2C{=}CFOC_2H_5 \xrightarrow{KMnO_4} (CF_3)_2C(OH){-}COOC_2H_5 \xrightarrow{NaOH} (CF_3)_2C(OH){-}COONa \rightarrow$$

$$\rightarrow CF_3C({=}O){-}CF_2H$$

Ethyl bis(trifluoromethyl)glycolate. In a 500 ml three-necked flask equipped with a stirrer and a reflux condenser, are placed a solution of 53.8 g (0.238 mol) of ethylfluoroisobutenyl ether (Note 1) and 5 ml of water in 110 ml of acetone, then 26.2 g (0.165 mol) of $KMnO_4$ is added gradually, in small portions, with vigorous stirring at room temperature. The mixture is stirred for 3 h, then SO_2 is bubbled to decolorization of the mixture, whereupon 150 ml of water is added. The lower layer is separated and washed with water. The yield of crude ethyl bis(trifluoromethyl)glycolate, containing about 10% of acetone, is 55.9 g; the yield of the pure product is about 50 g (87%). The crude product is used (without purification) for the preparation of sodium bis(trifluoromethyl)glycolate.

Sodium bis(trifluoromethyl)glycolate. In a 500-ml four-necked flask provided with a stirrer, a reflux condenser, a thermometer, and a dropping funnel, are placed 100 ml of water and 32,5 g of crude ethyl bis(trifluoromethyl)glycolate, containing 29.2 g (0.122 mol) of pure ester. The mixture is heated to 70 °C, then 100 ml of 5% NaOH solution is added, with vigorous stirring during 4–6 h, to pH 9; the mixture is stirred at 70 °C for an additional 1 h (Note 2), whereupon CO_2 is passed to pH 8. The resulting solution is evaporated in a porcelain dish in a boiling water-bath, the residue is ground in a mortar and dried over P_2O_5 in a vacuum-desiccator till it gains constant weight. The yield of crude sodium bis(trifluoromethyl)glycolate is 25.2 g (88%) (Note 3).

Pentafluoroacetone. The distillation apparatus consists of a 30 ml Wurtz flask, a downward condenser, an adapter, a receiver, and a trap cooled to —78 °C. The flask is charged with a thoroughly mixed mixture of 11 g (0.047 mol) of crude sodium bis(trifluoromethyl)glycolate and 10 g of pure calcinated sand. The mixture is heated to 300 °C on a metal bath; pure pentafluoroacetone is collected in a trap (Note 4). The yield of pentafluoroacetone is 6.2 g (90%); bp 14°–15 °C.

Notes

1. Ethylperfluoroisobutenyl ether is prepared by dehydrofluorination of ethyloctafluorobutyl ether [109].
2. If, after stirring is complete, pH ~9, it is advisable to add a small amount of an alkaline solution to pH 9.
3. Sodium bis(trifluoromethyl)glycolate may be obtained also by neutralization of a free acid. Bis(trifluoromethyl)glycolic acid [110, 153] is neutralized with an equal amount of 10% NaOH solution till it is neutral to phenolphthalein; the resulting solution is evaporated, the residue is ground and dried. The yield of the salt is 93%.
4. A small amount of a mixture of unidentified products is condensed in the trap.

Pentafluorobromoacetone

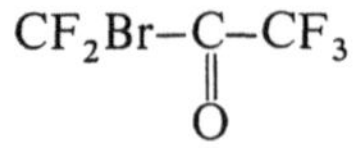

C_3BrF_5O MW 227

Pentafluorobromoacetone is a highly volatile colourless liquid with a strong unpleasant odour, soluble in common organic solvents; a lachrymator; dissolves in water, forming a hydrate. Bp 31 °C [111].

Method of Preparation

Pentafluorobromoacetone is prepared by the reaction of bromine with pentafluoropropenol-2 with subsequent β-dehydrobromination of pentafluoro-1,2-dibromopropanol-2 by heating it in N-methylpyrrolidone (53% yield) [111].

Procedure

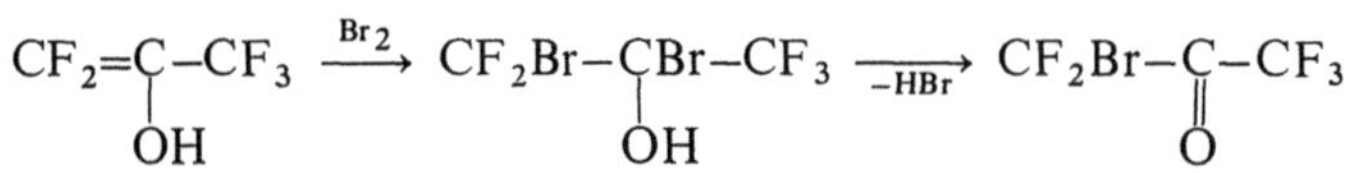

In a 50-ml three-necked flask provided with a stirrer, a dropping funnel, and a reflux condenser, is placed 2.8 g (0.0175 mol) of bromine, then 2.6 g (0.0175 mol) of perfluoropropenol-2 is added with stirring and cooling of the flask in an ice-water bath. Pentafluoro-1,2-dibromopropanol-2 formed in the reaction (as shown by [^{19}F] NMR and PMR) is shaken with metallic mercury to remove traces of bromine, then filtered and mixed with 3 ml of N-methylpyrrolidone. The mixture is heated in a water-bath, the distillate is collected in an ice-water-cooled receiver. Redistillation gives 2.1 g (52.6%) of pentafluorobromoacetone; bp 31 °C.

Likewise, pentafluoropropenol-2 and iodine chloride give pentafluoroiodoacetone (bp 54°–55 °C, yield 46%) [111].

Hexafluoroacetone

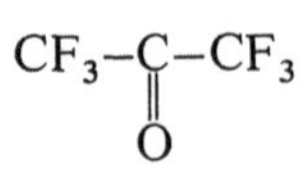

C_3F_6O MW 166

Hexafluoroacetone is a colourless toxic gas with a specific strong odour. Bp —27.4 °C; mp —122 °C [112]; for the toxicity of hexafluoroacetone see [113].

Method of Preparation

Hexafluoroacetone is prepared by the reaction of hexachloroacetone with hydrogen fluoride in the presence of CrF_3 [114] or salts of pentavalent chromium [115]; by oxidation of perfluoroisobutylene with potassium permanganate [21, 24]; by isomerization of perfluoropropylene oxide in the presence of SbF_5 [116]; or by oxidation of hexafluorothioacetone dimer with potassium iodate (89% yield) [117].

Procedure

$$(CF_3)_2C{=}CF_2 \xrightarrow[H_2O]{KMnO_4;} (CF_3)_2C(OH)_2 \xrightarrow[-H_2O]{H_2SO_4} CF_3COCF_3$$

Hexafluoroacetone hydrate. In a 200-ml three-necked flask provided with a stirrer, dry-ice reflux condenser, and a gas inlet tube, which extends to the bottom of the flask, are placed 300 g (2 mol) of $KMnO_4$ and 1500 ml of water. The reflux condenser is connected to a trap cooled to −78 °C, then 300 g of crude perfluoroisobutylene, containing 60%–80% of perfluoroisobutylene, is passed during 10–12 h from a cylinder into the flask with vigorous stirring and heating of the mixture to 70°–80 °C (Note 1). The unchanged gas, condensed in the trap, is reoxidized. The reaction mixture is cooled with ice-water, then SO_2 is passed, with stirring, to clarification of the solution. The reaction mixture is then extracted with ether (5 × 200–250 ml), the ether extract is dried over $MgSO_4$ and the ether distilled off on a Widmer column (Note 2). The yield of crude hexafluoroacetone hydrate is 200–250 g.

Anhydrous hexafluoroacetone. Anhydrous hexafluoroacetone is prepared in a 1 l three-necked flask provided with a stirrer, a dropping funnel, and a reflux condenser (Note 3), which is connected to a gas-washing bottle with H_2SO_4, a column packed with P_2O_5, a trap for unchanged hexafluoroacetone hydrate cooled to −20 °C, and a trap for hexafluoroacetone cooled to −78 °C. In the flask is placed 300 g of concentrated H_2SO_4, then 250 ml of crude hexafluoroacetone hydrate is dropped in, with vigorous stirring, at such a rate that all the liberated gas should condense in the trap. The flask is then slowly heated to 135°–140 °C, during 2–3 h. The yield of anhydrous hexafluoroacetone is 130–140 g; bp −28 °C.

Notes

1. *Caution*! Perfluoroisobutylene is extremely toxic; it should be handled only under an efficient hood (see p. 2).
2. At the end of this procedure the still residue is heated on a boiling water-bath.
3. All parts of the apparatus for the preparation of hexafluoroacetone should be thoroughly dried.

Pentafluoronitroacetone

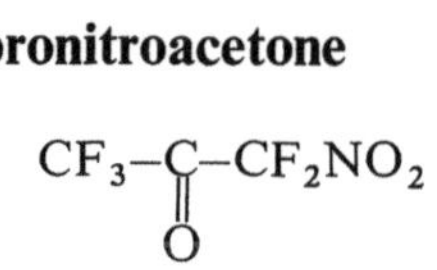

$C_3F_5NO_3$ MW 193

Pentafluoronitroacetone is a volatile colourless liquid; very hydroscopic, fumes in air; soluble in organic solvents; violently reacts with water to form a stable hydrate; toxic. Bp 32°–33 °C; n_D^{20} 1.2955, d_4^{20} 1.5350 [118].

Method of Preparation

Pentafluoronitroacetone is prepared by treatment of pentafluoronitroacetone hydrate with sulphuric acid (70% yield) [118].

Procedure

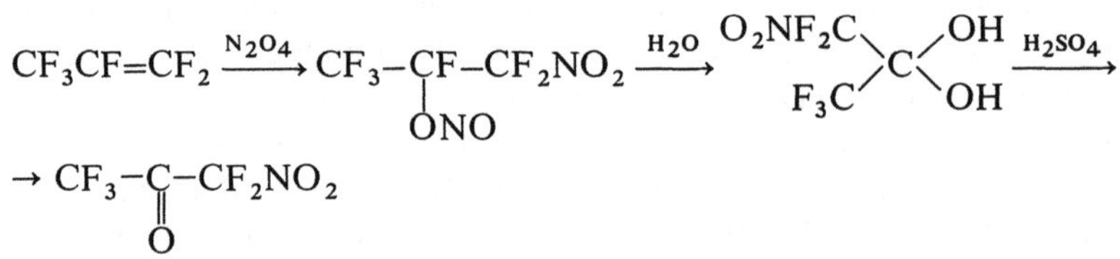

Pentafluoronitroacetone hydrate. In a 300-ml steel autoclave provided with a manometer, a valve, and evacuated to 3–5 mm Hg residual pressure, is placed 40 ml (64 g, 0.8 mol) of N_2O_4. The autoclave is cooled to —78 °C, then 18 l (0.7 mol) of perfluoropropylene is condensed into it. After the autoclave has warmed to room temperature, the pressure in it has increased to 6 atm; after 5–7 days the pressure in the autoclave has decreased to 1–1.5 atm. The autoclave is cooled to —78 °C, the pressure is released, cooling is stopped, and the reaction mixture is condensed under vacuum into a 100 ml trap cooled to —78 °C.

The product condensed in the trap is placed in a 500-ml three-necked flask fitted with a stirrer, a reflux condenser, and a dropping funnel (Note 1), and 100 ml of water is added dropwise with stirring and cooling with ice-water. The reaction mixture is then heated to 30 °C till nitrogen oxides are completely removed, the lower blue-coloured layer is separated (Note 2) and the aqueous one is extracted with ether. The organic layer and the extract are combined, dried over $MgSO_4$, the ether is removed and the residue distilled. The yield of pentafluoronitroacetone hydrate is 90 g (60%); bp 119 °C (Note 3).

Pentafluoronitroacetone. In a Claisen flask provided with a dropping funnel and a downward condenser connected to a receiver and an ice-cooled trap, is placed 70 g of concentrated H_2SO_4, then 18 g of pentafluoronitroacetone hydrate is dropped in with heating of the mixture to 80°–90 °C. The product condensed in the trap is combined with that from the receiver and redistilled. The yield of pentafluoronitroacetone is 11–12 g (up to 70%); bp 32°–33 °C.

Notes

1. The reaction products include volatile toxic compounds, therefore the reaction flask should be placed under an efficient hood.
2. The lower layer contains the by-product hexafluoro-1,2-dinitropropane; it may be isolated by distillation after the product was dried over $MgSO_4$.
3. Pentafluoronitroacetone and its hydrate are toxic. They must be handled with caution.

Bis(2-hydroperfluoroisopropyl)Ketone

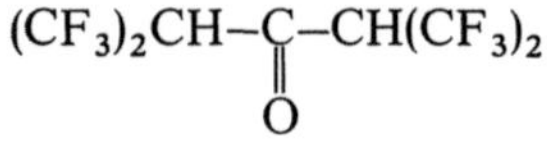

$C_7H_2F_{12}O$ MW 320

Bis(2-hydroperfluoroisopropyl)ketone is a highly volatile colourless crystalline substance with a specific strong odour, soluble in common organic solvents. Bp 102 °C, mp 71.5°–72 °C [119].

Method of Preparation

Bis(2-hydroperfluoroisopropyl)ketone is prepared by the reaction of hydration of tetrakis(trifluoromethyl)allene (82% yield) [119], or by the hydrolysis of bis(trifluoromethyl)ketene dimer (93%) [119].

Procedure

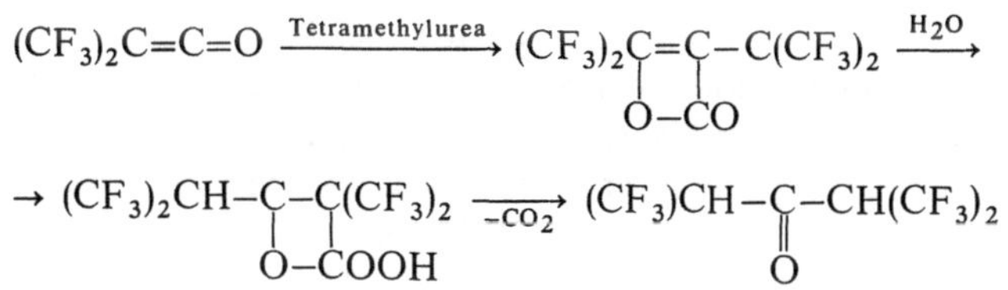

Bis(trifluoromethyl)ketene dimer [119]. A mixture of 42 g (0.135 mol) of bis(trifluoromethyl)ketene (Note 1) and 7 g (0.045 mol) of tetramethylurea, placed in a sealed glass ampoule, is allowed to stand at room temperature for one day (Note 2). The ampoule is then cooled to 0 °C, opened and the reaction mixture is distilled under reduced pressure to give 39.2 g (93%) of bis(trifluoromethyl)ketene dimer; bp 48 °C/149 mm Hg.

Bis(2-hydroperfluoroisopropyl)ketone. Bis(trifluoromethyl)ketene dimer (3.1 g, 0.009 mol) and an equal amount of water are placed in a 10-ml flask connected with a gas-washing bottle with H_2SO_4 and a trap cooled to −78 °C. The mixture is allowed to stand at room temperature for 2 days (till CO_2 has ceased to evolve), whereupon water is decanted off from the crystalline precipitate, which is pressed on a filter paper (Note 3) and distilled over concentrated H_2SO_4 (1 ml). Redistillation gives 2.6 g (93%) of bis(2-hydroperfluoroisopropyl)ketone; bp 102 °C, mp 71.5°–72 °C (in a sealed capillary).

Notes

1. For the preparation of bis(trifluoromethyl)ketene see p. 72.
2. In about 2 h the reaction mixture separates into layers.
3. During filtration, losses of the product are possible due to its volatility..

1,1,1,2-Tetrafluorohexanedione-3,5

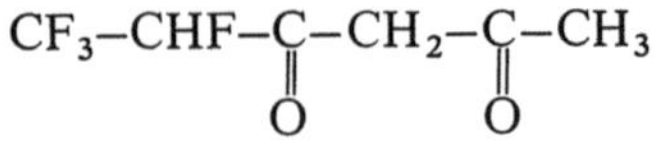

$C_6H_6F_4O_2$ MW 186

1,1,1,2-Tetrafluorohexanedione-3,5 is a colourless liquid with a specific faint odour, soluble in common organic solvents. Bp 136–138 °C [120].

Method of Preparation

1,1,1,2-Tetrafluorohexanedione-3,5 is prepared by the reaction of acetyl fluoride with trifluoroethylene in the presence of SbF_5 (40% yield) [120].

Procedure

$$CF_2{=}CHF + CH_3COF \xrightarrow{SbF_5} CF_3{-}CHF{-}CO{-}CH_2{-}CO{-}CH_3$$

In a 250-ml steel autoclave is placed 44 g (0.2 mol) of SbF_5 (Note). The autoclave is closed, cooled with liquid nitrogen and evacuated, then 70 g (1.1 mol) of acetyl fluoride and 8.5 l (0.3 mol) of trifluoroethylene are condensed into it at −196 °C under vacuum. The autoclave is shaken at room temperature during 5 h, the reaction mixture is poured onto ice, the organic layer is separated, washed with a saturated solution of $NaHCO_3$, then with water, dried over $MgSO_4$ and distilled to give 22 g (40%) of 1,1,1,2-tetrafluorohexanedione-3,5; bp 136–138 °C.

A similar reaction of benzoyl fluoride with vinylidene fluoride in the presence of SbF_5 (liquid SO_2; −20 °C) affords 1,1,1-trifluoropropiophenone (bp 105°–106 °C/4 mm Hg; mp 38 °C (from hexane); yield 33%) [120]; the reaction of benzoyl fluoride with trifluoroethylene gives 1,1,1,2-tetrafluoropropiophenone (bp 79°–81 °C/10 mm Hg, mp 41 °C (from hexane); yield 39%) [120].

Note. Antimony pentafluoride should be handled with utmost care, since in contact with the skin it causes painful, slow-healing wounds. For treatment of the skin burnt with SbF_5 see p. 133.

Trifluoromethyl-β-diethylaminovinylketone

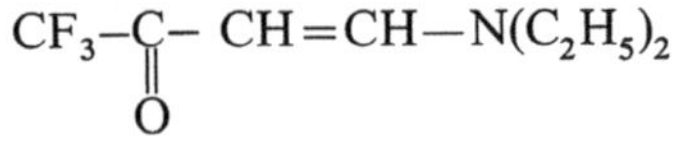

$C_8H_{12}F_3NO$ MW 195

Trifluoromethyl-β-diethylaminovinylketone is a colourless liquid with a faint odour, soluble in common organic solvents. Bp 97°–98 °C/2 mm Hg, mp 12°–13 °C, n_D^{23} 1.4867 [121].

Method of Preparation

Trifluoromethyl-β-diethylaminovinylketone is prepared by the reaction of triethylamine with trifluoroacetyl chloride (35% yield) [121].

Procedure

$$CF_3COCl + CH_3CH_2N(C_2H_5)_2 \rightarrow CF_3COCH{=}CHN(C_2H_5)_2$$

A 200-ml steel autoclave is charged with 48 g (0.47 mol) of anhydrous triethylamine and 50 ml of anhydrous benzene, and cooled to —78 °C (air moisture should be kept out); then 35 g (0.24 mol) of trifluoroacetyl chloride is quickly poured in. The autoclave is closed and allowed to stand at 0 °C for 12 h. The cooled reaction mixture is poured into water, the organic layer is separated, acidified with 10% HCl, washed with water and dried over $MgSO_4$. The benzene is distilled off and the residue distilled on a fractionating column with a glass packing. The yield of trifluoromethyl-β-diethylaminovinylketone is 8.2 g (35%); bp 97°–98 °C at 2 mm Hg.

Hexafluorothioacetone

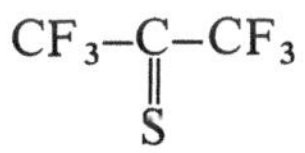

C_3F_6S MW 182

Hexafluorothioacetone is a gas; deep blue as a liquid. Bp 8 °C [122].

Method of Preparation

Hexafluorothioacetone is prepared by the reaction of bis(perfluoroisopropyl)mercury with boiling sulphur (62% yield) [122]; or by the pyrolysis of hexafluorothioacetone dimer at 600 °C in a platinum tube (92% yield) [122]. Hexafluorothioacetone dimer [2,2,4,4-tetrakis(trifluoromethyl)-1,3-dithiethane] is prepared by the reaction of bis(perfluoroisopropyl) disulphide with triphenylphosphine [122]; or by the reaction of perfluoroisopropylene with sulphur, in the presence of KF, in sulpholane or nitrobenzene in an autoclave (70% yield) [123], or in dimethylformamide under atmospheric pressure [117, 124].

Procedure [122, 123]

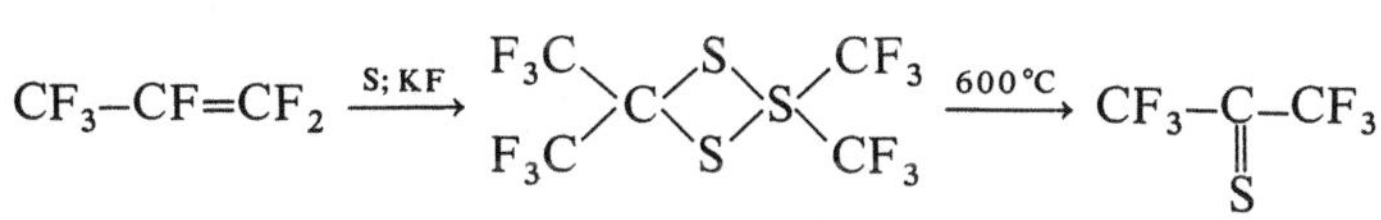

Hexafluorothioacetone dimer [123]. In a 100-ml steel autoclave are placed 25 ml of anhydrous sulpholane, 1.72 g (0.053 mol) of finely divided sulphur, 0.2 g (0.0034 mol) of calcinated KF. The autoclave is closed, cooled to —196 °C and evacuated; then

8 g (0.053 mol) of perfluoropropylene is condensed into it. The autoclave is heated over a period of 10 h in a tilting furnace at 130 °C, then cooled to room temperature, the contents are transferred into a distillation flask and distilled. The yield of hexafluorothioacetone dimer is 6.8 g (70%); bp 110°–111 °C; n_D^{20} 1.3342.

Hexafluorothioacetone [122]. Hexafluorothioacetone dimer (88 g, 0.24 mol) is placed in a dropping funnel connected with a platinum tube (510 mm long, 12.5 mm i.d., heated zone 300 mm), placed at 30° to the horizon. The tube is heated to 600 °C and hexaflurothioacetone dimer is introduced at that temperature dropwise, during 5 h; simultaneously, dry nitrogen is let in at the rate of 100 ml/min. The pyrolysis products are collected in a trap kept cooled to —78 °C, then quickly distilled under reduced pressure (Note). The yield of hexafluorothioacetone is 81 g (90%); bp —20 °C/200 mm Hg.

Note. Distillation of hexafluorothioacetone at the atmospheric pressure, as well as storing the monomer at the temperature close to boiling point, result in spontaneous dimerization. Hexafluorothioacetone should be stored at —78 °C.

Hexafluoroacetone Imine

$$(F_3C)_2C{=}NH$$

C_3HF_6N MW 165

Hexafluoroacetone imine is a colourless gas with a strong unpleasant odour, soluble in common organic solvents, insoluble in water. Bp 16°–16.5 °C [125], 16.5°–18 °C [126]; mp —47 °C; d_4^0 1.47 [125].

Method of preparation

Hexafluoroacetone imine is prepared by treatment of hexafluoroacetone with ammonia, then with $POCl_3$ in pyridine (70% yield) [125]; by decomposition of 2-amino-2-phenylaminohexafluoropropane by heating (86% yield) [126] or by the reaction with acetic oxide (72% yield) [127]; or by the reaction of hexafluoroacetone with triphenylphosphine imide [126].

Procedure

1. Dehydration of 2-Hydroxy-2-aminohexafluoropropane [125]

$$(CF_3)_2C{=}O \xrightarrow{NH_3} (CF_3)_2C(OH){-}NH_2 \xrightarrow{POCl_3;\,C_5H_5N} (CF_3)_2C{=}NH$$

In a 1-l four-necked round-bottom flask provided with a stirrer, a thermometer, a gas inlet tube, and a dry-ice reflux condenser, is placed 500 ml of anhydrous pyridine. The flask is cooled to —25 °C, whereupon first hexafluoroacetone (200 g, 1.2 mol), then dry ammonia (20.5 g, 25 ml at —78 °C, 1.2 mol) are gradually added with stirring and cooling of the mixture to —25 °C to —30 °C. Then the reaction mixture is heated to 25 °C, whereupon the gas inlet tube and the reflux condenser are replaced with a dropping funnel and a short reflux condenser cooled with running water; the condenser is connected with a trap cooled to —78 °C. Phosphorus oxychloride (184 g, 1.2 mol) is then added dropwise, with stirring, at such a rate as to maintain gentle refluxing of the mixture (~30 min). The reaction mixture is heated to 100 °C and kept at that temperature for 30 min. The liquid condensed in the trap is distilled on a fractionating column with a low-temperature distilling head. The yield of hexafluoroacetone imine is 138 g (70%); bp 16°–16.5 °C.

2. Thermolysis of 2-Amino-2-phenylaminohexafluoropropane [126]

$$(CF_3)_2C(NH_2){-}NHC_6H_5 \rightarrow (CF_3)_2C{=}NH + C_6H_5NH_2$$

In a 100-ml round-bottom flask provided with a reflux condenser, is placed 44 g of 2-amino-2-phenylaminohexafluoropropane. The condenser is connected to a trap, which has been cooled to —40° ÷ —50 °C, and the contents of the flask is refluxed gently for 3 h. To ensure complete removal of the product from the system, cooling of the reflux condenser is stopped at the end of the reaction. The crude product collected in the trap is distilled. The yield of hexafluoroacetone imine is 24 g (86%); bp 16.5°–18 °C.

Hexafluoroacetone Anil

$$(CF_3)_2C{=}NC_6H_5$$

$C_9H_5F_6N$ MW 241

Hexafluoroacetone anil is a yellow-green liquid with a specific odour, soluble in common organic solvents. Bp 135°–137 °C, 82 °C/120 mm Hg, 70°–73 °C/73 mm Hg, 48°–49 °C/21 mm Hg; n_D^{20} 1.4150; d_4^{20} 1.3782 [128].

Method of Preparation

Hexafluoroacetone anil is prepared by heating hexafluoroacetone with phenylisocyanate in the presence of triphenylphosphine oxide (90%–93% yield) [126]; by the reaction of hexafluoroacetone with triphenylphosphiophenylimide (65% yield) [128]; or by heating perfluoroisobutylene with nitrosobenzene (57% yield) [129].

Procedure [126]

$$(CF_3)_2C{=}O + C_6H_5N{=}C{=}O \xrightarrow[-CO_2]{(C_6H_5)_3PO} (CF_3)_2C{=}NC_6H_5$$

A 100-ml steel autoclave is charged with 4.2 g (0.015 mol) of triphenylphosphine oxide and 23.6 g (0.2 mol) of phenylisocyanate, then cooled to —78 °C (air moisture should be kept out) and 36 g (22 ml at —78 °C, 0.215 mol) of hexafluoroacetone is poured in. The autoclave is closed and heated during 17 h at 200 °C. After it has been cooled, the volatile products are passed through a gas-washing bottle with concentrated H_2SO_4 into a trap cooled to —78 °C, where unchanged hexafluoroacetone is collected. The contents of the autoclave are distilled in vacuum. The yield of hexafluoroacetone anil is 44 g (92%); bp 70°–73 °C/73 mm Hg.

Hexafluoroacetone Benzoylimine

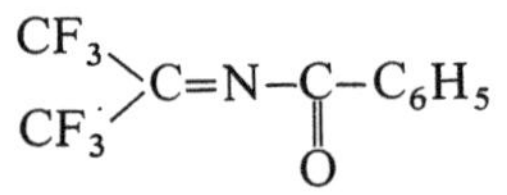

$C_{10}H_5F_6NO$ MW 269

Hexafluoroacetone benzoylimine is a hygroscopic liquid with a strong irritant odour, soluble in common organic solvents. Bp 71°–72 °C/15 mm Hg, 49°–50 °C/5 mm Hg; n_D^{20} 1.4298 [130].

Method of Preparation

Hexafluoroacetone benzoylimine is prepared by thermolytic decomposition of 2,2-bis(benzoylamino)hexafluoropropane (52% yield) [130], or by the reaction of hexafluoroacetone with triphenylphosphine benzoylimide (57% yield) [130].

Procedure [130]

$$(CF_3)_2C{=}O + 2\,C_6H_5CONH_2 \xrightarrow{(CH_3CO)_2O} (CF_3)_2C(NHCOC_6H_5)_2 \rightarrow$$
$$\rightarrow (CF_3)_2C{=}NCOC_6H_5$$

2,2-Bis(benzoylamino)hexafluoropropane. In a 150 ml glass ampoule are placed 20 g (0.165 mol) of finely ground benzamide and 50 ml of acetic oxide. The ampoule is cooled to —78 °C, then 15 g (8.8 ml at —78 °C, 0.09 mol) of hexafluoroacetone is poured in. The ampoule is sealed and gradually heated, with intermittent shaking, to room temperature, whereupon it is heated during 4 h in a boiling water-bath, the contents are cooled and poured into 200 ml of water. The precipitate is filtered off on a Büchner funnel, washed with water, dried first in air, then over P_2O_5 in a vacuum-desiccator, and crystallized from heptane. The yield of 2,2-bis(benzoylamino)hexafluoropropane is 28 g (85%); mp 131°–133 °C.

Hexafluoroacetone benzoylimine. In a 50-ml flask fitted with a downward condenser, is placed 27 g (0.07 mol) of 2,2-bis-(benzoylamino)hexafluoropropane. The flask is heated to 150 °C in 15 mm Hg vacuum, whereupon the temperature of the oil-bath is raised gradually (during 2 to 2.5 h) to 220 °C (Note 1). A fraction boiling at 80° to 145 °C is collected (Note 2), which is immediately redistilled. The yield of hexafluoroacetone benzoylimine is 5.6 g; bp 49°–50 °C/5 mm Hg (Note 3).

Notes.

1. At a higher temperature the reaction product contains benzonitrile as an impurity.
2. Heating is stopped when a compound that has been solidified in the condenser starts to distil off.
3. The still residue is treated with 10% aqueous alkali, the insoluble part is distilled off and the filtrate acidified with diluted HCl. The precipitate contains 11.5 g of crude 2,2-bis-(benzoylamino)hexafluoropropane, which is recycled after drying. Based on the starting bis-amide, the yield of benzoylimine is 52%.

Hexafluoroacetone Oxime

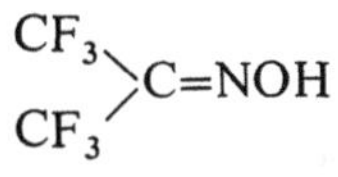

C_3HF_6NO MW 181

Hexafluoroacetone oxime is a highly volatile colourless liquid with a strong odour. Bp 72°–73 °C; n_D^{20} 1.2980 [131].

Method of Preparation

Hexafluoroacetone oxime is prepared by the reaction of potassium α-hydroperfluoroisobutirate [131] or of a free acid [132] with nitrosyl chloride in pyridine (53% yield); or by the reduction of 2-nitroperfluoropropane (70% yield) [133].

Procedure [132]

$$(CF_3)_2CH{-}COOH \xrightarrow{NOCl;\,C_5H_5N} (CF_3)_2C{=}NOH$$

In a three-necked flask provided with a stirrer, a gas inlet tube, and a reflux condenser, is placed 130 ml of anhydrous pyridine and 87 g (0.44 mol) of α-hydroperfluoroisobutyric acid. The flask is cooled to —5°–0 °C and 33 g (0.5 mol) of nitrosyl chloride is added at that temperature, with continuous stirring during 2 h. Towards the end of the addition the temperature has risen to 5 °C. The reaction mixture is stirred at room temperature for 2 h, then heated to 80 °C. Heating is continued till the gas has ceased to evolve (Note 1). The flask is cooled to room temperature, the reaction mixture is diluted with 70 ml of water and poured into 400 ml of cooled 15% HCl. The lower organic layer is separated and distilled twice in vacuum at 90–100 mm Hg, over an equal volume of H_2SO_4. The yield of hexafluoroacetone oxime is 50 g (53%); bp 72°–74 °C (Note 2).

Notes

1. Gas evolution usually stops 30 min after heating has started.
2. In order to increase the yield of the product, the aqueous layer should be extracted with ether, part of the ether distilled off, and the residue distilled in vacuum over H_2SO_4.

Perfluorocyclobutanone Oxime

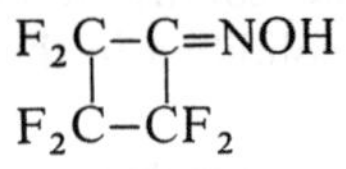

C_4HF_6NO MW 193

Perfluorocyclobutanone oxime is a colourless liquid with a faint pleasant odour, soluble in common organic solvents. Bp 42.5°–43.5 °C/103 mm Hg; forms a distillable complex with ether (4:3), bp 60 °C/90 mm Hg. Perfluorocyclobutanone oxime is a relatively strong acid: $pK_a = 5.5$ (in water) [134].

Method of Preparation

Perfluorocyclobutanone oxime is prepared by the reduction of nitrosoperfluorocyclobutane with hydrogen over palladium black in ether with subsequent treatment of perfluorocyclobutanone oxime etherate with concentrated H_2SO_4 (52% yield) [134].

Procedure

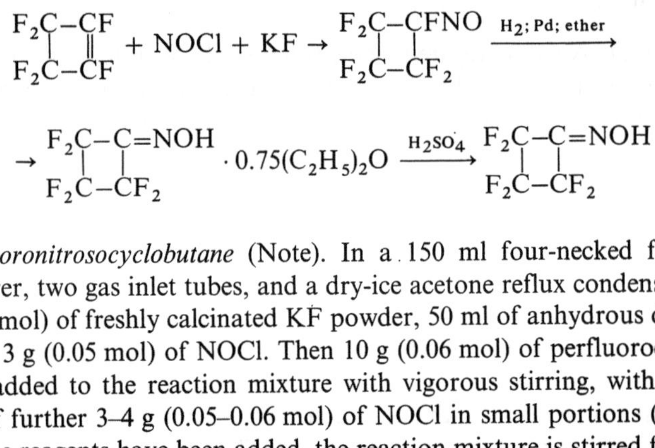

Heptafluoronitrosocyclobutane (Note). In a 150 ml four-necked flask provided with a stirrer, two gas inlet tubes, and a dry-ice acetone reflux condenser, are placed 11.6 g (0.2 mol) of freshly calcinated KF powder, 50 ml of anhydrous dimethylformamide and 3 g (0.05 mol) of NOCl. Then 10 g (0.06 mol) of perfluorocyclobutene is gradually added to the reaction mixture with vigorous stirring, with simultaneous addition of further 3–4 g (0.05–0.06 mol) of NOCl in small portions (of about 1 g). After all the reagents have been added, the reaction mixture is stirred for 1 h, whereupon the volatile products are distilled off in slight vacuum via the gas-washing bottle with concentrated H_2SO_4 into a trap cooled to −78 °C. The condensed product is distilled on a fractionating column with a low-temperature distilling head to give 5.6 g of the starting perfluorocyclobutene and 6.8 g (51%, based on the converted perfluorocyclobutene) of heptafluoronitrosocyclobutane; bp 24°–25 °C.

Perfluorocyclobutanone oxime. In a 250-ml stainless steel autoclave, previously cooled to 0°–5 °C, are placed 25.6 g (0.12 mol) of heptafluoronitrosocyclobutane, 30 ml of anhydrous ether and the catalytic amount (0.1–0.15 g) of palladium black. The autoclave is closed, cooled to −78 °C, flushed with hydrogen, the pressure is brought to 100 atm and the autoclave is shaken at room temperature during 10 h. The reaction mixture is then poured out of the autoclave, shaken with 5–7 g of calcinated KF, the ether is distilled off and the residue distilled in vacuum to give 15.7 g (52%) of perfluorocyclobutanone oxime etherate; bp 60 °C/90 mm Hg.

Distillation of the product in vacuum over concentrated H_2SO_4 gives perfluorocyclobutanone oxime in a quantitative yield; bp 42.5°–43.5 °C/103 mm Hg.

Note. Caution! Perfluorocyclobutene is toxic. The safety measures used in handling C_4F_6 is the same as for perfluoroisobutene (p. 2). The method suggested is a modification of the one described in [135].

Monofluoroacetic Acid

$CH_2F{-}COOH$

$C_2H_3FO_2$ MW 78

Monofluoroacetic acid is a colourless crystalline substance without odour; mp 31°–32 °C [136], 33°–34 °C [262], bp 167°–168.5 °C [136].

Method of Preparation

Monofluoroacetic acid is prepared by the hydrolysis of methyl or ethyl monofluoroacetate (97% yield) [136]. Monofluoroacetates are prepared from the corresponding monocloroacetates and KF in the absence of solvents, in an autoclave (50%–70% yield) [136], or in a high boiling polar solvent (e.g. acetamide) at atmospheric pressure (50% yield) [137]. Monofluoroacetic acid is also prepared by the hydrolysis of 2-fluoronitroethane (97% yield), obtained by nitration of ethylene in the presence of hydrogen fluoride (57% yield) [138].

Procedure [138]

$$CH_2{=}CH_2 \xrightarrow{HNO_3;\,HF} CH_2F{-}CH_2NO_2 \xrightarrow{H_2SO_4} CH_2F{-}COOH$$

Caution! Monofluoroacetic acid is highly toxic (see p. 22).

2-*Fluoronitroethane.* In a steel reactor provided with a stirrer, a gas inlet and outlet tubes, are placed 130 ml (3 mol) of anhydrous HNO_3 (Note 1) and 300 ml of anhydrous HF (Note 2). Ethylene is passed at −60 °C from a cylinder into the reactor through a gas meter and a column packed with calcinated $CaCl_2$, with vigorous stirring during 1 h, at the rate of 40 l/h. After 1 h, 1.46 mol of ethylene has been absorbed. The gas flow is cut off and the reaction mixture is poured onto ice. The resulting solution is extracted with ether (Note 3), the ether extract is washed with a saturated aqueous solution of urea and dried over $CaCl_2$. The ether is distilled off and the residue distilled twice in vacuum (Note 4). The yield of 2-fluoronitroethane is 78 g (57%, based on the converted ethylene); bp 44°–46 °C/6 mm Hg; n_D^{20} 1.3941.

Monofluoroacetic acid. In a three-necked flask equipped with a dropping funnel, a thermometer, and a reflux condenser, is placed 12 ml of 80% H_2SO_4. The flask is heated to 150 °C and 9.3 g (0.1 mol) of 2-fluoronitroethane is added dropwise at that temperature, at such a rate that the temperature of the reaction mixture should not rise above 175 °C. After all 2-fluoronitroethane has been added, the reaction mixture is heated at 175 °C for 30 min, and monofluoroacetic acid is distilled off under vacuum. The yield of 81.7% aqueous monofluoroacetic acid is 9.3 g (97%). Redistillation of that product affords 5.4 g of 67% aqueous acid (bp 110°–164 °C) and 3.9 g anhydrous crystalline monofluoroacetic acid, bp 164°–166 °C, mp 32 °C.

Notes

1. It is preferable to pour nitric acid, with stirring, to anhydrous HF cooled to −78 °C. Anhydrous HNO_3 should be used (d 1.51); otherwise, 2-nitroethanol nitrate will be formed as a by-product.
2. For the safety measures, see p. 1
3. Extraction of the aqueous solution containing HF should be conducted in a polyethylene or Teflon bottle. It is also possible to use a thick-walled glass separatory funnel with a Teflon stopcock; but it should be borne in mind that this funnel fails when used for a long time.
4. Distillation of 2-fluoronitroethane should be executed with care, since under heating nitroethanol nitrate, a possible by-product (bp 108°–110 °C/6 mm Hg), may decompose with an explosion.

Difluoronitroacetic Acid

$O_2NCF_2{-}COOH$

$C_2HF_2NO_4$ MW 141

Difluoronitroacetic acid is a colourless crystalline substance which spreads in air; soluble in organic solvents and water; toxic. Mp 37 °C [139]; bp 80 °C/32 mm Hg [139]; 78 °C/30 mm Hg [140]; 63°–65 °C/15 mm Hg [141]; 80°–83 °C/32–34 mm Hg [142], 58 °C/11 mm Hg [143].

Method of Preparation

Difluoronitroacetic acid is prepared by the hydrolysis of tetrafluoro-2-nitroethyl nitrite (65% yield) [141, 139, 144, 145], or of difluoronitrothioacetic acid (95% yield) [140]; by alkaline hydrolysis of tetrafluoro-2-nitroso-1-nitroethane (58% yield) [143]; or by the reaction of tetrafluoro-2-nitroso-1-nitroethane with hydroxylamine in an aqueous solution (70% yield) [142].

Procedure [141]

$$CF_2{=}CF_2 \xrightarrow{N_2O_4} \underset{\displaystyle NO_2}{\underset{|}{C}}F_2{-}CF_2ONO \xrightarrow{H_2O} O_2NCF_2COOH$$

In a 16-l glass reactor provided with two tubes with stopcocks (Note 1) and evacuated to 5–7 mm Hg residual pressure (Note 2), are placed 26 g (0.28 mol) of nitrogen tetroxide (in one portion) and 7 l (0.3 mol) of tetrafluoroethylene. By the end of charging of the reagents, the pressure in the reactor rises to that of the atmosphere. The reactor is hermetically closed. After 4 h, 75 ml of water is placed in the reactor, which is vigorously shaken with intermittent opening of one of the stopcocks. The aqueous solution is removed from the reactor through the short tube and treatment with water is repeated. The combined water solutions are filtered, extracted with ether, the ether extract is dried over $MgSO_4$ and distilled. The yield of difluoronitroacetic acid is 25 g (65%); bp 63°–65 °C/15 mm Hg (Notes 3, 4).

The reaction of difluoronitroacetic acid with PCl_5 (mol ratio 1:1.2) gives difluoronitroacetyl chloride. Difluoronitroacetic acid is carefully added to PCl_5; a vigorous reaction occurs, during which the chloride distils off. After the reaction is complete, the mixture is heated to 90°–95 °C in a water-bath. The product is redistilled. The yield of difluoroacetyl chloride is 50%; bp 52 °C [138].

Notes

1. One of the tubes is an emergency one, since the main tube may be clogged by the resinous products formed in the reaction of nitrogen tetroxide with the rubber tubes.
2. Before evacuation, the reactor should be placed in a bag of thick fabric.
3. Difluoronitroacetic acid is toxic. It is essential that hands should be protected during work with this compound. Discharging of the reactor and distillation of difluoronitroacetic acid should be carried out under an efficient hood, behind a plastics screen. The glassware and the reactor, which have been in contact with difluoronitroacetic acid, should be washed with an alkaline solution.
4. Difluoronitroacetic acid is obtained wet. In order to obtain dry acid, it is distilled two or three times.

α-Hydroperfluoroisobutyric Acid

CF_3\
  >CH–COOH\
CF_3

$C_4H_2F_6O_2$ MW 196

α-Hydroperfluoroisobutyric acid is a hydroscopic colourless crystalline substance, soluble in water and common organic solvents. Bp 124°–126 °C/750 mm Hg [109], 126°–128 °C/748 mm Hg [146], 95 °C/200 mm Hg [25]; mp 49°–50 °C [25], 51° to 52 °C [146]; pK_a 2.35 (in water) [25].

Method of Preparation

α-Hydroperfluoroisobutyric acid is prepared by treatment of perfluoroisobutylene with aqueous acetone (90%–95% yield) [146] or aqueous tetrahydrofuran (77% yield) [25].

Procedure [146]

$$(CF_3)_2C{=}CF_2 \xrightarrow{H_2O;\ acetone} (CF_3)_2CH{-}COOH$$

Caution! Perfluoroisobutylene is extremely toxic; work with this compound should be performed under an efficient hood (see p. 2).

In a three-necked flask provided with a stirrer, a dry-ice acetone reflux condenser, and a gas inlet tube, are placed 40 ml of acetone and 20 ml of water. Then 106 g (0.53 mol) of perfluoroisobutylene is passed during 30 min through the vigorously stirred solution cooled in an ice-water bath. The reflux condenser is connected with a trap cooled to —78 °C and the mixture is allowed to stand overnight, whereupon it is heated to reflux and unchanged perfluoroisobutylene (66 g) is distilled off into the trap. The residue is distilled in a boiling water-bath to remove low-boiling products. The still residue is cooled, mixed with an equal volume of concentrated H_2SO_4 and distilled. The yield of α-hydroperfluoroisobutyric acid is 35–37 g (90%–95%, based on converted perfluoroisobutylene); bp 126°–128 °C (Note).

Note. Conversion of perfluoroisobutylene to α-hydroperfluoroisobutyric acid may be easily enhanced by taking such an amount of acetone that after addition of all the perfluoroisobutylene a homogeneous solution is obtained.

Perfluorosuccinic Acid

$$HOOC{-}CF_2{-}CF_2{-}COOH$$

$C_4H_2F_4O_4$ MW 190

Perfluorosuccinic acid is a hygroscopic colourless crystalline substance. Mp 86.4° to 87.4 °C [147], 115°–116 °C [148]; bp 150 °C/15 mm Hg [147].

Method of Preparation

Perfluorosuccinic acid is prepared by oxidation of perfluorocyclobutene or 1,2-dichlorotetrafluorocyclobutene with $KMnO_4$ in an alkaline medium (28% yield) [147, 148].

Procedure [147]

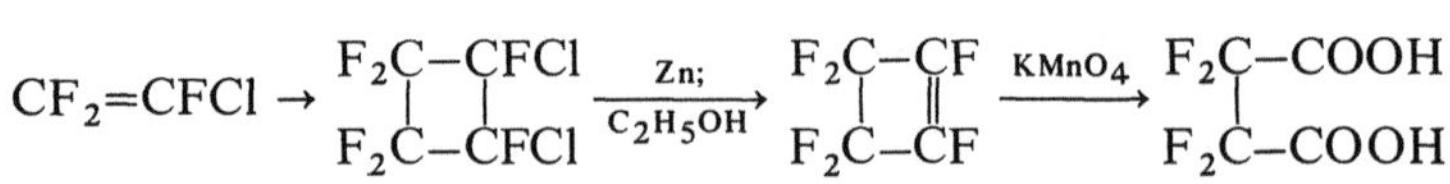

1,2-Dichlorohexafluorocyclobutane. Chlorotrifluoroethylene (140 g, 1.2 mol) and 1 g of hydroquinone are placed in a steel autoclave and stirred at 220°–230 °C for 8 h. The autoclave then cooled and its contents distilled. The yield of 1,2-dichlorohexafluorocyclobutane is 98 g (70%); bp 59 °C; n_D^{20} 1.3360.

Perfluorocyclobutene. Anhydrous ethanol (250 ml) is placed in a two-necked flask equipped with a column (500 × 30 mm), packed with granulated zinc, and a dropping funnel. The top of the column is connected with a reflux condenser. The flask is heated to boiling of ethanol and 100 g (0.42 mol) of 1,2-dichlorohexafluorocyclobutane is dropped in at that temperature during 5–6 h. The liberated gas is passed through a gas-washing bottle with H_2SO_4 and condensed in a trap cooled to −55 °C (Note 1). The condensed product is distilled. The yield of perfluorocyclobutene is 50 g (75%); bp 4°–6 °C, mp −60 °C.

Perfluorosuccinic acid. A mixture of 15.7 g of perfluorocyclobutene, 70 g of $KMnO_4$, and 300 ml of water is shaken in an autoclave at 105 °C for 14 h, then cooled. The reaction mixture is filtered, treated with SO_2 to decolorization, acidified with 50 ml of concentrated H_2SC_4 and extracted with ether. The ether is distilled off and the residue distilled in vacuum. The yield of perfluorosuccinic acid is 17.1 g (93%); bp 75 °C/15 mm Hg (Note 2).

Notes.

1. Perfluorocyclobutene is toxic (see p. 51).
2. Perfluorosuccinic acid thus obtained is dried over P_2O_5 in a vacuum-disiccator; this procedure gives anhydrous perfluorosuccinic acid, mp 115°–116 °C [148].

Bis(trifluoromethyl)acrylic Acid

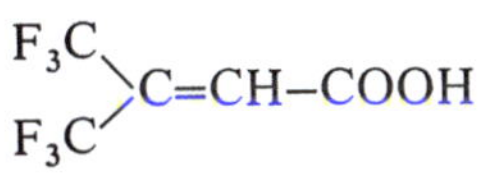

$C_5H_2F_6O_2$ MW 208

Bis(trifluoromethyl)acrylic acid is a low-melting crystalline substance, spreading in air, soluble in water and common organic solvents. Bp 158°–158.5 °C/762 mm Hg [149], 90 °C/51 mm Hg [150]; mp 28.5°–29.5 °C [149].

Method of Preparation

Bis(trifluoromethyl)acrylic acid is prepared by the reaction of β,β-bis(trifluoromethyl)-β-propiolactone with anhydrous sulphuric acid (94% yield) [149]; by heating β-hydroxy-β,β-bis(trifluoromethyl)propionic acid with anhydrous sulphuric acid (90% yield) [149]; or by heating hexafluoroacetone with acetic anhydride at 200 °C [150].

Procedure [149]

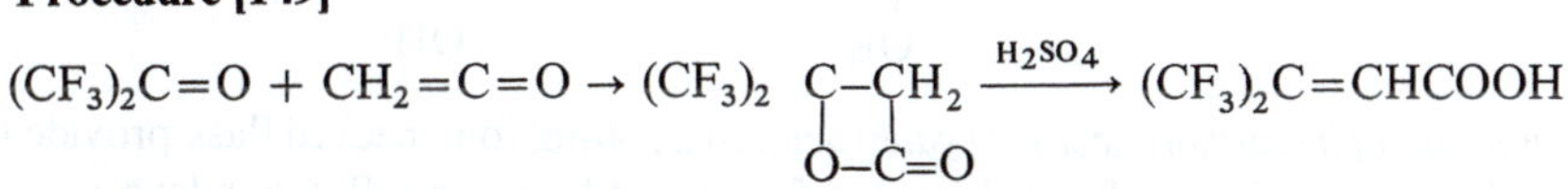

β,β-Bis(trifluoromethyl)-β-propiolactone. Anhydrous ether (50 ml) is placed in a four-necked flask provided with a stirrer, a dry-ice reflux condenser, and two gas inlet tubes extending to the bottom of the flask. The flask is cooled to −78 °C and 123 g (75 ml at −78 °C, 0.79 mol) of hexafluoroacetone is passed during 2 h, together with 40 g (0.95 mol) of ketene (Notes 1, 2). The mixture is slowly heated to room temperature and 10 g of P_2O_5 is added to the solution. The following day the solution is decanted from the precipitate and distilled on a fractionating column with a glass packing (Note 3). The yield of β,β-bis(trifluoromethyl)-β-propiolactone is 130 g (85%); bp 94°–95 °C; n_D^{20} 1.3240.

Bis(trifluoromethyl)acrylic acid. β, β-Bis(trifluoromethyl)-β-propiolactone (31.2 g, 0.15 mol) is added dropwise, with stirring and cooling, to 20 ml of 2%–3% oleum. After the exothermic reaction is complete, the mixture is heated under a downward condenser. The yield of bis(trifluoromethyl)acrylic acid is 29.2 g (94%); bp 157° to 163 °C; after redistillation bp 158°–158.5 °C/762 mm Hg.

Notes

1. The ketene is obtained by the pyrolysis of acetone.
2. The reaction time depends on the output of the ketene generator [151].
3. It is difficult to separate the ether from the reaction product.

Bis(trifluoromethyl)glycolic Acid

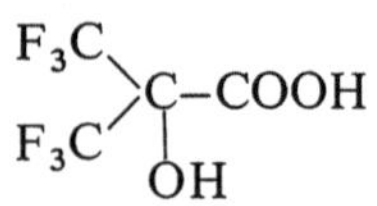

$C_4H_2F_6O_3$ MW 212

Bis(trifluoromethyl)glycolic acid is a hygroscopic colourless crystalline substance. Bp 153°–155 °C [152]; mp 78°–82 °C [152], 82°–85 °C [153]; pKa 2.1 [153].

Method of Preparation

Bis(trifluoromethyl)glycolic acid is prepared by heating hexafluoroacetone cyanhydrin [152] or its potassium salt [153] with concentrated sulphuric acid (60% yield in either case), or by the reaction of perfluoroisobutylene oxide with water in acetone (39% yield) [154].

Procedure [153]

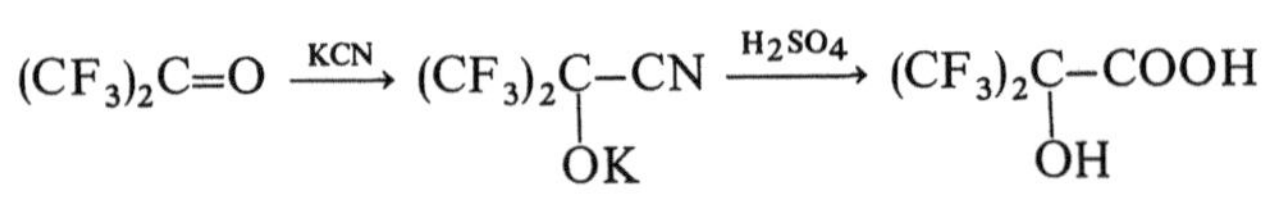

Potassium salt of hexafluoroacetone cyanhydrin. In a 500-ml four-necked flask provided with a stirrer, a gas inlet tube, a dropping funnel, and dry-ice reflux condenser, are placed 190 ml of anhydrous ether, 10 ml of anhydrous tetrahydrofuran and 9 g (0.14 mol) of KCN in an atmosphere of dry nitrogen. The mixture is cooled to −25 °C, then 23 g (0.14 mol) of hexafluoroacetone is slowly bubbled in with stirring. The reaction mixture is slowly heated to room temperature and allowed to stand at that temperature for 1 h, whereupon it is evaporated to remove the solvent. The yield of crude potassium salt of hexafluoroacetone cyanhydrin is 30 g (95%). The salt is used without further purification.

Bis(trifluoromethyl)glycolic acid. Potassium salt of hexafluoroacetone cyanhydrin (30 g, 0.13 mol) is added in portions, with efficient cooling, to 150 ml of concentrated H_2SO_4. The mixture is refluxed during 24 h (the temperature of the oil-bath 160° to 180 °C), then distilled. The yield of bis(trifluoromethyl)-glycolic acid is 13 g (60%); bp 153°–156 °C.

β-Amino-β,β-bis(trifluoromethyl)propionic Acid

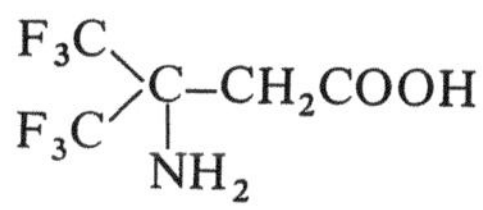

$C_5H_5F_6NO_2$ MW 225

β-Amino-β,β-bis(trifluoromethyl)propionic acid is a colourless crystalline substance, soluble in water, ether, acetone and ethanol. Mp 69°–70 °C [155].

Method of Preparation

β-Amino-β,β-bis(trifluoromethyl)propionic acid is prepared by hydrolysis of N-(p-toluenesulphonyl)-β,β-bis(trifluoromethyl)-β-propiolactam (82% yield) [155], or by deacylation of various β-acylamino-β,β-bis(trifluoromethyl)propionic acids (78% to 80% yield) [156].

Procedure [155]

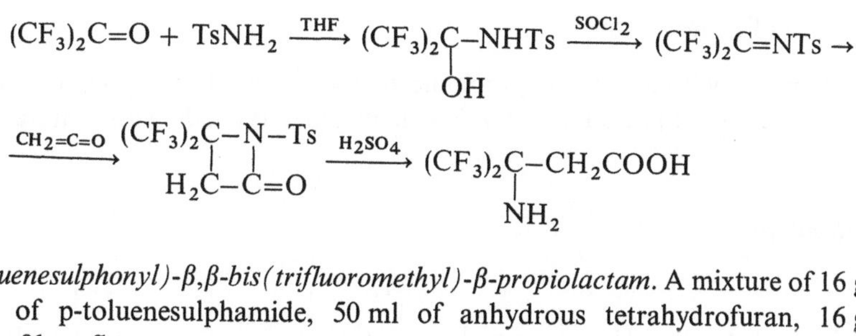

N-(p-Toluenesulphonyl)-β,β-bis(trifluoromethyl)-β-propiolactam. A mixture of 16 g (0.094 mol) of p-toluenesulphamide, 50 ml of anhydrous tetrahydrofuran, 16 g (0.096 mol) of hexafluoroacetone and several drops of pyridine are shaken in a sealed tube to complete dissolution of the precipitate, and allowed to stand at room temperature for 2 days. The contents of the tube are then transferred to a distillation flask and tetrahydrofuran is distilled off in vacuum of a water jet pump at a temperature not higher than 20 °C. To the solid residue are added 40 ml of benzene and 30 ml of thionyl chloride and the mixture is boiled till HCl has ceased to evolve (~2 h). Benzene and excess thionyl chloride are then distilled off in vacuum. The resulting crude hexafluoroacetone N-p-toluenesulphonylimine is dissolved in 100 ml of anhydrous ether, whereupon 5 ml (at —78 °C) of ketene is bubbled in, with intermittent cooling of the flask on an ice water bath. The mixture is allowed to stand at room temperature during 1 h, then 10 ml of ethanol is added with cooling. The resulting solution is shaken with a saturated $NaHCO_3$ solution in a separatory funnel, the ether layer is separated, washed with water and dried over $MgSO_4$. The ether is evaporated and the solid residue recrystallized from CCl_4. The yield of N-(p-toluenesulphonyl)-β,β-bis(trifluoromethyl)-β-propiolactam is 30 g (87%); mp 94°–96 °C.

β-Amino-β,β-bis(trifluoromethyl)propionic acid. N-(p-Toluenesulphonyl)-β,β-bis-(trifluoromethyl)-β-propiolactam (20 g) is dissolved, with heating, in 30 ml of concentrated H_2SO_4. The mixture is refluxed during 2 h in a boiling water-bath, then cooled and poured onto 120 g of ice. The resulting mass is extracted with ether (4 × 75 ml). The combined ether extracts are dried over $MgSO_4$, the ether is evaporated in vacuum, the solid residue is dissolved in hot benzene, the solution is filtered while still hot, the filtrate is cooled and the precipitate is separated. The yield of β-amino-β,β-bis(trifluoromethyl)propionic acid is 10 g (82%); mp 69°–70 °C.

Trifluoro-2-hydroxy-2-phenylpropionic Acid

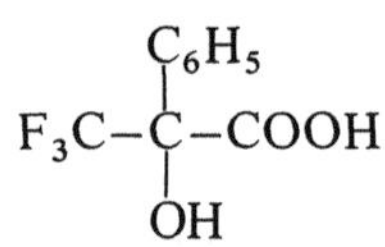

$C_9H_7F_3O_3$ MW 220

Trifluoro-2-hydroxy-2-phenylpropionic acid is a white crystalline substance, soluble in common organic solvents. Mp 110 °C [157], 110.5°–111 °C [158].

Method of Preparation

Trifluoro-2-hydroxy-2-phenylpropionic acid is prepared by saponification of pentafluöro-1-chloro-2-phenylpropanol-2 with an aqueous alkali, with heating (75% yield) [157].

Procedure

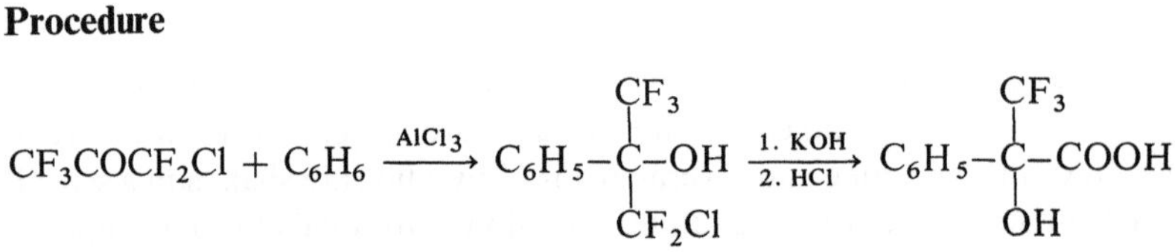

Pentafluoro-1-chloro-2-phenylpropanol-2 [157, 159]. In a 100-ml four-necked flask fitted with a stirrer, a gas inlet tube, a thermometer, and dry-ice reflux condenser, are placed 30 ml of anhydrous benzene and 0.5 g of $AlCl_3$, then 15 g (0.082 mol) of pentafluorochloroacetone is added to the stirred mixture; the temperature of the mixture should not rise above 15 °C. The reaction mixture is stirred during 0.5 h, then washed with water and the organic layer is dried over $MgSO_4$. Benzene is distilled off and the residue distilled in vacuum to give 14.2 (66.3%) of pentafluoro-1-chloro-2-phenylpropanol-2; bp 90 °C/30 mm Hg [157], 190 °C/760 mm Hg [159].

Trifluoro-2-hydroxy-2-phenylpropionic acid [157]. In a 150 ml flask fitted with a stirrer and a reflux condenser, are placed 5.2 g (0.02 mol) of pentafluoro-1-chloro-2-phenylpropanol-2 and a solution of 4.48 g (0.08 mol) of KOH in 70 ml of water. The mixture is refluxed for 3 h, then slightly acidified with HCl and extracted with ether. The ether extract is dried over $MgSO_4$, the ether is evaporated and the residue recrystallized from hexane to give 3.3 g (75%) of trifluoro-2-hydroxy-2-phenyl-propionic acid; mp 110 °C.

3-Fluoro-2-trifluoromethylhexen-2-oic Acid

$$CH_3-CH_2-CH_2-CF=\underset{\displaystyle CF_3}{\underset{|}{C}}-COOH$$

$C_7H_8F_4O_2$ MW 200

3-Fluoro-2-trifluoromethylhexen-2-oic acid is a colourless liquid with a specific odour, soluble in common organic solvents, parially soluble in water. Bp 98°–101 °C/10 mm Hg, n_D^{20} 1.3930, d_4^{20} 1.3151.

Method of Preparation

3-Fluro-2-trifluoromethylhexen-2-oic acid is prepared by the reaction of α-hydroperfluoroisobutyric acid with n-propylmagnesium bromide (the ratio of the reagents 1:3) (45% yield).

Procedure

$$(CF_3)_2CHCOOH \xrightarrow{n\text{-}C_3H_7MgBr} n\text{-}C_3H_7-CF=\underset{\displaystyle CF_3}{\underset{|}{C}}-COOH$$

In a 500-ml flask provided with a stirrer, a dropping funnel, and a reflux condenser, is placed a solution of 19.6 g (0.1 mol) of α-hydroperfluoroisobutyric acid (for the preparation see p. 53) in 40 ml of anhydrous ether. Then a solution of n-propylmagnesium bromide (0.3 mol), prepared from 7.5 g of magnesium and 37 g of n-propyl bromide, is dropped in during 1 h, with stirring; the reaction mixture is cooled so that its temperature should not rise above 20 °C. After the Grignard reagent has been added, the mixture is refluxed during 3 h, cooled and washed with 10% H_2SO_4. The ether layer is separated, the aqueous one is extracted three times with ether, the ethereal solutions are combined, dried over $MgSO_4$ and distilled. The yield of 3-fluoro-2-trifluoromethylhexen-2-oic acid is 9 g (45%); bp 98°–101 °C/10 mm Hg, n_D^{20} 1.3930.

Likewise, α-hydroperfluoroisobutyric acid and n-propylmagnesium bromide (1:6 ratio) give 2-trifluoromethyl-3-propyl-hexen-2-oic acid (bp 127°–130 °C/10 mm Hg, n_D^{20} 1.4205; d_4^{20} 1.1376, yield 35%).

Difluorophosgene

$F_2C{=}O$

CF_2O MW 66

Difluorophosgene is a colourless gas. Bp. —83 °C [160].

Method of Preparation

Difluorophosgene is prepared by the reaction of phosgene with NaF in acetonitrile (70%–80% yield) [160], or with SbF_3 (80%–85% yield) [161, 162].

Procedure [160]

$$COCl_2 \xrightarrow{\text{NaF; acetonitrile}} COF_2$$

In a 1-l three-necked flask provided with an efficient stirrer, a gas inlet tube and a dry-ice reflux condenser connected by rubber tubes with two traps cooled to —78 °C and —110 °C, are placed 100 g (2.4 mol) of finely powdered anhydrous NaF and 200 ml of anhydrous acetonitrile (Note 1). The flask is heated to 40°–50 °C, then 70–72 g (~0.73 mol) of phosgene is passed during 2.5–3 h from a cylinder into the vigorously stirred suspension. In the trap cooled to —110 °C, 27–30 ml (30–35 g) of liquid difluorophosgene is collected (Notes 2, 3).

Notes

1. All parts of the apparatus should be thoroughly dried. Before use, acetonitrile is distilled over P_2O_5.
2. This procedure gives a product containing 95% of pure COF_2, 4%–5% of CO_2 and traces of COFCl and phosgene [160].
3. *Caution*! Difluorophosgene is highly toxic. This substance is easily hydrolysed releasing HF (see p. 1), therefore skin (especially hands) should be protected from contact with COF_2. The poisonous effect of COF_2 is similar to that of phosgene. Inhalation of difluorophosgene vapours causes lung edema (high concentrations), irritation of the respiratory passage and eyes, coughing and sickness (small and average concentrations). First-aid treatment of poisoning: warmth, early inhalation of oxygen, caffeine (1 ml of 10% solution), camphor (1–2 ml of 20% solution), corasole (1ml of 10% solution), cordiamine (1 ml). Treatment of irritation of the respiratory passage: inhalation of the soda solution, hot milk with soda or alkaline mineral water. *The poisoned person should be hospitalized (even is in good condition)*!

2,2-Bis(trifluoromethyl)Propionyl Fluoride

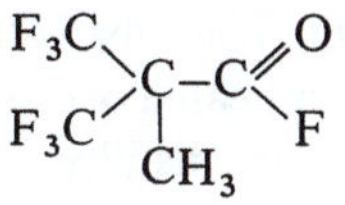

$C_5H_3F_7O$ MW 212

2,2-Bis(trifluoromethyl)propionyl fluoride is a low-boiling colourless liquid. Bp 47° to 47.5 °C/753 mm Hg; mp 7°–9 °C; n_D^{20} 1.284, d_4^{20} 1.473 [163].

Method of Preparation

2,2-Bis(trifluoromethyl)propionyl fluoride is prepared by heating methylperfluoroisobutenyl ether in the presence of triethylamine (70%–80% yield) [164], or by treatment of α-hydroperfluoroisobutyl fluoride-triethylamine complex with methyl iodide (76% yield) [163].

Procedure

$$(CF_3)_2CHCF_2OCH_3 \xrightarrow{KOH} (CF_3)_2C{=}CFOCH_3 \xrightarrow{(C_2H_5)_3N} (CF_3)_2C(CH_3){-}COF$$

$$(CF_3)_2C{-}CF_2 \xrightarrow{CH_3OH} (CF_3)_2CHCF_2OCH_3$$

Caution! Perfluoroisobutylene is extremely toxic. It should be handled only under an efficient hood (see p. 2).

Methyl-2-hydroperfluoroisobutyl ether [109]. In a three-necked flask equipped with a stirrer, a gas inlet tube, and a reflux condenser, is placed 50 ml of methanol, then 67 g (0.35 mol) of perfluoroisobutylene is passed in with vigorous stirring at room temperature. The gas is absorbed with slight evolution of heat. The reaction mixture is allowed to stand overnight at room temperature, then poured into cold water. The lower organic layer is separated (Note), washed with water, with a saturated solution of $NaHCO_3$ and once more with water, dried over $CaCl_2$ and distilled; a fraction boiling at 68.5°–70.5 °C/754 mm Hg is collected. The yield of methyl-2-hydroperfluoroisobutyl ether is 57 g (65%). Distillation gives the product boiling at 69.5°–70 °C; n_D^{20} 1.283.

Methylperfluoroisobutenyl ether [109]. In a 1-l three-necked rount-bottom flask provided with a stirrer, a dropping funnel, and a reflux condenser with a calcium chloride tube, are placed 68 g of KOH powder and 500 ml of anhydrous ether. Then a solution of 95 g (0.41 mol) of methyl-2-hydroperfluoroisobutyl ether in 50 ml of anhydrous ether is dropped in with vigorous stirring. The mixture is boiled during 3 h in a water-bath, with continuous stirring, the ethereal solution is decanted and distilled. After the ether has been removed, a fraction boiling at 100°–108 °C is collected, which is an almost pure methylperfluoroisobutenyl ether, as shown by GLC. The yield of the ether is 63 g (66%).

2,2-Bis(trifluoromethyl)propionyl fluoride [164]. A mixture of 22 g (0.104 mol) of methylperfluoroisobutenyl ether and 0.3–0.5 g of anhydrous triethylamine is heated during 8.5 h in a sealed glass tube, in a boiling water-bath. The tube is then cooled, further 0.2–0.3 g of triethylamine is added and the mixture is heated once more during 15 h in the sealed tube at 100 °C. The contents of the tube are distilled. The yield of 2,2-bis(trifluoromethyl)propionyl fluoride is 15 g (72%); bp 46°–49 °C.

Note. The lower organic layer contains a small amount of dissolved perfluoroisobutylene, therefore it should be handled with care.

Nonafluoropivalyl Fluoride and Hexafluoroisobutenyldimethylamine

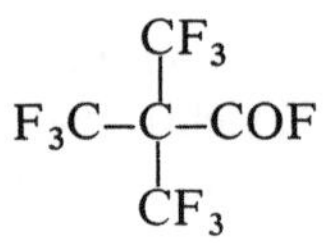

$C_5F_{10}O$ MW 266

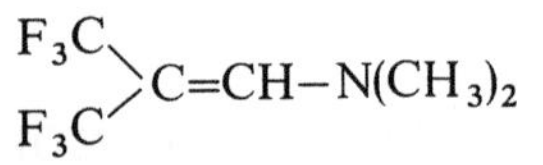

$C_6H_7F_6N$ MW 207

Nonafluoropivalyl fluoride is a highly volatile colourless crystalline substance with a strong odour, soluble in common organic solvents. Mp 34.5°–35.5 °C [165].

Hexafluoroisobutenyldimethylamine is a slightly coloured liquid, soluble in common organic solvents. Bp 171°–174 °C [165], 65 °C/19 mm Hg [166]; n_D^{20} 1.3842 [165].

Method of Preparation

Nonafluoropivalyl fluoride is prepared by the reaction of perfluoroisobutylene with dimethylformamide (64% yield); hexafluoroisobutenyldimethylamine is also formed in this reaction (87% yield) [165]. Nonafluoropivalyl fluoride is also obtained by the reaction of perfluoroisobutylene with phosgene in the presence of CsF (55% yield) [167]. Hexafluoroisobutenyldimethylamine is prepared by the reaction of bis(trifluoromethyl)ketene with dimethylformamide (72% yield) [168].

Procedure

1. Reaction of Perfluoroisobutylene with Dimethylformamide [165]

$$2\,(CF_3)_2C{=}CF_2 + (CH_3)_2N\overset{\overset{\displaystyle O}{\|}}{C}H \rightarrow (CF_3)_3C{-}COF + (CF_3)_2C{=}CHN(CH_3)_2$$

Caution! Perfluoroisobutylene is extremely toxic. It should be handled only under an efficient hood (see p. 2).

A mixture of 73 g (1 mol) of anhydrous dimethylformamide (Note 1) and 224 g (1.12 mol) of perfluoroisobutylene is maintained in a sealed glass tube for 2–3 days (Note 2). The tube is cooled with liquid nitrogen, opened, connected with a trap cooled to —78 °C and heated to 100 °C in a water-bath. The volatile products collected in the trap are distilled on a fractionating column with a glass packing and a low-temperature head. A low-boiling fraction (impurities, contained in the starting perfluoroisobutylene) is discarded. Towards the end of the distillation the distillate and the still residue (crude nonafluoropivalyl fluoride) have crystallized. The product is further purified by removing impurities under 0.01 mm Hg vacuum at a temperature below —35 °C and subsequent sublimation in a hermetically closed system of evacuated glass tubes connected in series and cooled with liquid nitrogen. The yield of nonafluoropivalyl fluoride is 96 g (64%); mp 34.5°–35.5 °C.

The residue after distilling of the fluoride from the reaction mixture is poured into ice-water, the oily layer is separated, washed with ice-water, dried over $MgSO_4$ and distilled in vacuum. The yield of hexafluoroisobutenyldimethylamine is 101 g (87%); bp 68°–70 °C/19 mm Hg.

2. Reaction of Perfluoroisobutylene with Phosgene in the Presence of CsF [167]

$$(CF_3)_2C{=}CF_2 + COCl_2 \xrightarrow{CsF} (CF_3)_3C{-}COF$$

Perfluoroisobutylene (25 g, 0.125 mol) is added to a suspension of 37.4 g (0.25 mol) of freshly calcinated CsF powder in 50 ml of anhydrous diglyme with stirring and cooling of the flask in an ice-water bath, then 12 g (0.12 mol) of $COCl_2$ is added to the mixture during 2 h. The mixture is stirred at 20 °C during 1 h, the volatile products are distilled off into a trap cooled to —78 °C and the condensate is distilled off from the trap to give 18 g (55%) of nonafluoropivalyl fluoride; mp 30°–35 °C.

Notes

1. Immediately before the reaction dry dimethylformamide is distilled over P_2O_5.
2. Repeated synthesis may in some cases lead to separation of the reaction mixture into layers and solidification of the lower layer in the tube cooled to —78 °C.

Ethyl Iododifluoroacetate

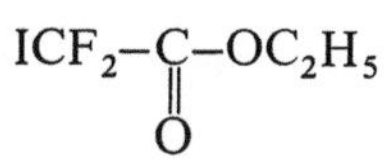

$C_4H_5IF_2O_2$ MW 250

Ethyl iododifluoroacetate is a slightly coloured liquid, soluble in common organic solvents. Bp 70°–72 °C/70 mm Hg [169].

Method of Preparation

Ethyl iododifluoroacetate is prepared by the addition of iodine monochloride to ethyltrifluorovinyl ether with subsequent heating of the addition product with concentrated H_2SO_4 [169].

Procedure

$$CF_2{=}CFOC_2H_5 \xrightarrow{ICl} ICF_2{-}CFCl{-}OC_2H_5 \xrightarrow{H_2SO_4} ICF_2{-}COOC_2H_5$$

Iodine monochloride. In a two-necked round-bottom flask provided with a calcium chloride tube and a wide gas inlet tube (7 mm i.d.) is placed 127 g (0.5 mol) of iodine (Note 1), then a flow of dry chlorine is let in with continuous shaking until a gain in weight of 37 g is obtained (Note 2). The resulting liquid is distilled with a small downward condenser; a fraction boiling at 98°–102 °C is collected. The yield of iodine monochloride is 132 g (81 %) (Note 3).

Ethyl-2-iodo-1-chlorotrifluoroethyl ether. Iodine monochloride (24.5 g, 0.15 mol) in 5 ml of methlyene chloride is added dropwise, with stirring and cooling to —78 °C, to a solution of 19 g (0.15 mol) of ethyltrifluorovinyl ether (for the preparation see p. 32) in 10 ml of dry methylene chloride. The mixture is heated to room temperature, allowed to stand for 1 h, then washed with 10% NaOH solution, with water and dried over $CaCl_2$. The solvent is distilled off in a slight vacuum and the residue distilled in vacuum. The yield of ethyl-2-iodo-1-chlorotrifluoroethyl ether is 24.1 g (55.5%); bp 62°–64 °C/45 mm Hg.

Ethyl iododifluoroacetate. A mixture of 16.5 g (0.057 mol) of ethyl-2-iodo-1-chlorotrifluoroethyl ether, 20 ml of concentrated H_2SO_4 and 5 g of glass powder is heated in a water-bath at 60°–70 °C with stirring during 1 h, then 10 ml of ethanol is added. Heating and stirring are continued for further 30 min. The reaction mixture is poured into ice-water, the organic layer is separated and the aqueous one extracted several times with ether. The organic layer and the ether extract are combined, washed with $NaHCO_3$ solution, then with the concentrated solution of $CaCl_2$, and dried over $CaCl_2$. The ether is distilled off and the residue distilled in vacuum. The yield of ethyliododifluoroacetate is 11.3 g (79%); bp 70°–72 °C/70 mm Hg.

Notes

1. The gas inlet tube should reach the surface of iodine.
2. When chlorine is passed into the flask, care should be taken to prevent clogging of the gas inlet tube by the precipitate.
3. Iodine monochloride in contact with the skin causes severe burns. One should wear rubber gloves and goggles during work which should be performed under an efficient hood.

Methyl Perfluoromethacrylate

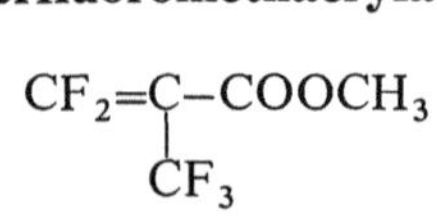

$C_5H_3F_5O_2$ MW 190

Methyl perfluoromethacrylate is a colourless liquid with a strong odour, soluble in common organic solvents. Bp 94°–95 °C [170], 92°–94 °C [171], 90°–92 °C [172]; n_D^{20} 1.3390 [171], 1.3288 [173].

Method of Preparation

Methyl perfluoromethacrylate is prepared by dehydrofluorination of methyl α-hydroperfluoroisobutyrate with $(C_2H_5)_3N \cdot BF_3$ (81% yield) [171]; by thermal dechlorination of methyl pentafluoro-1,2-dichloroisobutyrate with zinc [172]; or by elimination of methyl fluoride from bis(trifluoromethyl)ketene dimethylacetal in the presence of BF_3 etherate (89% yield) [170] (Note 1).

Procedure

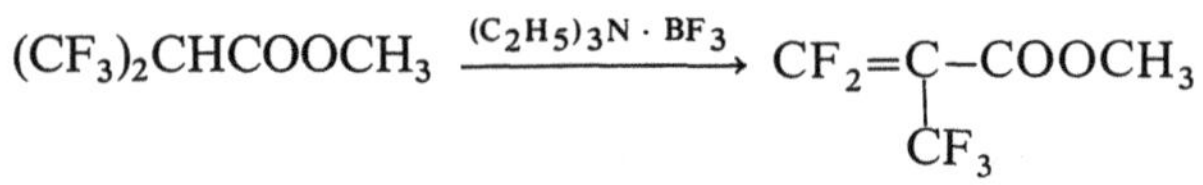

In a 30-ml round-bottom flask fitted with a reflux condenser are placed 11.73 g (0.07 mol) of $(C_2H_5)_3N \cdot BF_3$ [174] and 9.88 g (0.047 mol) of methyl α-hydroperfluoroisobutyrate (Note 2). The mixture is refluxed during 1 h, then cooled, distilled in vacuum at room temperature; the distillate is collected in a trap cooled to —78 °C. This procedure affords a mixture containing 98% of the major product and 2% of the starting ether. The yield of methyl perfluoromethacrylate is 7.25 g (81%); bp 92° to 94 °C.

Notes

1. It is impossible to obtain methyl perfluoromethacrylate directly from perfluoromethacrylyl fluoride and methanol; this reaction would give only 3,3-difluoro-3-methoxy-2-trifluoromethylpropanyl fluoride and methyl 3,3-difluoro-3-methoxy-2-trifluoromethylpropanate.
2. Methyl α-hydroperfluorisobutyrate can be easily prepared from methyl-2-hydroperfluorisobutyl ether [173]: a mixture of 121 g (0.52 mol) of methyl-2-hydroperfluoroisobutyl ether (for the preparation see p. 61) and 125 ml of concentrated H_2SO_4 are refluxed during 15 min, then cooled, the homogeneous solution is poured into 200 g of ice, the organic layer is separated, washed with water and distilled over P_2O_5. The yield of methyl α-hydroperfluoroisobutyrate is 102 g (93%); bp 90 °C.

(Perfluoropropenyl-2)acetate

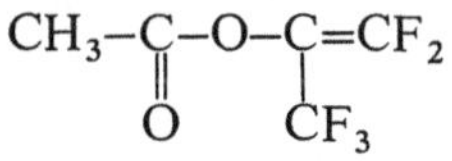

$C_5H_3F_5O_2$ MW 190

(Perfluoropropenyl-2)acetate is a colourless liquid with a pleasant ethereal odour, soluble in common organic solvents. Bp 75°–76 °C [176].

Method of Preparation

(Perfluoropropenyl-2)acetate is prepared by acylation of perfluoropropenol-2 with acetyl chloride (70% yield) [176].

Procedure

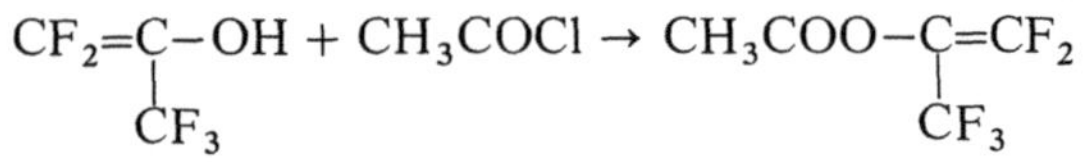

A mixture of 5.7 g (0.035 mol) of perfluoropropenol-2 and 3.1 g (0.0395 mol) of acetyl chloride is refluxed till hydrogen chloride has ceased to evolve. Distillation of the reaction mixture gives 5.2 g (70%) of (perfluoropropenyl-2) acetate; bp 75°–76 °C.

Likewise, perfluoropropenol-2 and benzoyl chloride give (perfluoropropenyl-2)-benzoate (bp 76 °C/20 mm Hg; yield 76.5%) [176]; perfluoropropenol-2 reacts with trifluoroacetic anhydride and perfluorobutyryl chloride in the presence of N-methyl-pyrrolidone to give perfluoro (propenyl-2)acetate (bp 45°–46 °C, 72.5% yield) and perfluoro (propenyl-2)butyrate (bp 85°–87 °C, 68.8% yield), respectively [176].

Perfluoroisopropenyl Perfluoromethacrylate

[Perfluoro(propenyl-2)methacrylate

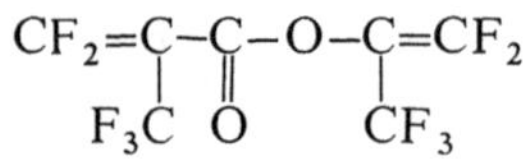

$C_7F_{10}O_2$ MW 306

Perfluoro(propenyl-2)methacrylate is a volatile colourless liquid with a pleasant ethereal odour. Bp 99°–100 °C [176].

Method of Preparation

Perfluoro(propenyl-2)methacrylate is prepared by dehydrofluorination of perfluoro-(propenyl-2)-α-hydroisobutyrate with triethylamine-boron trifluoride adduct (60% yield) [176].

Procedure

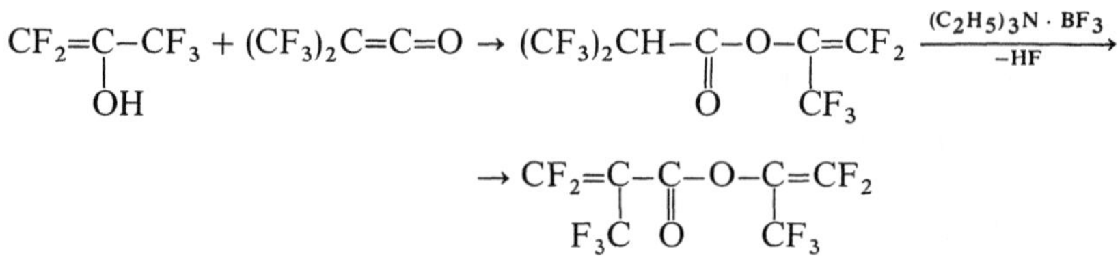

Perfluoro(propenyl-2)-α-hydroisobutyrate [176]. In a 50-ml four-necked flask provided with a stirrer, a dropping funnel, a gas inlet tube, and a dry-ice reflux condenser, is placed 4.6 g (0.031 mol) of perfluoropropenol-2 (for the preparation see p. 26), then 6.1 g (0.034 mol) of bis(trifluoromethyl)ketene (for the preparation see p. 72) is gradually added with stirring. The reaction mixture is cooled to 0 °C, whereupon several drops of anhydrous N-methylpyrrolidone are added. The mixture is allowed to warm to room temperature, stirred during 10 min and distilled. The yield of perfluoro(propenyl-2)-α-hydroisobutyrate is 7.5 g (74%); bp 99°–100 °C.

Perfluoro(propenyl-2)methacrylate [176]. A mixture of 4 g (0.0123 mol) of perfluoro(propenyl-2)-α-hydroisobutyrate and 4.7 g (0.028 mol) of $(C_2H_5)_3N \cdot BF_3$ [174] is refluxed at 80 °C during 45 min. The flask is then connected with a receiver cooled to −78 °C, whereupon the volatile reaction products are pumped off into the receiver at 1 mm Hg. The contents of the receiver are distilled. The yield of perfluoro(propenyl-2)methacrylate is 2.3 g (60%); bp 99°–100 °C.

Hexafluoro-2-chloroisopropyl Benzoate

[(2-chlorohexafluoropropyl)benzoate]

$$C_6H_5COO{-}\underset{\displaystyle Cl}{\underset{|}{C}}(CF_3)_2$$

$C_{10}H_5ClF_6O_2$ MW 306.5

(2-Chlorohexafluoroisopropyl)benzoate is a colourless liquid with a specific odour. Bp 41°–43 °C/2 mm Hg, 60°–62 °C/4 mm Hg, 76°–79 °C/10 mm Hg, 186°–192 °C/750 mm Hg [105].

Method of Preparation

(2-Chlorohexafluorisopropyl)benzoate is prepared by the reaction of benzoyl chloride with hexafluoroacetone in the presence of pyridine (63% yield) or triethylbenzylammonium chloride (45% yield) [105].

Procedure

$$C_6H_5COCl + (CF_3)_2CO \xrightarrow{C_5H_5N} C_6H_5COO{-}CCl(CF_3)_2$$

In a 20-ml glass tube is placed 4.08 g (0.029 mol) of benzoyl chloride. The tube is cooled to −78 °C (Note 1), then six drops of anhydrous pyridine are added, whereupon 6.05 g (0.036 mol) of hexafluoroacetone is quickly poured in. The tube is sealed, warmed to room temperature (Note 2) and maintained at that temperature for 23 h (Note 3). The tube is then again cooled to −78 °C, opened, excess hexafluoroacetone is distilled off and the residue dissolved in ether. The ethereal solution is washed with water, dried over $MgSO_4$, the ether is distilled off and the residue distilled in vacuum. The yield of (2-chlorohexafluoroisopropyl)benzoate is 5.6 g (63%); bp 60°–62 °C/4 mm Hg.

Notes

1. In order to keep air moisture out of the tube, the latter should be connected with a calcium chloride tube or a gas-washing bottle with concentrated H_2SO_4.
2. The reaction is exothermic, therefore when large quantities of starting compounds are charged the tube should be warmed very slowly; when the temperature rises above that of the room the tube should be immediately cooled with cold water. Due to the danger of an explosion in the tube, face should be protected with a mask and a screen of plastic; hands, with gloves. It is not recommended to use more than 20 g of hexafluoroacetone in one run.
3. Immediately after warming to room temperature the contents of the tube are a homogeneous solution. Two immiscible layers indicate that pyridine was taken in insufficient amount. In this case the tube should be cooled, opened and again sealed after addition of several drops of anhydrous pyridine.

Methyl Trifluoropyruvate

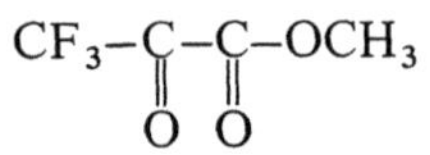

$C_4H_3F_3O_3$ MW 176

Methyl trifluoropyruvate is a hygroscopic colourless liquid with a specific odour, soluble in common organic solvents. Bp 84°–86 °C; n_D^{20} 1.3280 [177].

Method of Preparation

Methyl trifluoropyruvate is prepared by heating methyl α-methoxytetrafluoropropionate with concentrated H_2SO_4 (83% yield) [177].

Procedure

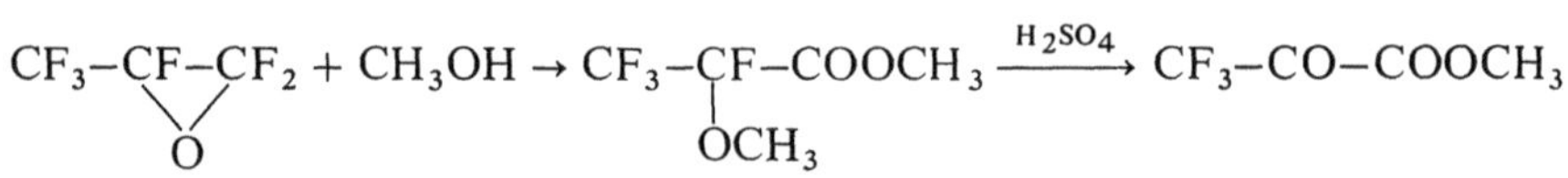

Methyl α-methoxytetrafluoropropionate [178]. Perfluoropropylene oxide (56.6 g) is bubbled with cooling during 4–5 h through 300 ml of anhydrous methanol (Note). The resulting mixture is poured into 1 l of water, the lower layer is separated, washed twice with water, dried over Na_2SO_4 and distilled. The yield of methyl α-methoxytetrafluoropropionate is 62 g (96%); bp 40°–41 °C/21 mm Hg.

Methyl trifluoropyruvate [177]. In a flask fitted with a reflux condenser with a calcium chloride tube, are placed 12 ml of concentrated H_2SO_4, 1 g of finely divided silica gel, and 9.5 g (0.05 mol) of methyl α-methoxytetrafluoropropionate. The mixture is heated to 140 °C in an oil-bath, then allowed to stand at 140°–150 °C for 20–30 min before vigorous boiling starts. The flask is cooled, connected with a downward condenser and the product is distilled in an oil-bath. A fraction boiling at 105°–110 °C is redistilled. The yield of methyl trifluoropyruvate is 6.5 g (83 %); bp 84°–86 °C.

Note. Perfluoropropylene oxide should be passed from the cylinder at such a rate that it is completely absorbed by the reaction mixture.

3,3,3-Trifluoroacetoacetic Ester

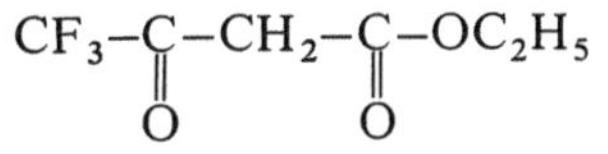

$C_6H_7F_3O_3$ MW 184

3,3,3-Trifluoroacetoacetic ester is a colourless liquid, soluble in common organic solvents. Bp 131.5 °C/757 mm Hg, 41 °C/22 mm Hg; $n_D^{15.5}$ 1.3783; $d_4^{15.5}$ 1.2586 [179].

Method of Preparation

3,3,3-Trifluoroacetoacetic ester is prepared by heating ethyl trifluoroacetate with ethyl acetate in the presence of metallic sodium (77% yield) [104], sodium hydride (84% yield) [180] or sodium alkoxide (54% yield) [101].

Procedure [104]

$$CF_3COOC_2H_5 + CH_3COOC_2H_5 \xrightarrow{Na} CF_3COCH_2COOC_2H_5$$

In a flask provided with an efficient water-cooled reflux condenser on the top of which is mounted a dry-ice reflux condenser, is placed 16.5 g (0.72 mol) of sodium wire; then a solution of 101.5 g (0.715 mol) of ethyl trifluoroacetate in 130 ml of dry ethyl acetate is quickly added. A violent reaction starts immediately (Note 1); sodium quickly dissolves (in about 10 min). After the reaction mixture has cooled, 500 ml of anhydrous ether is added and the resulting mixture is refluxed during 14 h (Note 2). The ether is evaporated in vacuum, the dark-brown residue is dissolved in 400 ml of ether (Note 3) and shaken with 200 ml of 15% H_2SO_4. The ether layer is separated and the aqueous one extracted with ether. The combined ethereal solutions are dried over MgSO and distilled. The yield of 3,3,3-trifluoroacetoacetic ester is 101 g (77%); bp 128°–132 °C.

Notes

1. It is not practical to effect this preparation on a large scale as in this case it is difficult to control the reaction.
2. If this procedure is omitted and the reaction mixture is hydrolysed immediately after the exothermic reaction, the yield of the product decreases to 61%.
3. Crude sodium 3,3,3-trifluoroacetoacetate may be used for the preparation of 1,1,1-trifluoroacetone. For that purpose it is boiled with 250 ml of 20% H_2SO_4; the volatile products are collected in a trap cooled to —78 °C. The yield of 1,1,1-trifluoroacetone is 63 g; bp 20°–22 °C [104].

α-Ketoperfluorovaleric Acid Dimethylamide

$$CF_3{-}CF_2{-}CF_2{-}\underset{\displaystyle \overset{\|}{O}}{C}{-}\underset{\displaystyle \overset{\|}{O}}{C}{-}N(CH_3)_2$$

$C_7H_6F_7NO_2$ MW 269

α-Ketoperfluorovaleric acid dimethylamide is a hygroscopic colourless liquid. Bp 82°–83 °C/10 mm Hg [181], 62°–66 °C/7 mm Hg [182]; n_D^{20} 1.3542 [181].

Method of Preparation

α-Ketoperfluorovaleric acid dimethylamide is prepared by the reaction of perfluorobutyryl chloride with dimethylformamide in the presence of triethylamine (33% yield) [182].

Procedure

$$C_3F_7COCl + HCON(CH_3)_2 \xrightarrow{(C_2H_5)_3N} C_3F_7COCON(CH_3)_2$$

In a glass tube are placed 24 g (0.33 mol) of anhydrous dimethylformamide and 6.5 g (0.064 mol) of dry triethylamine. The tube is cooled to −10° ÷ −20 °C, then 15 g (0.064 mol) of perfluorobutyryl chloride (for the preparation see p. 77) is added. The tube is sealed and allowed to stand at 0 °C for 15 h. Then the tube is cooled, its contents poured into ice-water, the organic layer is separated and the aqueous one extracted several times with ether. The organic layer and the ether extract are combined, washed with water, dried over $MgSO_4$, then maintained over P_2O_5 and distilled in vacuum. The yield of α-ketoperfluorovaleric acid dimethylamide is 5.7 g (33%); bp 62°–66 °C/7 mm Hg.

α-Hydroperfluoropropionitrile

$CF_3{-}CHF{-}CN$

C_3HF_4N MW 127

α-Hydroperfluoropropionitrile is a volatile colourless liquid. Bp 40°–41 °C; n_D^{20} 1.27; d_4^{20} 1.3620 [109].

Method of Preparation

α-Hydroperfluoropropionitrile is prepared by the reaction of perfluoropropylene with aqueous ammonia in dioxan (46% yield) [109].

Procedure

$$CF_3{-}CF{=}CF_2 \xrightarrow{NH_3} CF_3CHFCN$$

Perfluoropropylene (63 g, 0.42 mol) is passed with stirring and cooling into a mixture of 100 ml of dioxan and 20 ml of 30% aqueous ammonia. In the course of the reaction three 10 ml portions of 30% aqueous ammonia are added in the reaction flask. After perfluoropropylene has been added, the reaction mixture is quickly distilled into a cooled receiver. A fraction boiling at 36°–95 °C is poured into water, the organic layer is separated, thoroughly washed with cold water to remove dioxan, dried over $CaCl_2$, and distilled. The yield of α-hydroperfluoropropionitrile is 23.8 g (46%); bp 40°–41 °C; n_D^{20} 1.27; d_4^{20} 1.3620.

Trifluoroacrylonitrile

$CF_2{=}CF{-}CN$

C_3F_3N MW 107

Trifluoroacrylonitrile is a low-boiling colourless liquid. Bp 16.4°–16.5 °C, 16°–17 °C [183, 184]; 17.9°–18.5 °C [185].

Method of Preparation

Trifluoroacrylonitrile is prepared by dehydrofluorination of α-hydroperfluoropropionitrile (56% yield) [183, 184], or by dehalogenation of 1,2-dichloroperfluoropropionitrile (25%–32% yield) [185, 186].

Procedure

$$CF_3CHFCN \xrightarrow{600\,°C} CF_2{=}CFCN$$

α-Hydroperfluoropropionitrile (33 g, 0.26 mol) is passed during 25 min in a flow of dry nitrogen at 600 °C through a graphite-lined iron tube packed with granulated KCl (20 mm i.d.) (KCl layer 470 mm long). The exit gases are collected in a trap kept cooled to −78 °C. The condensate is distilled on a Podbielniak column to afford 9.1 g of trifluoroacrylonitrile together with 14 g of unchanged starting nitrile. The yield of trifluoroacrylonitrile is 56% (based on the converted starting nitrile).

Purification. Purification of trifluoroacrylonitrile from hydrogen fluoride and other impurities is carried out by bromination of the pyrolysis gases under UV-irradiation. The resulting 1,2-dibromotrifluoropropionitrile (bp 98°–100 °C/750 mm Hg; n_D^{20} 1.4150) is easily dehalogenated with zinc.

1,2-Dibromotrifluoropropionitrile is added dropwise, with vigorous stirring and heating to 90°–100 °C, to a mixture of 10 g of zinc dust and 40 ml of anhydrous dibutyl ether (Note). The liberated trifluoroacrylonitrile is condensed in a trap cooled to

—78 °C and distilled. The yield of trifluoroacrylonitrile is 8.7 g; bp 16°–17 °C/ 750 mm Hg.

Note. In order to ensure smooth dehaloganation, it is necessary to add a little zinc dust from the previous procedure.

1,1-Dicyano-2,2-bis(Trifluoromethyl)Ethylene

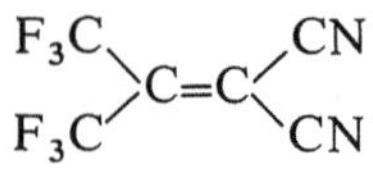

$C_6F_6N_2$ MW 214

1,1-Dicyano-2,2-bis(trifluoromethyl)ethylene is a colourless liquid, soluble in benzene, pentane, carbon tetrachloride. Bp 107 °C; n_D^{25} 1.3526; d_4^{25} 1.479 [187].

Method of Preparation

1,1-Dicyano-2,2-bis(trifluoromethyl)ethylene is prepared by the reaction of hexafluoroacetone with malononitrile in the presence of zinc chloride with subsequent heating of the adduct with P_2O_5 (50% yield) [187].

Procedure

$$(CF_3)_2C{=}O + CH_2(CN)_2 \rightarrow (CF_3)_2\underset{\displaystyle OH}{\underset{|}{C}}{-}CH(CN)_2 \quad {-}CH(CN)_2 \rightarrow (CF_3)_2C{=}C(CN)_2$$

A mixture of 75 g (1.14 mol) of malononitrile, 4 g of anhydrous $ZnCl_2$ and 150 g (0.9 mol) of hexafluoroacetone is heated during 8 h in a 240 ml autoclave at 80 °C. The resulting 211 g of a solid product (Note) is mixed with 454 g of P_2O_5 in a 5 l flask. The mixture is distilled with vigorous heating till the distillate does not collect any longer. Then the distillate is filtered and distilled. The yield of 1,1-dicyano-2,2-bis(trifluoromethyl)-ethylene is 97 g (50%); bp 107 °C.

Note. Recrystallization of the sample from methylene chloride gives hygroscopic white crystalline 1-hydroxyhexafluoroisopropylmalononitrile, mp 59°–70 °C.

Bis(trifluoromethyl)Ketene

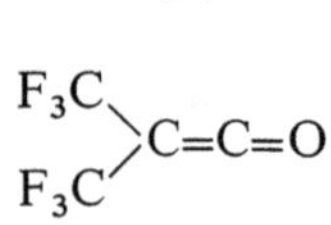

C_4F_6O MW 178

Bis(trifluoromethyl)ketene is a gas, stable at room temperature. Bp 5 °C [188].

Method of Preparation

Bis(trifluoromethyl)ketene is prepared by dehalogenation of 1-chloro- or 1-bromo-perfluoroisobutyryl chloride or bromide (70% yield) [189], or by dehydration of 1-hydroperfluoroisobutyric acid with P_2O_5, in a high yield [188, 190].

Procedure [188]

$$(CF_3)_2CH{-}COOH \xrightarrow{P_2O_5} (CF_3)_2C{=}C{=}O$$

1-Hydroperfluoroisobutyric acid (100 g, 0.5 mol) and 300 g of P_2O_5 are placed in a 2-l round-bottom flask provided with an efficient reflux condenser, the top of which is connected to a trap cooled to —78 °C. The contents of the flask are thoroughly mixed with a glass rod, then heated in an oil-bath; the temperature is slowly raised during 6–8 h from 170° to 200 °C. The liberated gas is condensed in a trap (Note) and distilled. The yield of bis(trifluoromethyl)ketene is 70 g (75%); bp 5°–7 °C. It may be stored in a sealed glass ampoule.

Note. Air moisture must be kept out of the trap in which bis(trifluoromethyl)ketene is collected, since this compound is extremely hygroscopic.

Bis(Trifluoromethyl)Thioketene

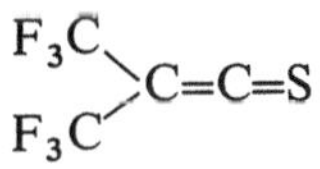

C_4F_6S MW 194

Bis(trifluoromethyl)thioketene is a volatile organe-red liquid with a specific strong odour. Bp 52 °C [191].

Method of Preparation

Bis(trifluoromethyl)thioketene is prepared by the pyrolysis of its dimer in a quartz tube at 650 °C (60%–70% yield) [191]. Bis(trifluoromethyl)thioketene dimer is obtained by the raction of perfluoroisobutylene with potassium 0,0-diethyldithiophosphate (78% yield) [192, 193], or with potassium sulphide (80% yield) [191]; or by the reaction of bis(trifluoromethyl)ketene with P_2S_5 (60% yield) [191].

Procedure

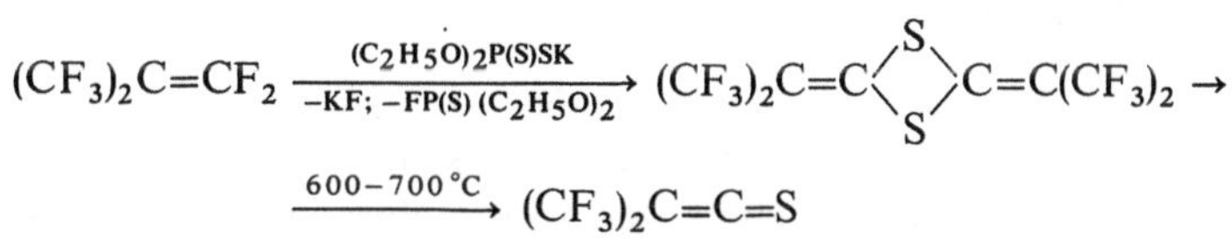

Potassium 0,0-diethyldithiophosphate [195]. In a 500-ml two-necked flask provided with a reflux condenser, and a dropping funnel, are placed 110 g (0.5 mol) of P_2S_5

and 150 ml of benzene. The reaction mixture is heated till benzene starts to reflux, then 92 g (2 mol) of anhydrous ethanol is gradually added during 2 h. The mixture is refluxed till hydrogen sulphide has ceased to evolve (10–14 h), then distilled in vacuum to give 110–135 g (60%–75%) of diethyldithiophosphoric acid.

A solution of the acid in the equal volume of anhydrous ethanol is neutralized with 8%–9% ethanolic KOH solution. After neutralization the solvent is partially removed on a rotary vacuum-evaporator and the residue distilled with a large amount of ether. The potassium diethyldithiophosphate precipitate is filtered off and dried in air. The yield is almost quantitative.

Bis(trifluoromethyl)thioketene dimer [192, 193]. In a 100-ml three-necked flask fitted with a stirrer, a gas inlet tube, and a dry-ice reflux condenser, are placed 18 g (0.08 mol) of potassium 0,0-diethyldithiophosphate and 50 ml of anhydrous acetonitrile; then 16 g (0.08 mol) of perfluoroisobutylene (Note) is gradually introduced with stirring. After the gas has been absorbed, the mixture is stirred during 20–30 min, filtered and cooled to —40° ÷ —50 °C. The crystalline precipitate is filtered off and recrystallized from petroleum ether (bp 70 °C). The yield of bis(trifluoromethyl)-thioketene dimer is 12 g (78%); mp 83.5°–84.5 °C [192], 85 °C [191].

Bis(trifluoromethyl)thioketene [191]. A quartz tube (500–700 mm long, 8–10 mm i.d.) is heated to 650°–700 °C in a tube furnace and bis(trifluoromethyl)thioketene dimer is passed, in a slow flow of an inert gas. Bis(trifluoromethyl)thioketene is collected in two receivers cooled to 0 °C and connected in series. Bis(trifluoromethyl)-thioketene may be stored for a long time in a glass container without being dimerized to any significant extent, but even traces of basic reagents may cause quick dimerization of this compound.

Note. Caution! Perfluoroisobutylene is extremely toxic. It should be handled only under an efficient hood (see p. 2).

N-Phenylbis(Trifluoromethyl)Keteneimine

$(CF_3)_2C=C=N-C_6H_5$

$C_{10}H_5F_6N$ MW 253

N-Phenylbis(trifluoromethyl)keteneimine is a light-yellow liquid, soluble in common organic solvents. Bp 70°–71 oC/14 mm Hg [196], 77°–78 °C/18 mm Hg [197].

Method of Preparation

N-Phenylbis(trifluoromethyl)keteneimine is prepared by dehydrofluorination of N-phenyl-2-hydroperfluoroisobutylimine with KOH (90% yield) (Note 1) [196]; by the reaction of perfluoroisobutylene with aniline in the presence of catalytic amounts of triethylamine (90% yield) [196], or with triphenylphosphinephenylimine (34% yield) [198]; or by the reaction of N,N-diphenyliminosulphurane with bis(trifluoromethyl)thioketene (38% yield) [197].

Procedure

1. Dehydrofluorination of N-phenyl-2-hydroperfluoroisobutylimine

$$(CF_3)_2C{=}CF_2 \xrightarrow{C_6H_5NH_2} (CF_3)_2CHCF{=}NC_6H_5 \xrightarrow{KOH} (CF_3)_2C{=}C{=}NC_6H_5$$

Caution! Perfluoroisobutylene is extremely toxic. It should be handled only under an efficient hood.

N-Phenyl-2-hydroperfluoroisobutylimine. Anhydrous ether (40 ml) is placed in a 200-ml four-necked flask equipped with a stirrer, a dropping funnel, a thermometer, and a dry-ice reflux condenser. The flask is cooled to —8° ÷ —10 °C, 22 g (0.11 mol) of perfluoroisobutylene is condensed into it and several drops of pyridine are added; then a solution of 4.65 g (0.05 mol) of freshly distilled aniline in 20 ml of anhydrous ether is added, with stirring, at such a rate that the temperature of the reaction mixture should not rise above —5 °C. The mixture is stirred at that temperature during 1 h and allowed to stand at room temperature for 2.5–3 h. The volatile products (Note 2) and the ether are distilled off on a fractionating column, the residue is distilled in vacuum. The yield of N-phenyl-2-hydroperfluoroisobutylimine is 12.7 g (93%); bp 68°–69 °C/18 mm Hg.

N-Phenylbis(trifluoromethyl)keteneimine. In a 200-ml flask provided with a stirrer and a reflux condenser, is placed a solution of 13.65 g (0.05 mol) of N-phenyl-2-hydroperfluoroisobutylimine in 50 ml of anhydrous ether; then 14.2 g (0.25 mol) of KOH powder is gradually added with continuous stirring and cooling of the flask with ice-water. The reaction mixture is stirred at room temperature for 1.5–2 h (completeness of the reaction is controlled by GLC). After the reaction is complete, the precipitate is filtered off, washed with anhydrous ether, which is then combined with the filtrate and distilled to give 11.4 g (90%) of N-phenylbis(trifluoromethyl)-keteneimine; bp 70°–71 °C/14 mm Hg.

2. Reaction of Perfluoroisobutylene with Aniline

$$(CF_3)_2C{=}CF_2 + C_6H_5NH_2 \rightarrow (CF_3)_2C{=}C{=}NC_6H_5 + (CF_3)_3CH$$

In a 200-ml three-necked flask provided with a stirrer, a dropping funnel, and a dry-ice reflux condenser, are placed 50 ml of anhydrous ether and 0.1–0.2 ml of triethylamine. The flask is cooled with ice-water, 20 g (0.1 mol) of perfluoroisobutylene is condensed into it and a solution of 3 g (0.032 mol) of freshly distilled aniline in 10 ml of anhydrous ether is gradually added with continuous stirring and cooling. The solution is allowed to warm to room temperature and to stand for 2.5–3 h. The volatile products and the ether are distilled off and the residue is distilled in vacuum to give 7.3 g (90%) of N-phenylbis(trifluoromethyl)keteneimine; bp 70°–71 °C/14 mm Hg.

Notes

1. This procedure may be used for the preparation of N-alkyl keteneimines.
2. Perfluoroisobutylene acts as a dehydrofluorinating agent forming monohydroperfluoroisobutane.

1,1-Bis(Dimethylamino)-2,2-bis(Trifluoromethyl)Ethylene

$(CF_3)_2C{=}C[N(CH_3)_2]_2$

$C_8H_{12}F_6N_2$ MW 250

1,1-Bis(dimethylamino)-2,2-bis(trifluoromethyl)ethylene is a white crystalline substance, well soluble in polar solvents (ether, tetrahydrofuran, dimethylformamide), partially soluble in alkanes and aromatic hydrocarbons. Mp 69°–71 °C (from CCl_4); bp 75°–80 °C/6 mm Hg [199].

Method of Preparation

1,1-Bis(dimethylamino)-2,2-bis(trifluoromethyl)ethylene is prepared by amination of perfluoroisobutylene with dimethylamine in ether (50% yield) [199].

Procedure

$$(CF_3)_2C{=}CF_2 + HN(CH_3)_2 \rightarrow (CF_3)_2C{=}C[N(CH_3)_2]_2$$

Caution! Perfluoroisobutylene is extremely toxic. It should be handled only under an efficient hood (see p. 2).

A solution of 18 g (0.4 mol) of anhydrous dimethylamine in 100 ml of anhydrous ether is placed in a 250-ml flask provided with a stirrer, a gas inlet tube, and a dry-ice reflux condenser. The flask is cooled to —30° ÷ —50 °C and 20 g (0.1 mol) of perfluoroisobutylene (Note) is gradually added to the reaction mixture with vigorous stirring. The reaction mixture is then stirred till it has reached room temperature and allowed to stand for 3 days. The precipitate is filtered off, the filtrate is allowed to stand at 0° to —5 °C overnight to ensure complete removal of dimethylamine hydrofluoride, then filtered once more, the ether is evaporated and the residue recrystallized from CCl_4. The yield of 1,1-bis(dimethylamino)-2,2-bis(trifluoromethyl)ethylene is 12.5 g (50%); mp 69°–71 °C; bp 75°–80 °C/6 mm Hg.

Interaction of perfluoroisobutylene with piperidine (—8 °C to 0 °C, without a solvent) gives 1,1-bis(piperidino)-2,2-bis(trifluoromethyl)ethylene (68% yield); mp 83°–85 °C (from hexane) [199]. Diethylaminoperfluoroisobutylene reacts with dimethylamine in ether (in a sealed tube, at 50 °C) to form 1-dimethylamino-1-diethylamino-2,2-bis(trifluoromethyl)ethylene (68% yield); mp 28°–30 °C; bp 90° to 94 °C/13 mm Hg [199]. Likewise, the reaction of (N-perfluoroisobutenyl)piperidine and dimethylamine (in a sealed tube, at room temperature) gives 1-dimethylamino-1-piperidino-2,2-bis(trifluoromethyl)ethylene (85% yield); bp 108°–110 °C/4 mm Hg, n_D^{24} 1.4508 [199].

Note. The gas inlet tube should be of a sufficiently large diameter; otherwise it may be clogged by dimethylamine hydrofluoride.

Perfluoropropylisocyanate

$$CF_3{-}CF_2{-}CF_2{-}N{=}C{=}O$$

C_4F_7NO MW 199

Perfluoropropylisocyanate is a low-boiling colourless liquid with a strong unpleasant odour, soluble in ether, partially soluble in benzene and heptane. Bp 24.5 °C [200], 27°–29 °C [201].

Method of Preparation

Perfluoropropylisocyanate is prepared by heating perfluorobutyryl azide in toluene (70%–80% yield) [200–201], or by the pyrolysis of the silver salt of perfluorobutyric acid N-bromoamide (83% yield) [202].

The following precedure is a modification of the one described in [201].

Procedure

$$C_3F_7COCl \xrightarrow{NaN_3} C_3F_7CON_3 \xrightarrow[-N_2]{} C_3F_7NCO$$

Perfluorobutyryl chloride. In a 150-ml two-necked flask provided with a dropping funnel and a dephlegmator (Note 1) connected with a downward condenser and a trap cooled to −78 °C, is placed 50 g (0.24 mol) of PCl_5; then 40 g (0.17 mol) of perfluorobutyric acid, previously distilled over concentrated H_2SO_4, is added dropwise. A vigorous reaction follows; when it is complete, the mixture is heated in a water-bath. The distillate and the condensate from the trap are combined and distilled on a fractionating column with a glass packing. The yield of perfluorobutyryl chloride is 42 g (96%); bp 38°–40 °C.

Perfluoropropylisocyanate. In a 250-ml four-necked flask provided with a stirrer, a thermometer, a dropping funnel, and a reflux condenser, are placed a solution of 8 g (0.12 mol) of previously activated sodium azide (Note 2) in 20 ml of water and 40 ml of toluene. Then a solution of 25 g (0.108 mol) of perfluorobutyryl chloride in 40 ml of toluene is dropped in with vigorous stirring and cooling of the mixture in an ice-water bath, at such a rate that the temperature of the mixture should not rise above 10 °C. Stirring is continued for 1 h with cooling, then for 6 h at room temperature. The organic layer is separated and the aqueous one extracted with 10 ml of toluene. The toluene solutions are combined, washed with 20 ml of water and dried, first over $MgSO_4$, then during 2 h over metallic sodium. The dried toluene solution is placed in a flask with a reflux condenser connected to a trap cooled to −78 °C and via a calcium chloride tube to a gasometer. The solution is slowly heated to 80 °C and allowed to stand at 80°–100 °C till the gas has ceased to evolve. The mixture is then refluxed (Note 3). The condensed product is distilled from the trap on a fractionating column with a glass packing. The yield of perfluoropropylisocyanate is 16 g (70%); bp 27–29 °C.

Notes

1. Instead of a dephlegmator, a water-cooled reflux condenser may be used.
2. The sodium azide activation is as follows: 10 g of the azide and 1 ml of hydrazine hydrate are ground in a porcelain mortar and allowed to stand at room temperature for 24 h, then dissolved in 10 ml of hot water and allowed to settle by adding 1 to 1.5 l of acetone. The precipitate is filtered off, washed on a filter with acetone and dried in air. The yield of the activated NaN_3 is 8–9 g.
3. To ensure complete removal of the product from the reaction mixture at the end of heating, cooling of the reflux condenser is stopped.

Perfluorotrimethylenediisocyanate

$$O{=}C{=}N{-}CF_2{-}CF_2{-}CF_2{-}N{=}C{=}O$$

$C_5F_6N_2O_2$ MW 234

Perfluorotrimethylenediisocyanate is a colourless liquid with a strong unpleasant odour, soluble in common organic solvents. Bp 84°–85 °C [203], 64°–65 °C [204].

Method of Preparation

Perfluorotrimethylenediisocyanate is prepared by decomposition of perfluoroglutaryl diazide obtained by nitrosation of perfluoroglutaryl dihydrazide (72% yield) [203], or by the reaction of perfluoroglutaryl dichloride with trimethylsilyl azide (79% yield) [204].

Procedure

$$C_2H_5OOC(CF_2)_3COOC_2H_5 \xrightarrow{N_2H_4} H_2NNHCO(CF_2)_3CONHNH_2 \xrightarrow{HNO_2}$$

$$\rightarrow N_3CO(CF_2)_3CON_3 \xrightarrow[-N_2]{heating} OCN(CF_2)_3NCO$$

Perfluoroglutaryl dihydrazide [205]. Hydrazine hydrate (3 g, 0.06 mol) is added to a stirred solution of 9 g (0.03 mol) of diethyl perfluoroglutarate in 25 ml of anhydrous ethanol. The mixture is stirred at room temperature for 1–1.5 h. The crystalline precipitate is filtered off. The yield of perfluoroglutaryl dihydrazide is 7.4 g (85%); mp 168°–169 °C (from ethanol).

Similar procedures give perfluoroadipyl dihydrazide [mp 188°–189 °C (from ethanol)], perfluorosebacyl dihydrazide [mp 202°–204 °C (from ethanol)].

Perfluorotrimethylenediisocyanate. To a solution of 2.7 g (0.01 mol) of perfluoroglutaryl dihydrazide and 3.5 ml of concentrated HCl in 15 ml of water is added 30 ml of m-xylene; then a solution of 2.8 g (0.04 mol) of $NaNO_2$ in 5 ml of water is slowly added with continuous stirring and cooling of the mixture to 0° ÷ —5 °C, at such a rate that the temperature of the reaction mixture should not rise above 5°–8 °C. After mixing is complete, stirring is continued at the same temperature for 2 h, the xylene layer is then separated, washed with $NaHCO_3$ solution and with water, dried over $MgSO_4$, then over metallic sodium. The resulting xylene solution of perfluoroglutaryl diazide is heated at 90°–130 °C in nitrogen atmosphere. After nitrogen has ceased to evolve, the reaction mixture is distilled. The yield of perfluorotrimethylenediisocyanate is 1.7 g (72%); bp 84°–85 °C.

A similar procedure gives perfluorotetramethylenediisocyanate; bp 105°–106 °C; yield 78% [203].

2,2,2-Trifluoroethylamine Hydrochloride

$CF_3{-}CH_2NH_2 \cdot HCl$

$C_2H_5ClF_3N$ MW 135.5

2,2,2-Trifluoroethylamine is a colourless liquid. Bp 37 °C/740 mm Hg; n_D^{30} 1.295; d_4^{30} 1.245 [206].

Method of Preparation

2,2,2-Trifluoroethylamine is prepared by the reaction of trifluoroacetic amide with $LiAlH_4$ (80% yield) [207]; by the catalytic reduction of trifluoroacetonitrile with hydrogen on platinum (80% yield) [208]; by the reduction of trifluoronitroethane with hydrogen in statu nascendi; by amination of 2-chloro-1,1,1-trifluoroethane with ammonia [209]; or by the Curtius rearrangement of trifluoropropionyl azide (25% yield) [206].

Procedure [133, 210]

$$CF_2{=}CH_2 \xrightarrow{HNO_3;\,HF} CF_3{-}CH_2NO_2 \xrightarrow{Fe;\,HCl} CF_3{-}CH_2NH_2 \cdot HCl$$

2,2,2-Trifluoronitroethane [210]. Anhydrous hydrogen fluoride (600 ml, 30 mol) (Note 1) is placed in a steel reactor provided with a metallic stirrer, a dropping funnel, a gas inlet tube, and a dry-ice reflux condenser. The reactor is cooled to —78 °C and 240 ml (6 mol) of anhydrous HNO_3 is added at that temperature with stirring; then dry vinylidene fluoride is let into the reactor at —60 °C at the rate of 150 l/h. After 50 min, vinylidene fluoride (5.4 mol) is absorbed and the flow is cut off. The reaction mixture is poured onto 3 kg of ice (Note 2), the lower organic layer is separated and the aqueous one is extracted with three portions of ether (Note 3). The ether extract is combined with the organic layer, washed twice with water, dried over $CaCl_2$ and distilled on a 250 mm fractionating column. The yield of 2,2,2-trifluoronitroethane is 400 g (59%); bp 43 °C/106 mm Hg; n_D^{20} 1.3300.

2,2,2-Trifluoroethylamine hydrochloride [133]. In a three-necked flask provided with a stirrer, a dropping funnel, and a reflux condenser, are placed 56 g of iron turnings, 0.2 g of $FeCl_3$, 200 ml of water and 26 g (0.2 mol) of 2,2,2-trifluoronitroethane. The flask is heated to 90°–95 °C and 120 ml of concentrated HCl is dropped in at that temperature, with stirring during 2 h. Stirring is continued at the same temperature for 2 h, then cooled, neutralized with a solution of 8 g of NaOH in 250 ml of water and steam-distilled. The distillate is acidified with HCl and evaporated in vacuum. The yield of 2,2,2-trifluoroethylamine hydrochloride is 21 g (77%).

Notes

1. For the safety measures to be taken while handling hydrogen fluoride see p. 1.
2. During decomposition of the reaction mixture with ice, the reactor should be cooled, e.g. with a mixture of carbon tetrachloride and chloroform (1:1 v/v) with dry-ice.
3. Extraction from aqueous solutions containing HF may be conducted in a thick-wall glass dropping funnel provided with a Teflon stopcock. However, it should be borne in mind that such a funnel does not with stand long employment.

2,2,3,3,3-Pentafluoropropylamine

$CF_3-CF_2-CH_2-NH_2$

$C_3H_4F_5N$ MW 149

2,2,3,3,3-Pentafluoropropylamine is a low-boiling colourless liquid. Bp 49°–49.5 °C/735 mm Hg [211], 48°–48.5 °C/755 mm Hg [212]; n_D^{20} 1.297.

Method of Preparation

2,2,3,3,3-Pentafluoropropylamine is prepared by the reduction of perfluoropropionyl amide with $LiAlH_4$ (80% yield) [211], or by the reaction of tetrafluoroethylene with hexamethylenetetramine in anhydrous HF (44% yield) [212].

Procedure [212]

$$CF_2{=}CF_2 \xrightarrow{HF;(CH_2)_6N_4} CF_3-CF_2-CH_2NH_2$$

In a 250-ml steel autoclave are placed 120 g of HF (Note 1) and 21 g (0.15 mol) of urotropine. The autoclave is cooled with liquid nitrogen and evacuated, then 60 g (0.6 mol) of tetrafluoroethylene is condensed into it. The autoclave is placed in a tilting furnace and heated at 50 °C during 5 h, whereupon it is cooled to room temperature and the unconverted gas (48 g, 0.48 mol) is let out in a gasometer. The reaction mixture is poured onto ice, made alkaline with aqueous KOH, extracted with ether (4 × 75 ml) and the ether extract is dried over $MgSO_4$. The bulk of the ether is distilled off in a water-bath and residual ether is removed by distillation in vacuum of a water-jet pump; the volatile products are collected in a trap cooled to —78 °C (Note 2). The residue is distilled at 120 °C/2 mm Hg to give 7 g of N,N′,N″-tris-(2,2,3,3,3-pentafluoropropyl)hexahydrotriazine; bp 76°–96 °C/7 mm Hg; n_D^{20} 1.3498.

Hexahydrotriazine is hydrolysed in a mixture of 15 ml of concentrated HCl and 100 ml of water; the liberated formaldehyde is distilled off with steam. Distilling off formaldehyde is stopped when the condensate does not give the precipitate with a solution of 2,4-dinitrophenylhydrazine any longer. The residue is evaporated on a rotary vacuum evaporator to give 6.6 g of pentafluoropropylamine hydrochloride. The ether distillate obtained earlier is saturated with dry hydrogen chloride; the precipitate is filtered to give further 3 g of pentafluoropropylamine hydrochloride; mp 248 °C (from ethanol).

In a 250-ml flask connected with a straight condenser is placed 65 g (0.35 mol) of pentafluoropropylamine hydrochloride, then a little of ice and 40 ml of 50% NaOH are added with cooling. The liberated amine is distilled off. The yield of pentafluoropropylamine is 40 g; bp 48°–48.5 °C/755 mm Hg; n_D^{20} 1.297.

The total yield of pentafluoropropylamine is 44% (based on the converted olefin).

Notes

1. For the safety measures to be taken while handling anhydrous HF see p. 1.
2. Pentafluoropropylamine condenses in the trap together with other volatile products.

Perfluoro-*tert*-butylamine

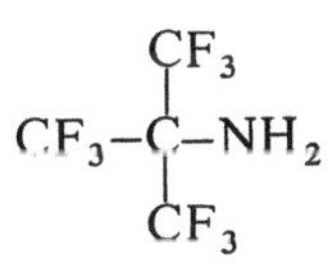

$C_4H_2F_9N$ MW 235

Perfluoro-*tert*-butylamine is a low-boiling colourless liquid. Bp 56.5°–57 °C; n_D^{20} 1.2795; d_4^{20} 1.6782 [213].

Method of Preparation

Perfluoro-*tert*-butylamine is prepared by the reduction of perfluoro-*tert*-butyl hydroxylamine with hydroiodic acid and red phosphorus (70% yield) [213].

Procedure

$$(CF_3)_3C{-}NO \xrightarrow{H_2;\,Pd} (CF_3)_3C{-}NHOH \xrightarrow{HI;\,P} (CF_3)_3CNH_2$$

Perfluoro-tert-butyl hydroxylamine. In a 250-ml stainless steel autoclave are placed 85 g (0.34 mol) of nitrosoperfluoro-*tert*-butane (for the preparation see p. 25), 15 ml of methanol and 0.3 g of palladium black. The autoclave is flushed with hydrogen, the pressure is brought to 100 atm and the autoclave is shaken at room temperature during 10 h. The reaction mixture is poured out of the autoclave, washed with water, with a saturated solution of $CaCl_2$, then with concentrated H_2SO_4, and distilled to give 69 g of compound containing, as shown by GLC, 87% perfluoro-*tert*-butyl hydroxylamine. Redistillation affords pure perfluoro-*tert*-butyl hydroxylamine. Bp 94°–95 °C; mp 40 °C; soluble in common organic solvents; slightly soluble in water.

Perfluoro-tert-butylamine. In a 50-ml Teflon-lined steel autoclave (Note) are placed 12 g of 87% perfluoro-*tert*-butyl hydroxylamine, 20 ml of 57% hydroiodic acid and 2 g of red phosphorus. The autoclave is placed in a tilting furnace and heated to 150°–160 °C during 15 h. Then the autoclave is cooled to room temperature, the reaction mixture is poured out of the autoclave and distilled. The yield of perfluoro-*tert*-butylamine is 5.75 g (70%); bp 56.5°–57 °C; n_D^{20} 1.2795.

Note. Reduction of perfluoro-*tert*-butyl hydroxylamine in a stainless steel autoclave leads to 2-hydroperfluoroisobutane.

2,2-Difluoro-3,3-bis(Trifluoromethyl)Aziridine

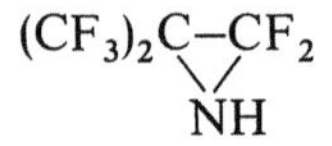

C_4HF_8N MW 215

2,2-Difluoro-3,3-bis(trifluoromethyl)aziridine is a low-boiling colourless liquid, soluble in common organic solvents. Bp 44°–45 °C; n_D^{20} 1.2797 [214].

Method of Preparation

2,2-Difluoro-3,3-bis(trifluoromethyl)aziridine is prepared by deacylation of 2,2-difluoro-1-benzoyl-3,3-bis (trifluoromethyl)-aziridine by heating it with concentrated H_2SO_4 (43.5% yield) [214].

The following procedure is a modification of the one described in [214].

Procedure

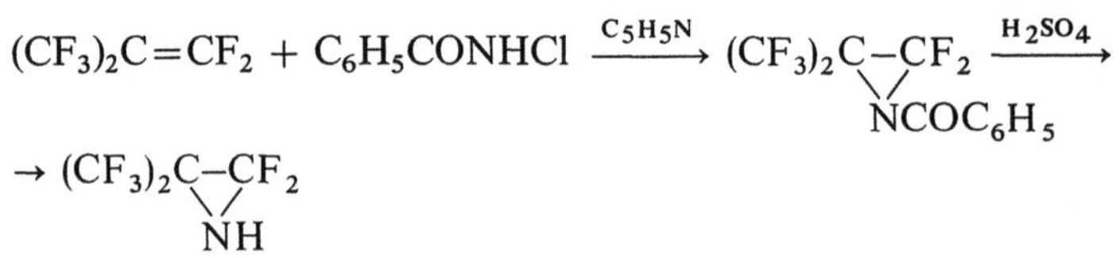

Caution! Perfluoroisobutylene is extremely toxic. It should be handled only under an efficient hood (see p. 2).

2,2-Difluoro-1-benzoyl-3,3-bis(trifluoromethyl)aziridine. In a 250-ml four-necked flask provided with a stirrer, a thermometer, a gas inlet tube, and a dry-ice reflux condenser, are placed 13.3 g (0.085 mol) of N-chlorobenzamide [215], 140 ml of anhydrous dimethylformamide and 3 ml of dry pyridine. Then 35 g (0.175 mol) of perfluoroisobutylene is bubbled during 1–1.5 h through the solution, with stirring and cooling of the flask with ice-water; the temperature of the reaction mixture should be maintained in the range of 8° to 12 °C. After the gas has been absorbed, the solution is allowed to warm to room temperature for 0.5 h, then it is allowed to stand for 30–40 min. The volatile products are then removed in vacuum (12–15 mm Hg) at 20 °C or below. The residue is poured into 500 ml of ice-water and acidified with diluted HCl. The oily layer is extracted with ether, the combined ether extracts are washed two or three times with cold water and dried over $MgSO_4$. The ether is evaporated in vacuum and the residue is distilled to give 17.7 g (65%) of crude 2,2-difluoro-1-benzoyl-3,3-bis(trifluoromethyl)aziridine, bp 60°–70 °C/6 mm Hg, which is used in the next stage without further purification.

2,2-Difluoro-3,3-bis(trifluoromethyl)aziridine. 2,2-Difluoro-1-benzoyl-3,3-bis(trifluoromethyl)aziridine (26.5 g, 0.083 mol) is placed in a 100-ml three-necked flask fitted with a thermometer, a dropping funnel, and a reflux condenser connected to a trap cooled to —40° ÷ —50 °C. Then 30 ml of concentrated H_2SO_4 is dropped in the flask at 80°–100 °C/100–110 mm Hg. The reaction mixture is then heated at 110 °C during 10 min and the product condensed in the trap is distilled to give 11.2 g (62%) of 2,2-difluoro-3,3-bis(trifluoromethyl)aziridine; bp 44°–46 °C.

Pentafluoro-2-azapropene

$CF_3{-}N{=}CF_2$

C_2F_5N MW 133

Pentafluoro-2-azapropene is a colourless gas. Bp —33.7 °C [216], —33° to —31 °C [217], —28° to —25 °C [218].

Method of Preparation

Pentafluoro-2-azapropene is prepared by the pyrolysis of perfluoro-2-methyl-1,2-oxazetidine (100% yield) [216], of trifluoronitrosomethane-tetrafluoroethylene copolymer (100% yield) [216], of perfluorodimethylcarbamoyl fluoride (96% yield) [217], or by the pyrolysis of pentafluoro-2-azapropene dimer over NaF (84% yield) [218].

Procedure [218]

$$ClCN \xrightarrow[250\,°C]{HF} (CF_3)_2NH \xrightarrow{KF} (CF_3)_2NCF{=}NCF_3 \xrightarrow{NaF} CF_3NCF_2$$

Bis(trifluoromethyl)amine. A mixture of 61 g (1 mol) of ClCN and 100 g (5 mol) of anhydrous HF is heated in a rotary autoclave at 75 °C for 3 h, then at 150 °C for 1 h and at 250 °C for 6 h. The autoclave is then cooled, the mixture of the products is passed through a column packed with granulated NaF (size of a granule ~3 mm) and condensed in a cooled metallic reactor designed for work under pressure and containing 250 g of granulated NaF. The procedure is rerun seven times (Note). The volatile products obtained in the eight runs are combined and maintained during

24 h over NaF at room temperature. Then the crude product is distilled on a fractionating column with a glass packing. The yield of bis(trifluoromethyl)amine is 141 g (92%); bp —8° to —6 °C.

Pentafluoro-2-azapropene. A mixture of 110 g of bis(trifluoromethyl)amine and 120 g of anhydrous KF is heated during 6 h in a 500 ml rocking autoclave at 200 °C. The products obtained in the two runs are combined and distilled to give 154 g (80% conversion) of pentafluoro-2-azapropene dimer; bp 33°–36 °C.

A nickel tube (18 mm in diameter) is packed with 90 g of granulated NaF (size of granule ~3 mm), heated to 700 °C in a flow of nitrogen and cooled to 525°–530 °C; then 194 g (0.73 mol) of pentafluoro-2-azapropene dimer is added during 8.5 h at the above temperature and atmospheric pressure. The pyrolysis products are collected in the cooled traps and distilled. The yield of pentafluoro-2-azapropene is 163 g (84%); bp —28 ° to —25 °C.

Note. The reactor is charged and discharged without being opened.

Trifluoronitrosomethane

CF_3—NO

CF_3NO MW 99

Trifluoronitrosomethane is a blue gas. Bp —84 °C [219].

Method of Preparation

Trifluoronitrosomethane is prepared by the reaction of iodotrifluoromethane with nitrogen oxide under UV irradiation (70%–90% yield) [216, 219], or by the pyrolysis of trifluoroacetyl nitrite (56% yield) [220].

Procedure [220, 221]

$$(CF_3CO)_2O \xrightarrow{N_2O_3} CF_3COONO \xrightarrow{90\ ^\circ C} CF_3NO$$

Trifluoroacetyl nitrite [221]. In a 250-ml three-necked flask provided with a dropping funnel, a gas inlet tube, and a dry-ice reflux condenser, is condensed 10.5 g (0.138 mol) of N_2O_3 at —78 °C. The temperature of the cooling bath is then raised to —10° to —5 °C and 28.7 g (0.138 mol) of trifluoroacetic anhydride is dropped into the flask at that temperature during 5 min. The reaction mixture is allowed to stand at —10° to —5 °C during 30 min, then distilled in vacuum. The yield of trifluoroacetyl nitrite is 36.5 g (quantitative); bp 45 °C/80 mm Hg (Note 1).

Trifluoronitrosomethane [220]. Trifluoroacetyl nitrite (24.7 g, 0.173 mol) is placed in a 100-ml pear-shaped flask provided with a gas inlet tube reaching to the bottom of the flask and a Pyrex capillary tube (280 mm long, 20 mm in dieameter, capillary diameter 1 mm). A thermocouple is attached to the capillary tube by means of aluminium foil to which is attached a heating coil. The flask is heated to 75°–80 °C in an oil-bath and a flow of nitrogen is let in at that temperature through the gas inlet tube at the rate of 150 ml/min (Note 2). Trifluoroacetyl nitrite vapours pass, together with nitrogen, through the capillary tube kept heated to 190°–192 °C, during 3.5 h. The pyrolysis products are collected in three traps connected in series. The first trap is cooled to —78 °C, the other two are cooled with liquid nitrogen.

After the pyrolysis is complete, the contents of the first trap are heated to —20 °C; the low-boiling volatile products are collected in two other traps cooled with liquid nitrogen, passed through a gas-washing bottle containing 5% KOH solution, then through a tube packed with Drierite, condensed and distilled. The yield of trifluoronitrosomethane is 9.6 g (56%); bp —84 °C (Note 3).

Notes

1. Trifluoroacetyl nitrile may explode upon heating. The explosive power of trifluoroacetyl nitrite is comparable to that of common explosives.
2. Trifluoroacetyl nitrite vapours explode upon heating unless diluted with an inert gas. Partial pressure of trifluoroacetyl nitrite vapours in the pyrolysis tube should not exceed 300 mm Hg, otherwise an explosion will occur.
3. Trifluoronitrosomethane is toxic, it should be handled only under an efficient hood.

Trifluoro-1-chloro-1-nitroso-2-nitroethane

$O_2N—CF_2—CFCl—NO$

$C_2ClF_3N_2O_3$ MW 192.5

Trifluoro-1-chloro-1-nitroso-2-nitroethane is a volatile dark-blue liquid. Bp 0 °C/33 mm Hg, n_D^{20} 1.3560, d_4^{20} 1.5952 [222]; bp 57 °C, n_D^{20} 1.3360, d_4^{20} 1.557 [223]; bp 10°–12 °C/90 mm Hg, n_D^{25} 1.349, d_4^{25} 1.5494 [224].

Method of Preparation

Trifluoro-1-chloro-1-nitroso-2-nitroethane is prepared by the reaction of trifluorochloroethylene with wet nitrogen dioxide (61% yield) [222], by the gas-phase reaction of trifluorochloroethylene with N_2O_3 (61% yield) [223], or by the reaction of NO with trifluorochloroethylene under pressure (3% yield) [224].

Procedure

$$N_2O_4 \xrightarrow{H_2O} N_2O_3 + HNO_3$$

$$CF_2{=}CFCl + N_2O_3 \rightarrow O_2NCF_2CFClNO\ (+\ O_2NCF_2CFClONO)$$

In a three-necked flask equipped with a stirrer, a gas inlet tube, and a reflux condenser, is placed a solution of 10 ml of wet N_2O_4 in 50 ml of tetrachloroethane, then trifluorochloroethylene is passed in with stirring till the nitrogen oxide vapours are no longer present over the reaction mixture. The mixture is washed with two 20–25 ml portions of cold water, the organic layer is separated and washed with concentrated H_2SO_4. Trifluoro-1-chloro-1-nitroso-2-nitroethane is distilled off under reduced pressure (at 20°–30 °C/30–40 mm Hg) into a trap cooled to —78 °C, then redistilled on a fractionating column with a low-temperature distilling head. The yield of trifluoro-1-chloro-1-nitroso-2-nitroethane is 39 g (61%); bp 0 °C/33 mm Hg; n_D^{20} 1.3560.

Tetrafluoro-2-nitrosoethyl Ether of Hexafluoroacetone Oxime

$$(CF_3)_2C{=}N{-}O{-}CF_2{-}CF_2{-}NO$$

$C_5F_{10}N_2O_2$ MW 310

Tetrafluoro-2-nitrosoethyl ether of hexafluoroacetone oxime is a volatile dark-blue liquid. Bp 32°–33 °C/330 mm Hg, n_D^{20} 1.2900, d_4^{20} 1.6210 [225].

Method of Preparation

Tetrafluoro-2-nitrosoethyl ether of hexafluoroacetone oxime is prepared by the addition of hexafluoroacetone oxime nitrite obtained by the reaction of bis(trifluoromethyl)ketene with N_2O_3, to tetrafluoroethylene with heating to 150 °C (35% yield) [225].

Procedure

$$(CF_3)_2C{=}C{=}O + N_2O_4 + NO \xrightarrow[-CO_2]{} (CF_3)_2C{=}N{-}ONO$$

$$(CF_3)_2C{=}N{-}ONO + F_2C{=}CF_2 \rightarrow (CF_3)_2C{=}N{-}O{-}CF_2CF_2NO$$

Hexafluoroacetone oxime nitrite [225]. In a 100-ml five-necked flask fitted with a stirrer, a thermometer, two gas inlet tubes, and a reflux condenser which is connected in series with a trap cooled to —78 °C and a gasometer (Note), is placed 12.2 g (0.132 mol) of N_2O_4, then NO is passed at —5° to —10 °C with gradual cooling of the flask to —60 °C. Bis(trifluoromethyl)ketene (46.3 g, 0.26 mol) (for the preparation see p. 72) is slowly added to the reaction mixture with continuous passing of NO. After all the bis(trifluoromethyl)ketene has been added, the temperature of the mixture is raised during 4 h to that of the room; by the end of that period 5.6 l of CO_2 has been liberated and the colour of the mixture has changed from blue to brown. The mixture is distilled on a fractionating column to give 38.8 g (74%) of hexafluoroacetone oxime nitrite; bp 46°–47 °C, n_D^{22} 1.3140.

Tetrafluoro-2-nitrosoethyl ether of hexafluoroacetone oxime. Tetrafluoroethylene (2.5 l, 0.11 mol) is bubbled during 1 h through 7 g (0.03 mol) of hexafluoroacetone oxime nitrite placed in a 30 ml gas-washing bottle. The vapours are passed through a glass tube 500 mm long, heated to 150 °C. The products are collected in two traps connected in series, cooled to 0° and —70 °C, respectively. The products condensed in the two traps are combined and distilled twice under reduced pressure to give

3.5 g (35 %) of tetrafluoro-2-nitrosoethyl ether of hexafluoroacetone oxime; bp 32° to 33 °C/330 mm Hg, n_D^{22} 1.2900.

Note. The starting compounds and the products are higly hygroscopic, therefore it is important that all parts of the apparatus be thoroughly dried and precautions be taken to keep out air moisture.

Ethyl α-Nitrosoperfluoropropionate

$$CF_3-\underset{\displaystyle NO}{\underset{|}{C}}F-COOC_2H_5$$

$C_5H_5F_4NO_3$ MW 203

Ethyl α-nitrosoperfluoropropionate is a deep-blue liquid. Bp 42 °C/100 mm Hg; n_D^{20} 1.3159 [226].

Method of Preparation

Ethyl α-nitrosoperfluoropropionate is prepared by decomposition of ethylperfluoropropenyl ether-nitrogen tetroxide adduct with water (69 % yield) [226].

Procedure

$$CF_3-CF=CF-OC_2H_5 \xrightarrow{N_2O_4;\,H_2O} CF_3-\underset{\displaystyle NO}{\underset{|}{C}}F-COOC_2H_5$$

Ethylperfluoropropenyl ether (9.6 g, 0.05 mol) is added dropwise at −78 °C, with intermittent shaking, to a suspension of 7.5 g (0.08 mol) of N_2O_4 in 15 ml of dichlorodifluoromethane. After 15 min, the resulting dark-blue solution is slowly poured with stirring into ice-water and stirring is continued for further 30 min. The organic layer, a dark-blue liquid, is separated, dried over P_2O_5 and distilled in vacuum. The yield of ethyl α-nitrosoperfluoropropionate is 7.7 g (69 %); bp 42 °C/100 mm Hg.

Difluoronitromethane

F_2CH-NO_2

CHF_2NO_2 MW 87

Difluoronitromethane is a colourless liquid, soluble in organic solvents, partially soluble in water. Bp 40°–41 °C [227], 42 °C [228, 118], 43°–43.5 °C [141], 2 °C/2 mm Hg [229]; n_D^{20} 1.3168 [230], 1.3200 [141]; d_4^{20} 1.4600 [228, 118], 1.4660 [227].

Method of Preparation

Difluoronitromethane is prepared by decomposition of pentafluoronitroacetone in an alkaline medium (50% yield) [118], by decarboxylation of difluoronitroacetic acid in an alkaline medium (46% yield) [141] or in the presence of KF (a small yield) [227], by oxidation of 1,1,4,4-tetrafluoro-1,4-dinitrobutene-2 with $KMnO_4$ in an alkaline medium (36% yield) [230], or by the pyrolysis of 5,5-difluoro-5-nitro-4-hydroxy-4-trifluoromethyl-2-methylpentene-1 at 100 °C (82% yield) [231].

Procedure [228, 118]

$$(HO)_2\underset{\displaystyle CF_3}{\underset{|}{C}}-CF_2NO_2 + NaOH \rightarrow CHF_2NO_2$$

In a 100-ml Claisen flask provided with a dropping funnel, a thermometer, and a downward condenser connected with a receiver and an ice-cooled trap, is placed 50 ml of 10% NaOH solution, then 22.8 g (0.1 mol) of pentafluoronitroacetone hydrate (for the preparation see p. 42) is cautiously (to avoid boiling up of the mixture) added dropwise at 80 °C. The resulting difluoronitromethane is removed, dried over $MgSO_4$ and distilled. The yield of difluoronitromethane is 6.5 g (62.5%); bp 42 °C.

Perfluoro(2-methylpropane)thiol-2

(Perfluoro-*tert*-butylmercaptan)

$(CF_3)_3C-SH$

C_4HF_9S MW 252

Perfluoro(2-methylpropane)thiol-2 is an extremely volatile white crystalline substance with an unpleasant odour. Mp 63.5°–64 °C [232].

Method of Preparation

Perfluoro(2-methylpropane)thiol-2 is prepared by treatment of its mercury salt with hydrochloric acid (92% yield) [232].

Procedure [233, 232]

$$(CF_3)_2C{=}CF_2 \xrightarrow[(CF_3COO)_2Hg]{KF;} [(CF_3)_3C]_2Hg \xrightarrow{S;\,KF} [(CF_3)_3CS]_2Hg \xrightarrow{HCl}$$

$$\rightarrow (CF_3)_3CSH$$

Caution! Perfluoroisobutylene is extremely toxic. It should be handled only under an efficient hood (see p. 2).

Bis(perfluoro-tert-butyl)mercury [233]. In a 100-ml three-necked flask provided with a stirrer, a gas inlet tube, and a dry-ice reflux condenser, are placed 17 g (0.04 mol) of mercury trifluoroacetate (Note 1), 7 g (0.12 mol) of freshly calcinated KF powder, 50 ml of anhydrous dimethylformamide, then 15 g (0.08 mol) of perfluoroisobutylene is gradually added with vigorous stirring. After the exothermic reaction is complete, the reaction mixture is stirred during 15 min, then poured into water, the lower layer is separated and distilled from concentrated H_2SO_4; the distillate (a fraction boiling at 120°–140 °C) is collected in a water-cooled receiver. The yield of bis(perfluoro-tert-butyl)mercury is 19 g (80%); mp 64°–65 °C (Note 2).

Bis(perfluoro-tert-butylthio)mercury [232]. In a 100-ml round-bottom flask fitted with a stirrer and a reflux condenser, are placed 19.1 g (0.03 mol) of bis(perfluoro-tert-butyl)mercury, 6.1 g (0.18 mol) of sulphur, 1 g (0.017 mol) of freshly calcinated KF powder and 60 ml of anhydrous dimethylformamide. The mixture is heated at 70°–80 °C with vigorous stirring for 20–25 min, during which time it turns black. The completeness of the reaction is checked by [^{19}F] NMR (Note 3). The reaction mixture is then cooled, filtered, and the filtrate washed with water. The crystalline precipitate is filtered off and recrystallized from hexane to give 17.3 g (82%) of bis-(perfluoro-*tert*-butylthio)mercury as a 2:1 complex with dimethylformamide; mp 78.6°–79.5 °C.

Perfluoro(2-methylpropane)thiol-2 [232]. To 17.3 g (0.023 mol) of bis(perfluoro-*tert*-butylthio)mercury-dimethylformamide complex is added 35 ml of concentrated HCl. The mixture is cautiously heated in vacuum of a water-jet pump till the crystalline thioxide has completely disappead. The volatile products are passed through the gas-washing bottle with concentrated H_2SO_4 and collected in a U-tube cooled to —78 °C. The yield of perfluoro(2-methylpropane)thiol-2 is 11.4 g (92%); mp 63.5°–64 °C (in a sealed capillary).

Notes

1. For the preparation of mercury trifluoroacetate see p. 97.
2. Bis(perfluoro-*tert*-butyl)mercury is extremely volatile; its vapours are toxic. It should be handled under a hood; hands should be protected with rubber gloves.
3. Instead of the signal of bis(perfluoro-*tert*-butyl)mercury (singlet at 21.4 ppm from CF_3COOH as external reference, $J_{Hg-C-C-F}$ 145 Hz) there appears the signal of thioxide (singlet at 10 ppm).

1,2-Difluoro-2-chlorovinylethyl Sulphide

$CFCl{=}CF{-}S{-}C_2H_5$

$C_4H_5ClF_2S$ MW 158.5

1,2-Difluoro-2-chlorovinylethyl sulphide is a transparent liquid. Bp 106°–106 °C, n_D^{20} 1.4385, d_4^{20} 1.282 [234].

Method of Preparation

1,2-Difluoro-2-chlorovinylethyl sulphide is prepared by dehydrofluorination of 1,1,2-trifluoro-2-chlorodiethyl sulphide with dry alkali (90% yield) [234].

Procedure

$$CF_2{=}CFCl + C_2H_5SH \xrightarrow{KOH} CHFCl{-}CF_2{-}S{-}C_2H_5 \xrightarrow[-HF]{KOH} CFCl{=}CF{-}S{-}C_2H_5$$

1,1,2-Trifluoro-2-chlorodiethyl sulphide. In a 100-ml stainless steel autoclave are placed 17 g (0.15 mol) of trifluorochlorethylene, 7 g (0.11 mol) of ethyl mercaptan, and 0.3 g (0.007 mol) of KOH. The autoclave is heated at 115°–125 °C during 6 h. After the unconverted trifluorochloroethylene has been distilled off, the reaction mixture is washed with an aqueous alkaline solution and extracted with ether. The ether extract is dried over $MgSO_4$, the ether is distilled off and the residue distilled. The yield of 1,1,2-trifluoro-2-chlorodiethyl sulphide is 16 g (80%); bp 120 °C.

1,2-Difluoro-2-chlorovinylethyl sulphide. In a 30-ml flask fitted with a dephlegmator are placed 10 g (0.056 mol) of 1,1,2-trifluoro-2-chlorodiethyl sulphide and 6.5 g (0.114 mol) of dry KOH. The mixture is heated at 120°–130 °C with simultaneous removal of 1,2-difluoro-2-chlorovinylethyl sulphide formed in the reaction. After redistillation the yield of 1,2-difluoro-2-chlorovinylethyl sulphide is 8 g (90%); bp 106°–107 °C.

Ethylperfluoroisobutenyl Sulphide

$(CF_3)_2C{=}CF{-}S{-}C_2H_5$

$C_6H_5F_7S$ MW 242

Ethylperfluoroisobutenyl sulphide is a colourless liquid with a faint sweet odour, soluble in common organic solvents. Bp 66 °C/88 mm Hg, n_D^{23} 1.3759, d_4^{23} 1.4193 [235].

Method of Preparation

Ethylperfluoroisobutenyl sulphide is prepared by the reaction of perfluoroisobutylene with sodium ethanthioxide in ether at 5°–6 °C (55% yield) [235].

Procedure

$$(CF_3)_2C{=}CF_2 + NaSC_2H_5 \rightarrow (CF_3)_2C{=}CF{-}S{-}C_2H_5$$

Caution! Perfluoroisobutylene is extremely toxic. It should be handled only under an efficient hood (see p. 2).

In a 100-ml four-necked flask equipped with a stirrer, a thermometer and a dry-ice reflux condenser, are placed, with cooling of the flask to —30 °C, 20 g (0.1 mol) of perfluoroisobutylene and 50 ml of anhydrous ether. Then sodium ethanthioxide (8.5 g, 0.1 mol) (Notes 1, 2) is gradually added with vigorous stirring, at such a rate that the temperature of the mixture should not rise above 5°–6 °C (the reaction flask is intermittently cooled in an ice-bath). The mixture is stirred for 1.5 h, whereupon the ethereal solution is filtered, the ether is distilled off and the residue is distilled in vacuum. The yield of ethylperfluoroisobutenyl sulphide is 13.3 g (55%); bp 66 °C /80 mm Hg, n_D^{23} 1.3759.

Likewise, perfluoroisobutylene and sodium butanethioxide give butylperfluoroisobutenyl sulphide (bp 86°–87 °C/50 mm Hg, n_D^{21} 1.3970; yield 48%) [235]; tetrafluoroethylene and sodium ethanthioxide (autoclave, ether, 20 °C) afford ethyltrifluorovinyl sulphide (bp 73 °C, yield 31%) [236].

Notes

1. Sodium ethanthioxide should be thoroughly purified from ethyl mercaptan. For that purpose the thioxide obtained is heated at 120°–150 °C in vacuum of an oil pump till ethyl mercaptan, condensed in a trap cooled to —78 °C, is no longer distilled off.
2. Addition of perfluoroisobutylene to a suspension of sodium ethanthioxide in ether at 20°–25 °C results in the formation of bis(trifluoromethyl)thioketene S,S-diethylacetal (bp 85°–86 °C at 10 mm Hg, yield 41%) [235]; likewise, perfluoroisobutylene and sodium butanethioxide give bis(trifluoromethyl)thioketene butylacetal (bp 120 °C/10 mm Hg, yield 45%) [235].

Bis(trifluoromethyl)Disulphide

$CF_3—S—S—CF_3$

$C_2F_6S_2$ MW 202

Bis(trifluoromethyl) disulphide is a low-boiling liquid, partially soluble in ether, dioxan, dimethylformamide, better soluble in halogenated alkanes (chloroform, reons). Bp 34 °C [237], 35 °C [238].

Method of Preparation

Bis(trifluoromethyl) disulphide is prepared by heating trifluoroiodomethane with sulphur (in an autoclave at 260 °C, yield 60%) [238, 239]; by fluorination of thiophosgene with sodium fluoride in sulpholane [240]; by fluorination of CS_2 with IF_5 at 200 °C (76% yield) [241]; by the reaction of chloroform with sulphur and KF in sulpholane (19% yield) [237]; or by the reaction of trichloromethylsulphenyl bromide with KF (58% yield) [242].

Procedure

1. Reaction of Chloroform with Sulphur and KF [237]

$$CHCl_3 + S + KF \xrightarrow[-CS_2]{} CF_3\text{–}S\text{–}S\text{–}CF_3$$

In a 250-ml three-necked flask provided with a stirrer, a thermometer, and a reflux condenser connected with a trap cooled to —78 °C, are placed 13 g (0.11 mol) of chloroform, 6.5 g (0.21 mol) of sulphur, 40 g (0.67 mol) of freshly calcinated KF powder and 120 ml of dry sulpholane. The mixture is heated at 140°–150 °C with vigorous stirring during 20 h. The reflux condenser is then replaced with a downward one and the mixture is heated to 205°–215 °C; the volatile products are condensed in a receiver cooled to —10 °C. The products collected in the trap and the receiver are combined (Note 1) and distilled on a fractionating column. The yield of bis(trifluoromethyl) disulphide is 0.96 g (19.6%); bp 34 °C (Note 2).

2. Reaction of Trifluoroiodomethane with Sulphur [239]

$$CF_3I + S \rightarrow CF_3\text{—}S\text{—}S\text{—}CF_3$$

In a 100-ml steel autoclave is placed 25 g (0.8 mol) of sulphur. The autoclave is closed, cooled with liquid nitrogen and evacuated. Trifluoroiodomethane (30 g, 0.15 mol) is condensed into the autoclave and heated at 220 °C for 30 h. The temperature is then raised to 265 °C and the autoclave is heated for further 24 h. After the reaction is complete, the liquid part of the reaction mixture is distilled. The yield of bis(trifluoromethyl) disulphide is 9 g (60%) (Note 3).

Likewise, heptafluoro-1-iodopropane is converted to bis(heptafluoropropyl) disulphide (bp 123 °C; yield 53%) [243]; tridecafluoro-2-iodohexane: to bis(perfluoro-1-methylpentyl) disulphide (bp 94 °C/8 mm Hg) [244]; octafluoro-1,4-diiodobutane: to perfluoro(tetrahydrothiophene) (bp 40.7 °C; yield 80%) [245].

Notes

1. The mixture (6.6 g) contains, as shown by GLC, bis(trifluoromethyl) disulphide (15%), CS_2 (15%), $CHCl_3$ (68%), and CCl_4 (2%).
2. Bis(trifluoromethyl) disulphide is more toxic than phosgene; it should be handled only under an efficient hood.
3. The by-product of the reaction is bis(trifluoromethyl) trisulphide (bp 86.5 °C).

Bis(perfluoroisopropyl)Disulphide

$$(CF_3)_2CF\text{—}S\text{—}S\text{—}CF(CF_3)_2$$

$C_6F_{14}S_2$ MW 402

Bis(perfluoroisopropyl) disulphide is a colourless liquid, soluble in halogenated alkanes. Bp 105°–107 °C [246], 119 °C/773 mm Hg [247].

Method of Preparation

Bis(perfluoroisopropyl) disulphide is prepared by heating perfluoroisopropylmercuric derivatives with sulphur at 200 °C (25% yield) [247]; by the reaction of perfluoroisopropyl iodide with sulphur at 240 °C (34% yield) [248] (Note 1); by the reaction of hexafluoropropene with sulphur and SbF_5 (40% yield) [249]; or by chlorination of bis(perfluoroisopropylthio)mercury (61% yield) [246].

Procedure

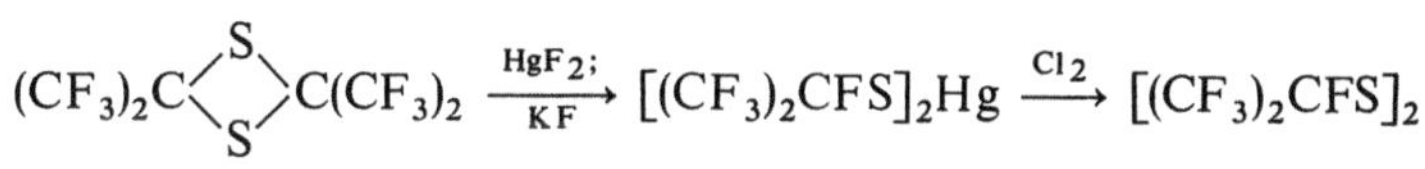

In a 250-ml three-necked flask provided with a stirrer, a reflux condenser, and a gas inlet tube, are placed 8.6 g (0.024 mol) of 2,2,4,4-tetrakis(trifluoromethyl)-1,3-dithiethane (for the preparation see p. 45) (Note 2), 11.6 g (0.049 mol) of HgF_2, 4 g (0.068 mol) of freshly calcinated KF, and 50 ml of anhydrous dimethylformamide. The mixture is stirred at 80°–100 °C for 15–20 min (Note 3), during which time it acquires a deep black colour. The completeness of the reaction is checked by [^{19}F] NMR (Note 4). Through the resulting solution cooled in an ice-water bath is bubbled 10 g (0.14 mol) of chlorine. The reaction mixture is then filtered, the lower layer of the filtrate is separated and distilled. The yield of bis(perfluoroisopropyl) disulphide is 5.8 g (61%); bp 105°–105 °C.

Notes

1. In these reactions bis(perfluoroisopropyl) disulphide is formed together with mono-, tri-, and tetrasulphides.
2. 2,2,4,4-Tetrakis(trifluoromethyl)-1,3-dithiethane is highly toxic. It should be handled only under an efficient hood.
3. Bis(perfluoroisopropylthio)mercury is highly hygroscopic; the reaction mixture should be safely protected from air moisture.
4. Instead of the signal of 2,2,4,4-tetrakis(trifluoromethyl)-1,3-dithiethane (singlet at 7.57 ppm from CF_3COOH), there appears the signal of bis(perfluoroisopropylthio)mercury (doublet at 1.43 ppm, $J_{F-C-C-F}$ 12.9 Hz); it is difficult to detect the CF signal due to its low intensity and splitting to a heptet, as well as to the low resolution resulting from the presence of solid particles in the reaction mixture.

Bis(perfluoro-*tert*-butyl)Disulphide

$(CF_3)_3C—S—S—C(CF_3)_3$

$C_8F_{18}S_2$ MW 502

Bis(perfluoro-*tert*-butyl) disulphide is a volatile colourless liquid. Bp 143°–145 °C [249].

Method of Preparation

Bis(perfluoro-*tert*-butyl) disulphide is prepared by the reaction of perfluoroisobutylene with sulphur and SbF_5 (47% yield) [249].

Procedure

$$(CF_3)_2C{=}CF_2 + S \xrightarrow{SbF_5} (CF_3)_3C{-}S{-}S{-}C(CF_3)_3$$

Caution! Perfluoroisobutylene is extremely toxic. It should be handled only under an efficient hood (see p. 2).

In a 250-ml steel autoclave are placed 10 g (0.3 mol) of finely divided sulphur and 97 g (0.45 mol) of SbF_5. The autoclave is closed, cooled with liquid nitrogen, evacuated and 24 g (0.12 mol) of perfluoroisobutylene is condensed into it. The autoclave is heated, while constantly rocking, at 100 °C during 6 h. After the reaction the autoclave is connected via a trap, cooled to —78 °C, to a vacuum system, then heated to 80 °C and the volatile products are condensed in the trap. The gaseous products are allowed to escape from the condensate, the residue is washed with water, dried over $MgSO_4$ and distilled. The yield of bis(perfluoro-*tert*-butyl) disulphide is 14 g (47%); bp 143°–145 °C (Note).

Likewise, the reaction of hexafluoropropylene with sulphur and SbF_5 affords bis-(perfluoroisopropyl) disulphide; bp 105°–107 °C, yield 40% [249].

Note. Bis(perfluoro-*tert*-butyl) disulphide is volatile; it is stored at 0 ÷ —5 °C.

α-Hydrotetrafluoroethanesulphonyl Fluoride

$CF_3{-}CHF{-}SO_2F$

$C_2HF_5O_2S$ MW 184

α-Hydrotetrafluoroethanesulphonyl fluoride is a colourless liquid. Bp 62 °C [250], 62°–63 °C [251]; n_D^{20} 1.3000; d_4^{20} 1.6496 [250].

Method of Preparation

α-Hydrotetrafluoroethanesulphonyl fluoride is prepared by the hydrolysis of hexafluoropropane-2-β-sultone (90%–93% yield) [250, 251].

Procedure

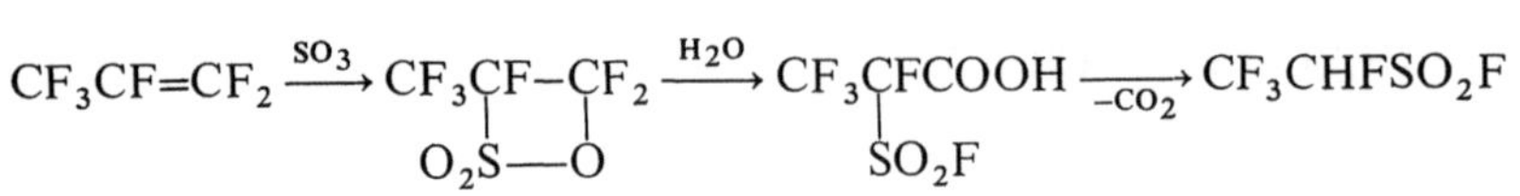

Hexafluoropropane-2-β-sultone [252]. In a 100-ml steel autoclave cooled to —78 °C are placed 14 g (0.175 mol) of freshly distilled liquid SO_3 and 30 g (0.2 mol) of hexafluoropropylene. The autoclave is closed and heated to 0 °C during 1.5–2 h, then at 60°–80 °C during 2 h, whereupon it is cooled to —78 °C and opened. The reaction mixture is distilled. The yield of hexafluoropropane-2-β-sultone is 37 g (92%); bp 42°–43 °C, n_D^{20} 1.3000, d_4^{20} 1.667.

Likewise, tetrafluoroethylene reacts with SO_3 (at 25°–35 °C, in an autoclave) to give tetrafluoroethane-β-sultone (bp 40°–41 °C, yield 93%–97%); at 90°–100 °C β-sultone is quantitatively isomerized to fluorosulphonyldifluoroacetyl fluoride (bp 30°–31 °C) [253].

α-Hydrotetrafluoroethanesulphonyl fluoride [250]. Water (10 ml) is slowly added dropwise to 23 g (0.1 mol) of hexafluoropropane-2-β-sultone with vigorous stirring and cooling of the flask with ice. A violent reaction follows, during which a gas is liberated. After the reaction has finished the lower layer is separated, washed with ice-water, dried over $MgSO_4$ and distilled. The yield of α-hydrotetrafluoroethanesulphonyl fluoride is 17 g (93%); bp 60°–62 °C.

Hexafluoro-2-chloropropylfluorosulphate

$$CF_3{-}CFCl{-}CF_2{-}O{-}SO_2F$$

$C_3ClF_6O_3S$ MW 314.5

Hexafluoro-2-chloropropylfluorosulphate is a colourless liquid, soluble in organic solvents, insoluble in water; its vapours produce a choky effect. Bp 76 °C, n_D^{20} 1.3070, d_4^{20} 1.701 [254].

Method of Preparation

Hexafluoro-2-chloropropylfluorosulphate is prepared by the reaction of chlorine fluorosulphate with perfluoro propylene (95% yield) [254].

Procedure

$$CF_3CF{=}CF_2 + ClOSO_2F \rightarrow CF_3CFClCF_2OSO_2F$$

In a 100-ml flask fitted with a reflux condenser and a gas inlet tube, is placed 26.9 g (0.2 mol) of freshly distilled chlorine fluorosulphate (Note 1), then 5 l (0.22 mol) of dry perfluoropropylene (Note 2) is slowly let in. The reaction mixture is distilled to give 54 g (95%) of hexafluoro-2-chloropropyl-fluorosulphate; bp 76 °C.

Notes

1. Chlorine fluorosulphate is a colourless liquid, bp 44 °C; in the presence of traces of moisture acquires a dark-brown colour; vigorously reacts with most organic and inorganic compounds. The most suitable solvents are perfluoroalkanes.
2. Perfluoropropylene is toxic; it should be handled under a hood.

Trifluoro-1,2-dichloroethylfluorosulphate

$CF_2Cl{-}CFCl{-}O{-}SO_2F$

$C_2Cl_2F_4O_3S$ MW 251

Trifluoro-1,2-dichloroethylfluorosulphate is a colourless liquid with a choky odour, soluble in organic solvents, insoluble in water. Bp 87°–88 °C, n_D^{20} 1.3511, d_4^{20} 1.722 [255].

Method of Preparation

Trifluoro-1,2-dichloroethylfluorosulphate is prepared by the reaction of chlorine fluorosulphate with 1,1,2-trifluorotrichloroethane (84% yield) [255].

Procedure

$$CF_2Cl{-}CFCl_2 + ClOSO_2F \xrightarrow[-Cl_2]{} CF_2Cl{-}CFCl{-}OSO_2F$$

In a three-necked flask provided with a thermometer, a dropping funnel, and a reflux condenser with a gas outlet tube (Note 1), are placed 18.75 g (0.09 mol) of freshly distilled 1,1,2-trifluorotrichloroethane and 0.45 g (0.45 mol) of HSO_3F, then 20.2 g (0.15 mol) of chlorine fluorosulphate (Note 2) is cautiously added dropwise. The reaction mixture is allowed to stand at 45°–50 °C for 8 h, then distilled. The yield of trifluoro-1,2-dichloroethylfluorosulphate is 21 g (84%); bp 87°–88 °C.

Notes

1. During the reaction chlorine is liberated. The exit gases should be led off into a trap.
2. For the properties of chlorine fluorosulphate see the preceding preparation.

Bis(Trifluoromethylthio)Mercury

$CF_3{-}S{-}Hg{-}S{-}CF_3$

$C_2F_6HgS_2$ MW 402

Bis(trifluoromethylthio)mercury is a white crystalline substance with a strong sweetish odour, soluble in water and polar organic solvents; sublimes at atmospheric pressure. Bp 80°–81 °C/21 mm Hg [256]; mp 37.5 °C [239].

Method of Preparation

Bis(trifluoromethylthio)mercury is prepared by the reaction of bis(trifluoromethyl) disulphide with mercury under UV-irradiation (53% yield) [239], or by heating CS_2 with HgF_2 in an autoclave at 250 °C (70% yield) [256, 257].

Procedure

$$CS_2 + HgF_2 \rightarrow (CF_3S)_2Hg$$

A 1-l steel autoclave is charged with 715 g (3 mol) of HgF_2 and 455 g (6 mol) of CS_2, and heated at 250 °C for 4 h. The autoclave is then opened, the reaction mixture is filtered to remove mercury sulphide, CS_2 is evaporated and the residue is vacuum-distilled to give 250–390 g of bis(trifluoromethylthio)mercury (yield ~70%); bp 80°–81 °C/21 mm Hg (Note).

Likewise, heating of silver fluoride with CS_2 in an autoclave (at 140 °C, for 12 h) gives silver trifluoromethylmercaptide (79% yield) [256].

Note. Bis(trifluoromethylthio)mercury causes skin abscess and injures the gastro-enteric canal. Due to high volatility it should be handled only under an efficient hood; hands should be protected with rubber gloves.

Bis(Trifluoromethyl)Mercury

$CF_3—Hg—CF_3$

C_2F_6Hg MW 338

Bis(trifluoromethyl)mercury is a white crystalline substance with a strong unpleasant odour, mildly volatile, soluble in water, ethers, dimethylformamide, and benzene. Mp 163°–166 °C [8, 258] (Note 1).

Method of Preparation

Bis(trifluoromethyl)mercury is prepared by the pyrolytic decomposition of mercury trifluoroacetate in the presence of anhydrous K_2CO_3 (92.5% yield) [8]; by the reaction of trifluoromethylmercuric iodide with cadmium amalgam (80%–90% yield) [258]; or by the reaction of tris(trifluoromethyl)phosphine with HgO (96% yield) [259].

Procedure

$$CF_3COOH \xrightarrow{HgO} (CF_3COO)_2Hg \xrightarrow[-CO_2]{K_2CO_3} (CF_3)_2Hg$$

Mercury trifluoroacetate. In a 250-ml beaker are placed 14 g (0.12 mol) of trifluoro-acetic acid and 45 ml of water, then 12 g (0.056 mol) of HgO is added in small portions, with stirring of the mixture with a glass rod. The mixture is heated during 3–5 min in a water-bath at 50°–60 °C, then evaporated in vacuum and the residue is dried in vacuum at 100 °C; yield is quantitative.

Bis(trifluoromethyl)mercury. A mixture of 14.1 g (0.03 mol) of mercury trifluoro-acetate and 10 g (0.07 mol) of calcinated K_2CO_3 is ground in a mortar, then placed in a 50-ml flask with a wide side-arm held cool, and a thermometer. The flask is connected, via a U-tube held at −78 °C, with a water-jet pump. The mixture is heated to 180°–200 °C (Note 2). Bis(trifluoromethyl)mercury, formed in the reaction, sublimes and crystallizes in a cooled side arm and a U-tube (Note 3). The yield of bis(trifluoromethyl)mercury is 10.3 g (92.5%); mp 163°–166 °C (Note 4).

Likewise, phenylmercuric trifluoroacetate prepared from trifluoroacetic acid, benzene, and HgO (50°–60 °C; 5 min; quantitative yield), gives phenyltrifluoromethylmercury (mp 137°–138 °C; yield 70%) [8].

Notes

1. Bis(trifluoromethyl)mercury vapours cause a strong headache. In contact with the skin this compound makes it temporarily insensitive. Bis(trifluoromethyl) mercury should be handled under an efficient hood. Hands should be protected with rubber gloves.
2. It is convenient to do the heating in a metallic bath.
3. Walls and neck of the flask should be intermittently heated with burners to avoid crystallization of bis(trifluoromethyl)mercury.
4. Bis(trifluoromethyl)mercury obtained in the reaction may contain a small amount of metallic mercury as an impurity. In order to purify bis(trifluoromethyl)mercury, it is dissolved in benzene, the solution is filtered, and benzene evaporated.

Tetrakis(Trifluorovinyl)Silane

$(CF_2{=}CF)_4Si$

$C_8F_{12}Si$ MW 352

Tetrakis(trifluorovinyl)silane is a colourless liquid, soluble in common organic solvents. Bp 119°–120 °C, n_D^{20} 1.3621, d_4^{20} 1.6182 [260].

Method of Preparation

Tetrakis(trifluorovinyl)silane is prepared by the reaction of trifluorovinylmagnesium iodide with $SiCl_4$ (42% yield) [260].

Procedure

$$SiCl_4 \xrightarrow{(CF_2=CF)MgI} (CF_2{=}CF)_4Si$$

In a 500-ml three-necked flask provided with a stirrer, a dropping funnel, and a reflux condenser, is placed a freshly prepared solution of trifluorovinylmagnesium iodide (from 5.2 g of Mg and 45 g of iodotrifluoroethylene) (Note 1) in 220 ml of anhydrous ether. The flask is cooled to —15° ÷ —18 °C and 7.4 g (0.044 mol) of freshly distilled $SiCl_4$ is added with stirring. The mixture is stirred at that temperature for 6–7 h (till a precipitate has been formed) and allowed to stand overnight at —40 °C. The next day the ethereal solution is decanted from the precipitate, the precipitate is washed with anhydrous ether, the ethereal solutions are combined and distilled (Note 2). The yield of tetrakis(trifluorovinyl)silane is 4.8 g (42% based on the converted $SiCl_4$); bp 119°–120 °C; n_D^{20} 1.3621.

Notes

1. The apparatus should be thoroughly dried; air moisture should be kept out.
2. The ethereal solution contains a certain amount of trifluorovinylmagnesium iodide, which can detonate upon heating. Therefore, it is preferable first to distil off the volatile products in vacuum of an oil pump and then to isolate tetrakis(trifluorovinyl)silane by distillation of the residue.

2-(Difluoroamino)-2-methylpropane

$(CH_3)_3C{-}NF_2$

$C_4H_9F_2N$ MW 109

2-(Difluoroamino)-2-methylpropane is an explosive colourless liquid. Bp 57.5° to 58.5 °C/742 mm Hg, n_D^{20} 1.3470, d_4^{20} 0.952 [263].

Method of Preparation

2-(Difluoroamino)-2-methylpropane is prepared by the reaction of *tert*-butanol with difluoroamine in the presence of sulphuric acid (63% yield) [263], or by alkylation of difluoroamine with isobutylene [264].

Procedure [263]

$$(CH_3)_3C{-}OH + HNF_2 \xrightarrow{H_2SO_4} (CH_3)_3C{-}NF_2$$

Caution! 2-(Difluoroamino)-2-methylpropane is explosive.

In a 1-l long-necked round-bottom flask provided with a manometer (Note 1), are placed 2.86 g of 72% H_2SO_4, 4.45 g (0.06 mol) of *tert*-butanol, then 3.2 g (0.06 mol) of difluoroamine (Note 2) is condensed into the flask, with cooling to —78 °C. The flask is then hermetically closed and cooling is stopped. After H_2SO_4 has melted, eight–ten drops of 100% H_2SO_4 are introduced from a syringe into the reaction flask. The volatile products are vacuum-distilled into a trap held at —78 °C. The condensate is distilled in a flow of nitrogen (Note 3). The yield of 2-(difluoroamino)-2-methylpropane is 3.9 g (63%); bp 57.5°–58.5 °C/742 mm Hg.

A similar reaction of 2-methylbutanol-2 with difluoroamine affords 2-difluoroamino-2-methylbutane (bp 87.5–88 °C, n_D^{20} 1.3690, d_4^{20} 0.958; yield 75%) [263].

Notes

1. A U-tube, filled with mercury to the height of 400 mm, is used as a manometer. In order to prevent ejection of mercury in an emergency case (a jump of pressure in the reactor, etc.), the open end of the tube is connected to the trap.
2. Difluoroamine is a colourless gas with a specific odour, bp —23 °C; toxic; it can spontaneously explode. Its mixture with vapours of organic compounds is also explosive, especially in air.

The least dangerous method for the preparation of difluoroamine is hydrolysis of an aqueous solution of N,N-difluorourea with concentrated H_2SO_4 [265, 266]. A disadvantage of this method is a low stability of N,N-difluorourea. On a small scale difluoroamine may be prepared by the reaction of N_2F_4 with thiophenol [272], or, which is still more convenient, by decomposition of trityldifluoroamine (triphenyldifluoroaminomethane) with concentrated H_2SO_4 in vacuum [267]. Handling difluoroamine requires taking special precautions. All the procedures should be carried out in an armoured box. Remote control of the process should be provided. Difluoroamine should be introduced through vacuum rubber tubes; their service life should not exceed four or five runs.

3. Organic compounds containing difluoroaminogroups are toxic and liable to explode, therefore the products should be handled with the same precautions as for highly sensitive explosives.

The glassware should be degassed by washing it successively with water (ethanol, acetone, CCl_4) and an aqueous (or ethanolic) alkali. The still residue should be dissolved in a large amount of an organic solvent, then cautiously decomposed with 10% alcoholic solution of an alkali.

Distillation of organic difluoroamino-derivatives should be carried out at 90° to 100 °C or below.

During any manipulation with organic difluoroamino-derivatives face should be protected with a screen of plastic, hands with leather gloves. The compounds should be stored and transported in special containers.

Do not use glass stoppers. Bottles (test tubes) containing organic difluoroamino-derivatives should be closed with rubber stoppers wrapped in thin Teflon films.

1-Difluoroaminoethanol

$$CH_3{-}\underset{\displaystyle OH}{\underset{|}{C}H}{-}NF_2$$

$C_2H_5F_2NO$ MW 97

1-Difluoroaminoethanol is a colourless liquid, soluble in organic solvents. Bp 52° to 55 °C/150 mm Hg, n_D^{20} 1.3584, d_4^{20} 1.223 [268].

Method of Preparation

1-Difluoroaminoethanol is prepared by the addition of difluoroamine to acetaldehyde (82% yield) [268, 269].

Procedure

$$CH_3CHO + HNF_2 \rightarrow CH_3CH(OH)NF_2$$

Caution! 1-Difluoroaminoethanol is explosive.

In a 25-ml ampoule is placed 5.2 g (0.12 mol) of acetaldehyde. The ampoule is evacuated with cooling to −78 °C, then 6 g (0.11 mol) of difluoroamine (Notes 1, 2) is condensed into it, whereupon cooling is stopped. After the reaction mixture has warmed to room temperature, the ampoule is flushed with nitrogen and the reaction mixture is distilled in vacuum, in a flow of nitrogen (Note 3). The yield of 1-difluoroaminoethanol is 9 g (82%); bp 52°–53 °C/150 mm Hg.

Similar reactions of the appropriate aldehydes and difluoroamine give difluoroaminomethanol (48°–50 °C/150 mm Hg, n_D^{20} 1.3470, d_4^{20} 1.3771; yield 84%), 1-difluoroaminopropanol (bp 44°–46 °C at 50 mm Hg, n_D^{20} 1.3752, d_4^{20} 1.1674, yield 62%), 1-difluoroaminobutanol (bp 52°–56 °C/50 mm Hg, n_D^{20} 1.3875, d_4^{20} 1.0763, yield 80%) [269].

Notes

1. To close the ampoule hermetically, a thick-walled rubber tube with a clamp may be used.
2. For the properties of difluoroamine and the safety measures, see p. 100.
3. For the safety measures to be taken while working with organic difluoroamino-derivatives, see p. 100.

Decafluoro-1,2-epoxyhexene-5

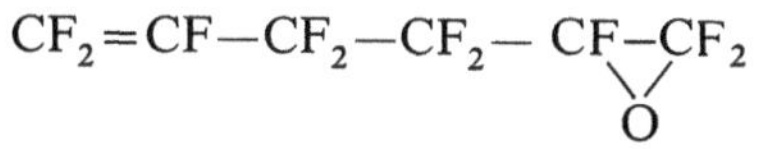

$C_6F_{10}O$ MW 278

Decafluoro-1,2-epoxyhexene-5 is a colourless liquid. Bp 54 °C, n_D^{20} 1.2950, d_4^{20} 1.6138 [270].

Method of Preparation

Decafluoro-1,2-epoxyhexene-5 is prepared by telomerization of heptafluoro-1,2-dichloro-4-iodobutane with trifluoroethylene with subsequent dehydroiodination, oxidation of olefin and dechlorination of decafluoro-5,6-dichloro-1,2-epoxyhexane. The total yield is 5% [270, 271].

Procedure

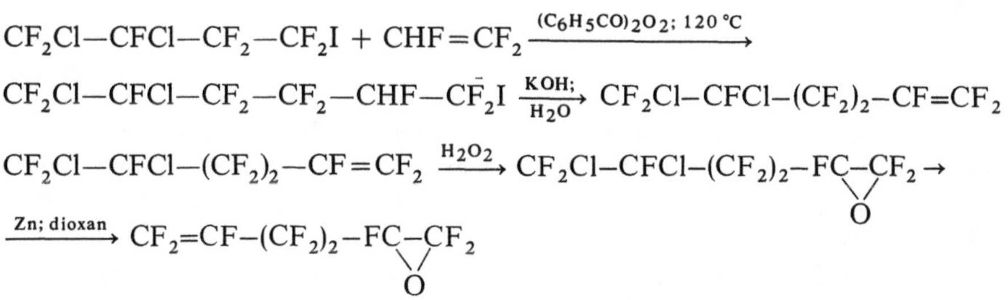

5-Hydroundecafluoro-1,2-dichloro-6-iodohexane. In a 0.5-l steel autoclave are placed 247 g (0.65 mol) of heptafluoro-1,2-dichloro-4-iodobutane, 3.5 g of benzoyl peroxide and 70 g (1.17 mol) of trifluoroethylene. The autoclave is placed in a tilting furnace and heated at 100°–120 °C during 3–4 h. Then it is cooled, the gaseous products are allowed to escape in the atmosphere and the reaction mixture is distilled. The yield of 5-hydroundecafluoro-1,2-dichloro-6-iodohexane is 93 g (44%, based on the converted amount of heptafluoro-1,2-dichloro-4-iodobutane); bp 185 °C, n_D^{20} 1.3975, d_4^{20} 2.0930 (Note 1).

Decafluoro-5,6-dichlorohexene-1. In a 250-ml three-necked flask provided with a short dephlegmator with a downward condenser, a stirrer, and a dropping funnel, are placed 22 g (0.4 mol) of KOH and 150 ml of water. The solution is heated to reflux and to it is gradually added (during 2 h) 65 g (0.14 mol) of a mixture of isomers of 5-hydroundecafluoro-1,2-dichloro-6-iodohexane and 6-hydroundecafluoro-1,2-dichloro-6-iodohexane; the product is simultaneously distilled off with steam. The organic layer is separated, dried over $CaCl_2$ and dissolved. The yield of decafluoro-5,6-dichlorohexene-1 is 31.3 g (85%); bp 103°–105 °C; n_D^{20} 1.3380, d_4^{20} 1.7342 (Note 2).

Decafluoro-5,6-dichloro-1,2-epoxyhexane. In a 500-ml three-necked flask fitted with a stirrer, a dropping funnel and a thermometer, are placed 100 ml of methanol and 80 ml of 30% H_2O_2. The flask is cooled to —20 °C and 24 g (0.07 mol) of decafluoro-5,6-dichlorohexene-1 is poured into a vigorously stirred solution held at —20 °C, whereupon a solution of 12 g of KOH in 40 ml of water is gradually added during 1 h. Stirring is continued for 3 h; the temperature of the mixture is then slowly raised to that of the room. The reaction mixture is poured into ice-water, the organic layer is separated and dried over $CaCl_2$ to afford a mixture of unconverted olefin and epoxide. The mixture is placed into a quartz tube and bromine is gradually added under UV-irradiation to complete decolorization (the amount of bromine is previously calculated from the GLC data). The products are separated by distillation. The yield of decafluoro-5,6-dichloro-1,2-epoxyhexane is 10.7 g (41%); bp 103 °C, n_D^{20} 1.3275, d_4^{20} 1.7350 (Note 3).

Decafluoro-1,2-epoxyhexene-5. In a 100-ml four-necked flask equipped with a stirrer, a dephlegmator with a downward condenser, a dropping funnel, and a thermometer, are placed 4 g (0.07 mol) of activated zinc dust (Note 4) and 40 ml of dry dioxan. The mixture is heated to reflux, then 14 g (0.04 mol) of decafluoro-5,6-dichloro-1,2-epoxyhexane is slowly added, with vigorous stirring. The product is distilled off with dioxan. The distillate is poured into water, the organic layer is separated, washed with water, dried over $CaCl_2$ and distilled. The yield of decafluoro-1,2-epoxyhexene-5 is 3 g (30%).

Notes

1. The reaction products contain about 17% of 6-hydroundecafluoro-1,2-dichloro-6-iodohexane isomers (it is impossible to separate them by distillation) and higher telomeric homologues. Unchanged heptafluorodichloroiodobutane (70 g, ~30% of the charged amount) may be isolated from the reaction mixture.
2. A highly boiling fraction, 6-hydroundecafluoro-1,2-dichloro-6-iodohexane, is isolated from the residue left in the flask.
3. Distillation gives also the bromination product, decafluoro-1,2-dichloro-5,6-dibromohexane [bp 99 °C/40 mm Hg, n_D^{20} 1.3990, d_4^{20} 2.1380, yield ~2.5 g (~7%)].

4. Zinc dust (49.2 g) is mixed during 1 min with 40 ml of 3% HCl, then washed by decanting with 3% HCl (3 × 50 ml), five times with distilled water, twice with 2% $CuSO_4$ solution (the total amount is 75 ml), five times with distilled water, four times with anhydrous ethanol, and five times with anhydrous ether, filtered on a Büchner funnel and dried over P_2O_5 in a desiccator.

References

1. Hauptschein, M., Braid, P.: J. Am. Chem. Soc. *83*, 2383 (1961)
2. Krespan, C.: J. Org. Chem. *27*, 1813 (1962)
3. Yarovenko, N. N., Raksha, M. A.: Zh. Obshch. Khim. *29*, 2159 (1959)
4. Desreux, V.: Bull. Sci. Acad. R. Belg. *44*, 8 (1934)
5. Hoffman, F.: J. Am. Chem. Soc. *70*, 2596 (1948)
6. Gryszkievisz-Trochimowski, E., Sporsynski, A., Vnuk, J.: Rec. Trav. Chim. *66*, 413 (1947)
7. Delyagina, N. I., Pervova, E. Ya., Knunyants, I. L.: Izv. Akad. Nauk SSSR Ser. Khim. *1972*, 376
8. Knunyants, I. L. et al.: Izv. Akad. Nauk SSSR Ser. Khim. *1973*, 943
9. Seyfer, T.: J. Org. Chem. *32*, 2980 (1967)
10. Franzen, V.: Chem. Ber. *95*, 1964 (1962)
11. Birchall, J., Cross, G., Haszeldine, R.: Proc. Chem. Soc. *1960*, 81
12. Birchall, J., Haszeldine, R., Roberts, D.: Chem. Commun. *1967*, 287
13. Park, I., Seff, R., Laecher, I.: J. Am. Chem. Soc. *78*, 59 (1956)
14. Sterlin, R. N.: D. Sc. Dissertation, Moscow, 1963
15. Booth, H., Burchfield, P., Bixey, E.: J. Am. Chem. Soc. *55*, 2231 (1933)
16. Locke, E., Brode, W., Henne, A.: J. Am. Chem. Soc. *56*, 1726 (1934)
17. Haszeldine, R., Steele, B. J.: Chem. Soc. (Lond.) *1954*, 923
18. Tarrant, P., Lovelace, A., Lilyquist, M.: J. Am. Chem. Soc. 77, 2783 (1955)
19. USSR Pat. 130, 895; C.A. *55*, 6372 (1961)
20. Sheppard, W., Sharts, C.: Organic Fluorine Chemistry. Benjamin, New York, 1969
21. Brice, T., Lazerte, J., Hais, L., Pearlson, W.: J. Am. Chem. Soc. *75*, 2698 (1953)
22. Haszeldine, R., Steele, B.: J. Chem. Soc. (Lond.) *1955*, 3005
23. Fokin, A. V., Kosyrev, Yu. M.: Khim. Nauka, Prom. *1960*, 183
24. Morse, A., Ayscough, P., Leitch, L.: Can. J. Chem. *33*, 453 (1955)
25. England, D., Krespan, C.: J. Am. Chem. Soc. *88*, 5582 (1962)
26. Tyuleneva, V. V. et al.: Izv. Akad. Nauk SSSR Ser. Khim *1975*, 1136
27. Delyagina, N. I.: Ph. D. Dissertation, Moscow, 1975
28. Mochalina, E. P. et al.: Dokl. Akad. Nauk SSSR *169*, 1346 (1966)
29. Gervits, L. et al.: J. Fluor. Chem. *9*, 45 (1977)
30. Haszeldine, R.: J. Chem. Soc. (Lond.) *1952*, 4423
31. Miller, W. T.: Preparation, Properties and Technology of Fluorine and Organic Fluorocompounds. Classer (ed), New York, 1951
32. Haszeldine, R.: J. Chem. Soc. (Lond.) *1954*, 4026
33. Henne, A., Postelnec, W.: J. Am. Chem. Soc. *77*, 2334 (1955)
34. Manler, W.: J. Am. Chem. Soc. *84*, 4600 (1962)
35. Krasnikova, G. S., German, L. S., Knunyants, I. L.: Izv. Akad. Nauk SSSR, Ser. Khim. *1973*, 459
36. Banks, R. et al.: Tetrahedron Lett. *1970*, 5215
42. Knunyants, I. L. et al.: Izv. Akad. Nauk SSSR Ser. Khim. *1964*, 384
38. Emeleus, H., Haszeldine, R.: J. Chem. Soc. (Lond.) *1949*, 2949
39. Knunyants, I. L. et al.: Izv. Akad. Nauk SSSR, Ser. Khim. *1961*, 1462
40. Hauptschein, M., Braid, M., Fainberg, A.: J. Am. Chem. Soc. *83*, 2495 (1961)
41. Aronov, Yu. E., Cheburkov, Yu. A., Knunyants, I. L.: Izv. Akad. Nauk SSSR, Ser. Khim. *1967*, 1758
42. Haszeldine, R., Steele, B.: J. Chem. Soc. (Lond.) *1951*, 588

43. Finnegan, W., Norriss, W.: J. Org. Chem. *28*, 1139 (1963)
44. Hasek, W., Smith, W., Engelhardt, V.: J. Am. Chem. Soc. *82*, 543 (1960)
45. Tullock, C. et al.: J. Am. Chem. Soc. 82, *539* (1960)
46. USSR Pat. 276,020; C.A. *74*, 55819 (1971)
47. Henne, A., Finnegan, W.: J. Am. Chem. Soc. *71*, 298 (1949)
48. Krespan, C., Harder, R., Drysdale, I.: J. Am. Chem. Soc. *83*, 3424 (1961)
49. Chambers, R., Gohes, C., Taylor, G.: J. Fluor. Chem. *18*, 407 (1981)
50. Henne, A., Trott, P.: J. Am. Chem. Soc. *69*, 1820 (1947)
51. Delyagina, N. I., Dyatkin, B. L., Knunyants, I. L.: Izv. Akad. Nauk SSSR, Ser. Khim. *1974*, 1666
52. Knunyants, I. L., Kildisheva, O. V., Petrov, I. P.: Zh. Obshch. Khim. *19*, 95 (1949)
53. Hoffman, F.: J. Org. Chem. *15*, 425 (1950)
54. Saunders, B., Stacey, G.: J. Chem. Soc. (Lond.) *1949*, 773
55. Campbell, K., Knoblock, J., Campbell, B.: J. Am. Chem. Soc. *72*, 4380 (1950)
56. Henne, A., Alm, R.: J. Am. Chem. Soc. *70*, 1968 (1948)
57. Gilman, H., Jones, R.: J. Am. Chem. Soc. *70*, 1281 (1948)
58. German, L. S., Knunyants, I. L.: Dokl. Akad. Nauk SSSR *166*, 602 (1966)
59. Belenky, G. G., German, L. S., Knunyants, I. L.: Dokl. Akad. Nauk SSSR *171*, 1333 (1966)
60. Belenky, G. G., German, L. S., Knunyants, I. L.: Izv. Akad. Nauk SSSR Ser. Khim. *1968*, 554
61. Middleton, W., Lindsey, R.: J. Am. Chem. Soc. *86*, 4948 (1964)
62. USSR Pat. 138,604; C.A. *56*, 8563 (1962)
63. Gambaryan, N. P., Knunyants, I. L.: Izv. Akad. Nauk SSSR Ser. Khim. *1965*, 722
64. Knunyants, I. L., Dyatkin, B. L.: Izv. Akad. Nauk SSSR, Ser. Khim. *1964*, 923
65. US Pat. 3,317,616; C.A. *67*, 63755 (1967)
66. US Pat. 3,385,904; C.A. *69*, 26753 (1968)
67. USSR Pat. 482,432; C.A. *83*, 147119 (1975)
68. Martynov, B. I.: Ph. D. Dissertation, Moscow, 1974
69. Becker, R. A. et al.: Dokl. Akad. Nauk SSSR *217*, 1320 (1974)
70. Becker, R. A., et al.: Zh. Org. Khim. *11*, 1370 (1975)
71. Gambaryan, N. P., Cheburkov, Yu. A., Knunyants, I. L.: Izv. Akad. Nauk SSSR Ser. Khim. *1964*, 1526
72. Knunyants, I. L., Fokin, A. V., Komarov, V. A.: Izv. Akad. Nauk SSSR Ser. Khim. *1966*, 466
73. USSR Pat. 161,717; C.A. *61*, 4218 (1964)
74. Fokin, A. V. et al.: Zh. Obshch. Khim. *37*, 152 (1967)
75. US Pat. 3,203,999; C.A. *63*, 14760 (1965)
76. Hill, M., Ross, L.: J. Org. Chem. *32*, 2595 (1967)
77. US Pat. 3,783,144; C.A. *80*, 70328 (1974)
78. Fokin, A. V. et al.: Zh. Org. Khim. *7*, 1165 (1971)
79. Fokin, A. V. et al.: Izv. Akad. Nauk SSSR Ser. Khim. *1975*, 1448
80. Pattiso, F. L. M.: Toxic aliphatic fluorine compounds. Elsevier, Amsterdam, pp. 208–209, 1959
81. Lovelace, A. M., Rausch, D. A., Postelnek, W.: Aliphatic fluorine compounds. Reinhold, New York, 1958
82. German, L. S., Knunyants, I. L.: Zh. Vses. Khim. Obshch. Im DI Mendeleeva, *11*, 354 (1966)
83. Weinmayer, V.: J. Org. Chem. *28*, 492 (1963)
84. Gervits, L. L. et al.: Izv. Akad. Nauk SSSR Ser. Khim. *1974*, 2256
85. Dyatkin, B. L. et al.: Zh. Org. Khim. *3*, 1006 (1967)
86. Okuhara, K., Baba, H., Kojima, R.: Bull. Chem. Soc. Jpn *35*, 534 (1962)
87. Fokin, A. V. et al.: Izv. Akad. Nauk SSSR Ser. Khim. *1976*, 461
88. Germ. Pat. 947,364; C.A. *53*, 4299 (1959)
89. Knunyants, I. L., Shokina, V. V., Galakhov, I. V.: Khim. Geterotsikl. Soedin. *1966*, 878
90. Sianesi, D., Pasetti, A., Corti, C.: Macromol. Chem. *86*, 308 (1965)
91. Brit. Pat. 1,036,174; C.A. *65*, 15322 (1966)
92. US Pat. 3,385,904; C.A. *69*, 26753 (1968)
93. Pavlik, F., Toren, P.: J. Org. Chem. *35*, 2054 (1970)
94. Becker, R. A. et al.: Dokl. Akad. Nauk SSSR *204*, 606 (1972)
95. Kausman, M., Braun. J.: J. Org. Chem. *31*, 3090 (1966)
96. USSR Pat. 394,371; C.A. *79*, 146364 (1973)

97. Becker, R. A., Asratyan, G. V., Dyatkin, B. L.: Zh. Org. Khim. *9*, 1635 (1973)
98. Henne, A., Pelley, R., Alm, R.: J. Am. Chem. Soc. *72*, 3370 (1950)
99. Shechter, H., Conrad, F.: J. Am. Chem. Soc. *72*, 3371 (1950)
100. Braid, M., Isersone, H., Lawlore, F.: J. Am. Chem. Soc. *76*, 4027 (1954)
101. Henne, A. et al.: J. Am. Chem. Soc. *69*, 1819 (1947)
102. Swarts, F.: Bull. Sci. Acad. R. Belg. *13*, 177 (1927)
103. Sykes, A., Tatlow, J., Thomas, C.: J. Chem. Soc. (Lond.) *1956*, 835
104. Burdon, J., McLoughlin, V.: Tetrahedron *20*, 2163 (1964)
105. Utebaev, U. et al.: Izv. Akad. Nauk SSSR Ser. Khim. *1974*, 387
106. Hynes, J. et al.: Can. J. Chem. *45*, 2278 (1967)
107. McBee et al.: J. Am. Chem. Soc. *72*, 3152 (1953)
108. Haszeldine, R.: J. Chem. Soc. (Lond.) *1954*, 1273
109. Knunyants, I. L., German, L. S., Dyatkin, B. L.: Izv. Akad. Nauk SSSR Ser. Khim. *1956*, 1353
110. Knunyants, I. L., Rokhlin, E. M., Gambaryan, N. P.: Khim. Nauka Prom. *4*, 802 (1959)
111. Becker, R. A. et al.: Zh. Org. Khim. *11*, 1604 (1975)
112. Murfhy, K.: J. Chem. Eng. Data *9*, 259 (1964)
114. Fr. Pat. 1,372,549; C.A. *62*, 6397 (1965)
115. Fr. Pat. 1,369,784; C.A. *62*, 1570 (1965)
116. US Pat. 3,213,134; C.A. *63*, 18029 (1965)
113. Krespan, C., Middleton, W., in: Fluorine Chemistry Reviews Vol. 1. Tarrant, P. (ed), 145 (1967)
117. Anello, L., Van Der Puy.: J. Org. Chem. *47*, 3377 (1982)
118. Knunyants, I. L., Fokin, A. V., Komarov, B. A.: Izv. Akad. Nauk SSSR Ser. Khim. *1966*, 466
119. Aronov, Yu. E., Cheburkov, Yu. A., Knunyants, I. L.: Izv. Akad. Nauk SSSR Ser. Khim. *1967*, 1758
120. Belenky, G. G., German, L. S.: Izv. Akad. Nauk SSSR Ser. Khim. *1974*, 942
121. Platoshkin, A. M., Cheburkov, Yu. A., Knunyants, I. L.: Izv. Akad. Nauk SSSR Ser. Khim. *1969*, 112
122. Middleton, W., Howard, E., Sharkey, W.: J. Org. Chem. *30*, 1375 (1965)
123. Sterlin, S. R., Dyatkin, B. L., Knunyants, I. L.: Izv. Akad. Nauk SSSR Ser. Khim. *1967*, 2583
124. England, D.: J. Org. Chem. *46*, 153 (1981)
125. Middleton, W., Krespan, C.: J. Org. Chem. *30*, 1398 (1965)
126. Zeifman, Yu. V., Gambaryan, N. P., Knunyants, I. L.: Dokl. Akad. Nauk SSSR *153*, 1334 (1963)
127. Zeifman, Yu. V., Gambaryan, N. P., Knunyants, I. L.: Izv. Akad. Nauk SSSR Ser. Khim. *1965*, 450
128. Knunyants, I. L. et al.: Izv. Akad. Nauk SSSR, Ser. Khim. *1962*, 684
129. Cheburkov, Yu. A., Knunyants, I. L.: Izv. Akad. Nauk SSSR Ser. Khim. *1967*, 829
130. Zeifman, Yu. V., Gambaryan, N. P., Knunyants, I. L.: Izv. Akad. Nauk SSSR Ser. Khim. *1965*, 2046
131. Dyatkin, B. L. et al.: Zh. Vses. Khim. Obshch. Im DI Mendeleeva *10*, 469 (1965)
132. Cheburkov, Yu. V., Mirzabekyants, N. S.: Vysokomol. Soedin. *9B*, 190 (1967)
133. Knunyants, I. L. et al.: Izv. Akad. Nauk SSSR Ser. Khim. *1966*, 250
134. Dyatkin, B. L., Sterlin, S. R., Knunyants, I. L.: Zh. Vses. Khim. Obshch Im DI Mendeleeva, *13*, 468 (1968)
135. Dyatkin, B. L. et al.: Tetrahedron *23*, 4291 (1967)
136. Saunders, B., Stacey, G.: J. Chem. Soc. (Lond.) *1948*, 1773
137. Beromann, E., Blank, I.: J. Chem. Soc. (Lond.) *1953*, 3786
138. German, L. S., Rozhkov, I. N., Knunyants, I. L.: Zh. Vses. Khim. Obshch Im DI Mendeleeva *10*, 599 (1965)
139. Knunyants, I. L., Fokin, A. V.: Dokl. Akad. Nauk SSSR *112*, 67 (1957)
140. Fokin, A. V. et al.: Zh. Obshch. Khim. *35*, 1662 (1965)
141. Fokin, A. V., Uzun, A. T., Kosyrev, Yu. M.: Zh. Obshch. Khim. *36*, 540 (1966)
142. Fokin, A. V., Uzun, A. T.: Zh. Obshch. Khim. *36*, 117 (1966)
143. Fokin, A. V., Kosyrev, Yu. M., Uzun, A. T.: Zh. Obshch. Khim. *36*, 119 (1966)
144. Knunyants, I. L., Fokin, A. V.: Dokl. Akad. Nauk SSSR *111*, 1035 (1956)
145. Knunyants, I. L., Fokin, A. V.: Izv. Akad. Nauk SSSR Ser. Khim. *1957*, 1439
146. USSR Pat. 129,653; C.A. *55*, 4364 (1961)

147. Henne, A., Zimmerschied, W.: J. Am. Chem. Soc. *69*, 281 (1947)
148. Buxton, M. et al.: J. Chem. Soc. (Lond.) *1952*, 3830
149. Knunyants, I. L., Cheburkov, Yu. A.: Izv. Akad. Nauk SSSR Ser. Khim. *1960*, 678
150. US Pat. 3,271,419; C.A. *66*, 2220 (1967)
151. Organic Reactions, vol. 3. R. Adams (ed). Wiley, New York, 1947
152. Knunyants, I. L., Rokhlin, E. M., Gambaryan, N. P.: Khim. Nauka Prom. *4*, 802 (1959)
153. Filler, R., Schure, R.: J. Org. Chem. *32*, 1217 (1967)
154. Knunyants, I. L., Shokina, V. V., Tyuleneva, V. V.: Izv. Akad. Nauk SSSR Ser. Khim. *1968*, 415
155. Zeifman, Yu. V., Knunyants, I. L.: Dokl. Akad. Nauk SSSR *173*, 1354 (1967)
156. Zeifman, Yu. V. et al.: Zh. Obshch. Khim. *37*, 2476 (1976)
157. Asratyan, G. V.: Ph. D. Dissertation, Moscow, 1973
158. Barna, P.: Aust. J. Chem. *21*, 1089 (1968)
159. Farah, B., Gilbert, E., Sibilia, J.: J. Org. Chem. *30*, 998 (1965)
160. Fawsett, F., Tullock, C., Coffman, D.: J. Am. Chem. Soc. *84*, 4275 (1962)
161. Haszeldine, R., Isersone, H.: J. Am. Chem. Soc. *79*, 5801 (1957)
162. Emeleus, H., Wood, J.: J. Chem. Soc. (Lond.) *1948*, 2183
163. Cheburkov, Yu. A., Bargamova, M. D., Knunyants, I. L.: Izv. Akad. Nauk SSSR Ser. Khim. *1964*, 367
164. Knunyants, I. L. et al.: Izv. Akad. Nauk SSSR Ser. Khim. *1967*, 110
165. Cheburkov, Yu. A., Knunyants, I. L.: Izv. Akad. Nauk SSSR Ser. Khim. *1967*, 346
166. Knunyants, I. L., Krasusskaya, M. P., Gambaryan, N. P.: Izv. Akad. Nauk SSSR Ser. Khim. *1965*, 723
167. Knunyants, I. L., Zeifman, Yu. A., Lantseva, L. T.: Dokl. Akad. Nauk SSSR *254*, 117 (1980)
168. Cheburkov, Yu. A., Mukhamadaliev, H., Knunyants, I. L.: Izv. Akad. Nauk SSSR Ser. Khim. *1965*, 1478
169. USSR Pat. 289,081; C.A. *74*, 111582 (1971)
170. Knunyants, I. L. et al.: Tetrahedron *29*, 595 (1973)
171. Knunyants, I. L. et al.: J. Fluor. Chem. *6*, 227 (1975)
172. Cheburkov, Yu. A., Knunyants, I. L.: Izv. Akad. Nauk SSSR Ser. Khim. *1963*, 1573
173. England, D., Solomon, L., Krespan, C.: J. Fluor. Chem. *3*, 63 (1973)
174. Kraus, S., Brown, E.: J. Am. Chem. Soc. *51*, 2690 (1929)
175. US Pat. 3,174,988; C.A. *63*, 494 (1965)
176. Becker, R. A. et al.: Zh. Org. Khim. *11*, 1600 (1975)
177. Knunyants, I. L., Shokina, V. V., Tyuleneva, V. V.: Dokl. Akad. Nauk SSSR *169*, 594 (1966)
178. Sianesi, D., Psetti, A., Tarli, F.: J. Org. Chem. *31*, 2312 (1966)
179. Swarts, F.: Bull. Sci. Acad. R. Belg. *12*, 679 (1926)
180. McBee, E. et al.: J. Am. Chem. Soc. *75*, 3152 (1953)
181. Knunyants, I. L., Cheburkov, Yu. A., Aronov, Yu. E.: Izv. Akad. Nauk SSSR Ser. Khim. *1966*, 1038
182. Cheburkov, Yu. A. et al.: Izv. Akad. Nauk SSSR Ser. Khim. *1970*, 2272
183. US Pat. 2,795,601; C.A. *51*, 16518 (1957)
184. Dyatkin, B. L.: Ph. D. Dissertation, Moscow, 1958
185. Paleta, O., Posta, A., Novotna, Z.: Collect. Czech. Chem. Commun. *33*, 2970 (1968)
186. Lazerte, J. et al.: J. Am. Chem. Soc. *78*, 5639 (1956)
187. Middleton, W.: J. Org. Chem. *30*, 1403 (1965)
188. Knunyants, I. L. et al.: Izv. Akad. Nauk SSSR Ser. Khim. *1966*, 1111
189. Cheburkov, Yu. A., Bargamova, M. D.: Izv. Akad. Nauk SSSR Ser. Khim. *1967*, 832
190. England, D. Krespan, C.: J. Am. Chem. Soc. *87*, 4019 (1965)
191. Raasch, M.: Chem. Commun. *1966*, 577
192. Sterlin, S. R. et al.: Izv. Akad. Nauk SSSR Ser. Khim. *1971*, 2517
193. USSR Pat. 246,508; C.A. *71*, 101317 (1969)
194. Krespan, C., England, D.: J. Org. Chem. *33*, 1850 (1968)
195. Fletcher, J. et al.: J. Am. Chem. Soc. *72*, 2461 (1950)
196. Zeifman, Yu. V. et al.: Izv. Akad. Nauk SSSR Ser. Khim. *1973*, 795
197. Raasch, M.: J. Org. Chem. *37*, 1347 (1972)
198. Rokhlin, E. M. et al.: Izv. Akad. Nauk SSSR Ser. Khim. *1965*, 749

199. Zeifman, Yu. V., Tyuleneva, V. V., Knunyants, I. L.: Dokl. Akad. Nauk SSSR *201*, 1359 (1971)
200. Barr, D., Haszeldine, R.: J. Chem. Soc. (Lond.) *1956*, 3428
201. Sander, M.: Monatsh. Chem. *95*, 608 (1964)
202. Barr, D., Haszeldine, R.: J. Chem. Soc. (Lond.) *1957*, 30
203. Deltsova, D. P.: Ph. D. Dissertation, Moscow, 1974
204. Peterson, V., Padell, J., Washburne, S.: J. Fluor. Chem. *2*, 437 (1972/1973)
205. Krasusskaya, M. P., Deltsova, D. P., Knunyants, I. L.: Izv. Akad. Nauk SSSR Ser. Khim. *1965*, 2039
206. Henne, A., Steward, J.: J. Am. Chem. Soc. *77*, 1901 (1955)
207. Bourne, E.: J. Chem. Soc. (Lond.) *1952*, 4015
208. Gilman, H., Jones, R.: J. Am. Chem. Soc. *65*, 1458 (1943)
209. US Pat. 2,348,321; C.A. *39*, 625 (1945)
210. Knunyants, I. L., German, L. S., Rozhkov, I. N.: Izv. Akad. Nauk SSSR Ser. Khim. *1963*, 1948
211. Husted, D., Ahlbrecht, A.: J. Am. Chem. Soc. *75*, 1605 (1953)
212. Podolsky, A. V., German, L. S., Knunyants, I. L.: Izv. Akad. Nauk SSSR Ser. Khim. *1966*, 1575
213. Knunyants, I. L., Dyatkin, B. L., Mochalina, E. P.: Izv. Akad. Nauk SSSR Ser. Khim. *1965*, 1091
214. Zeifman, Yu. V., Koshtoyan, S. O., Knunyants, I. L.: Dokl. Akad. Nauk SSSR *195*, 93 (1970)
215. Becwith, A., Goodrich, J.: Aust. J. Chem. *18*, 747 (1965)
216. Barr, D., Haszeldine, R.: J. Chem. Soc. (Lond.) *1955*, 1881
217. Young, J., Simmons, T., Hoffman, F.: J. Am. Chem. Soc. *78*, 5637 (1956)
218. Fawsett, F., Tullock, C., Coffman, D.: J. Chem. Eng. Data *10*, 398 (1965)
219. Banus, J.: J. Chem. Soc. (Lond.) *1953*, 3755
220. Rice, D., Crawford, C.: J. Org. Chem. *28*, 872 (1963)
221. Taylor, C., Brice, T., Wear, R.: J. Org. Chem. *27*, 1064 (1962)
222. Knunyants, I. L., Dyatkin, B. L., Gevorkyan, A. A.: Izv. Akad. Nauk SSSR Ser. Khim. *1965*, 1377
223. Ginsburg, V. A. et al.: Zh. Obshch. Khim. *30*, 2406 (1960)
224. Park, J. et al.: J. Org. Chem. *26*, 3316 (1961)
225. Isaev, V. L. et al.: Zh. Vses. Khim. Obshch. Im DJ Mendeleeva *18*, 713 (1973)
226. Dyatkin, B. L., Becker, R. A., Knunyants, I. L.: Dokl. Akad. Nauk SSSR *166*, 106 (1966)
227. Knunyants, I. L., Fokin, A. V., Komarov, V. A.: Zh. Vses. Khim. Obshch. Im DJ Mendeleeva *7*, 709 (1962)
228. Bissel, E.: J. Org. Chem. *28*, 1717 (1963)
229. Makarov, S. P. et al.: Dokl. Akad. Nauk SSSR *141*, 357 (1968)
230. Fokin, A. V. et al.: Zh. Obshch. Khim. *37*, 633 (1967)
231. Gambaryan, N. P., Rokhlin, E. M., Zeifman, Yu. V.: Izv. Akad. Nauk SSSR Ser. Khim. *1965*, 1466
232. Dyatkin, B. L. et al.: Tetrahedron Lett. *1971*, 345
233. Martynov. B. I., Sterlin, S. R., Dyatkin, B. L.: Izv. Akad. Nauk SSSR Ser. Khim. *1974*, 1642
234. Knunyants, I. L., Fokin, A. V.: Izv. Akad. Nauk SSSR Ser. Khim. *1952*, 261
235. Sterlin, S. R. et al.: Zh. Vses. Khim. Obshch. Im DJ Mendeleeva *18*, 711 (1973)
236. Yakubovich, A. Ya., et al.: Zh. Obshch. Khim. *37*, 847 (1967)
237. USSR Pat. 417,419; C.A. *80*, 145403 (1974)
238. Benett, F. et al.: Nature (Lond.) *166*, N 4190, 225 (1950)
239. Brandt, G., Emeleus, H., Haszeldine, R.: J. Chem. Soc. (Lond.) *1952*, 2198
240. Tullock, C., Coffman, D.: J. Org. Chem. *25*, 2016 (1960)
241. Haszeldine, R., Kidd, J.: J. Chem. Soc. (Lond.) *1953*, 3219
242. Dear, R., Gilbert, E.: Synthesis *1972*, 310
243. Tiers, G.: J. Org. Chem. *26*, 3515 (1961)
244. Hauptschein, M., Braid, M., Lawer, F.: J. Am. Chem. Soc. *79*, 6248 (1957)
245. Tiers, G.: J. Org. Chem. *26*, 2538 (1961)
246. Dyatkin, B. L. et al.: Tetrahedron *29*, 2759 (1973)
247. Middleton, W., Howard, E., Sharkey, W.: J. Org. Chem. *30*, 1375 (1965)

248. Chambers, R., Musgrave, W., Savody, J.: J. Chem. Soc. (Lond.) *1962*, 1993
249. Belenky, G. G., et al.: Dokl. Akad. Nauk SSSR *201*, 603 (1971)
250. Ragulin, L. I. et al.: Izv. Akad. Nauk SSSR Ser. Khim. *1967*, 1754
251. England, D., Dietrich, M., Lindsey, R.: J. Am. Chem. Soc. *82*, 6181 (1960)
252. Dmitriev, M. A., Sokolsky, G. A., Knunyants, I. L.: Izv. Akad. Nauk SSSR Ser. Khim. *1960*, 847
253. Belaventsev, M. A., Sokolsky, G. A., Knunyants, I. L.: Izv. Akad. Nauk SSSR Ser. Khim. *1965*, 1613
254. Fokin, A. V. et al.: Izv. Akad. Nauk SSSR Ser. Khim. *1974*, 471
255. Fokin, A. V. et al.: Izv. Akad. Nauk SSSR Ser. Khim. *1974*, 1892
256. Man, E., Coffman, D., Mutterties, E.: J. Am. Chem. Soc. *1959*, 3575
257. US Pat. 2,729,663; C.A. *50*, 11362 (1956)
258. Emeleus, H., Haszeldine, R.: J. Chem. Soc. (Lond.) *1949*, 2953
259. Griffiths, J., Burg, A.: J. Am. Chem. Soc. *82*, 5723 (1960)
260. Sterlin, S. R., Pinkina, V. N., Yatsenko, R. D.: Izv. Akad. Nauk SSSR Ser. Khim. 1959, 1492
261. McBee, E. T., Pierce, O. P., Higgins, J. F.: J. Am. Chem. Soc. *74*, 1736 (1952)
262. Gryszkiewicz-Trochimowski, E., Sporzynski, A., Wnuk, J.: Rec. Trav. Chim. *66*, 419 (1947)
263. USSR Pat. 232,979; C.A. *70*, 87011 (1969)
264. Steevens, F., Freeman, J.: J. Org. Chem. *29*, 2279 (1964)
265. Lawton, E., Cain, E., Sheftran, D.: J. Inorg. Nucl. Chem. *17*, 188 (1961)
266. Lawton, E., Pilipovich, D.: Inorg. Chem. *4*, 118 (1965)
267. Graham, W., Parker, C.: J. Org. Chem. *28*, 850 (1963)
268. USSR Pat. 245,750; C.A. *71*, 123524 (1965)
269. Fokin, A. V. et al.: Zh. Obshch. Khim. *39*, 2519 (1969)
270. Ankudinov, A. K., Ryazanova, R. M., Sokolov, S. V.: Zh. Org. Khim. *10*, 2503 (1974)
271. US Pat. 3,366,610; C.A. *67*, 109263 (1967)
272. Freeman, J., Kennedy, A., Colburn, C.: J. Am. Chem. Soc. *82*, 5304 (1960)

Fluoroaromatic Compounds

G. G. Furin, G. F. Grebenshchikova, A. Ya. Lvova, V. M. Vlasov, G. G. Yakobson

Toxicity of fluoroaromatic compounds is similar to that of the corresponding nonfluorinated hydrocarbons. Some polyfluoroaromatic compounds are considerably less toxic than their nonfluorinated analogues (e.g. hexafluorobenzene is just slightly toxic), whereas other compounds (e.g. octafluorotoluene and octafluoronaphthalene) are somewhat more toxic. Pentafluorobenzaldehyde is about ten times as toxic as benzaldehyde.

Polyfluorinated aromatic and heterocyclic compounds are nonflammable and nonexplosive.

Fluorobenzene

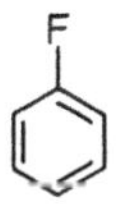

C_6H_5F MW 96

Fluorobenzene is a colourless liquid with a specific strong odour, soluble in common organic solvents. Bp 84°–85 °C [1, 2], 85°–85.5 °C/751 mm Hg [3]; n_D^{20} 1.4670; d_4^{20} 1.009 [3].

Method of Preparation

Fluorobenzene is prepared by the Balz-Schiemann reaction from phenyldiaznium tetrafluoroborate (51%–57% yield) [1, 2]; by co-pyrolysis of cyclopentadiene-1,3 or its dimer with difluorochloromethane or other fluorohalomethanes at 400°–600 °C (67% yield) [3]; or by the direct fluorination of benzene with fluorine in acetonitrile at −35 °C (~1%) or with XeF_2 (68% yield) [5].

The following procedure is a modification of the one described in [3].

Procedure

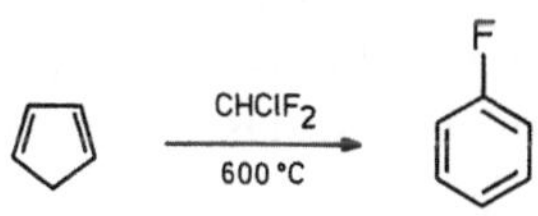

Freshly distilled cyclopentadiene-1,3 (12 g, 0.182 mol) and difluorochloromethane (29.7 g, 0.34 mol) are simultaneously passed during 20 min (flow rate 23 l/h) through

a quartz tube (250 × 20 mm) placed in a tube furnace at 600 °C and atmospheric pressure (the contact time 1.5 sec). The product is collected in a condensing system which consists of a water-cooled reflux condenser with a funnel and two dry-ice acetone traps arranged in series; it is washed with water (3 × 50 ml), dried over $CaCl_2$ and distilled through a fractionating column (~20 theoretical plates). A fraction boiling at 85°–85.5 °C is collected. The yield of fluorobenzene is 11.6 g (67%); n_D^{20} 1.4670, d_4^{20} 1.009.

1,3-Difluorobenzene

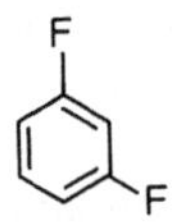

$C_6H_4F_2$ MW 114

1,3-Difluorobenzene is a colourless liquid with a specific odour, well soluble in common organic solvents and insoluble in water. Bp 82°–86 °C [6], 83°–84 °C [7]; n_D^{18} 1.4401 [7].

Method of Preparation

1,3-Difluorobenzene is prepared by decomposition of bis-m-phenylenediazonium borofluoride [6, 7].

Procedure

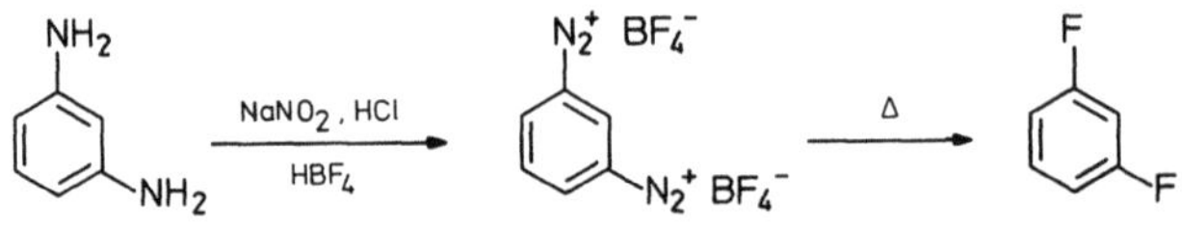

A cooled solution of $NaNO_2$ (120 g) in 240 ml of water is poured with stirring into a mixture of 400 ml of HCl (d 1.19) and 640 g of ice at −22 °C; the temperature should not rise above −15 °C. To the resulting blue-greenish solution is added at −12 to −17 °C, with stirring, a cooled to −5 °C suspension of m-phenylenediamine hydrochloride (72 g) in 100 ml of HCl (d 1.19). After all the amine has been added, diazo solution is stirred for another 5–10 min and quickly filtered. A solution of borofluorohydric acid, prepared by adding boric acid to 250 ml of 48% hydrofluoric acid to saturation, is then poured in. The mixture is colled to 0 °C with stirring for 15 min and filtered. The residue is washed with water (2 × 100 ml), alcohol (3 × 100 ml), and ether (3 × 100 ml), each of them cooled to 0 °C. After each washing the residue is thoroughly squeezed. The product is dried in air. The yield is 95–100 g (76%–81%), decomp. p. 204 °C.

In a 1-l Wurtz flask is placed 200 g of dry bis-m-phenylenediazonium borofluoride (Note). The Wurtz flask is connected to a long condenser, which is attached to three 1-l Erlenmeyer flasks arranged in series and cooled in ice-salt baths. Each supply tube should extend to the bottom of a flask. The flask with the product is carefully heated with an open smoky flame to complete decomposition of the salt. Two portions of meta-diazo fluorobenzene condensed in the first two flasks are combined, washed with water, then with 10% solution of NaOH (2 × 50 ml) and again with water, dried over $CaCl_2$ and distilled on a fractionating column (~20 theoretical plates). The yield of meta-difluorobenzene is 40–45 g (55%–60%); bp 83°–84 °C, n_D^{18} 1.4400.

Note. Diazonium borofluoride should be thoroughly dried, as decomposition of the wet product would sharply decrease the yield of the fluoro-derivative. It is convenient to dry diazonium borofluoride by placing it on filter paper under a hood or near a good hood.

Hexafluorobenzene and Chloropentafluorobenzene

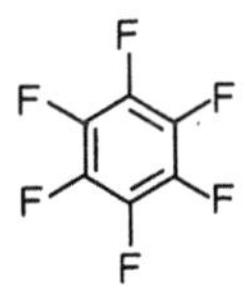

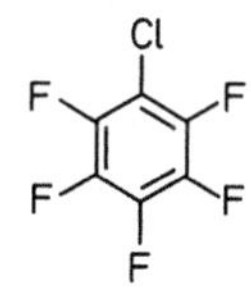

C_6F_6 MW 186 C_6F_5Cl MW 202.5

Hexafluorobenzene is a volatile colourless liquid with a strong odour, well soluble in common organic solvents, insoluble in water. Bp 79 °C [8], 80 °C [9–13], 80.1 °C [14], 80.26 °C [15], 80°–81 °C [16]; mp 5.2 °C [12], 5.29 °C [9]; d_4^{20} 1.6181 [9], 1.6188 [12]; n_D^{20} 1.3765 [8], 1.3768 [12], 1.3777 [9, 16].

Chloropentafluorobenzene is a volatile colourless liquid with a strong odour, well soluble in common organic solvents, insoluble in water. Bp 117°–118 °C [12], 122° to 123 °C [17]; mp −15 °C [12]; d_4^{20} 1.6425 [12]; d_4^{25} 1.5680 [17], 1.6403 [12]; n_D^{20} 1.4251 [12], 1.4256 [17].

Method of Preparation

Hexafluorobenzene is prepared by treatment of hexachlorobenzene with anhydrous KF at 430°–530 °C in the absence of a solvent (40%–45% yield; the by-products are chloropentafluorobenzene and a mixture of dichlorotetrafluoro- and trichloro-trifluorobenzenes) [12, 18]; by treatment of chloropentafluorobenzene with anhydrous CsF at 160°–190 °C in sulpholane (42% yield) [19]; by defluorination of a mixture of octafluorocyclohexadienes-1,3 and −1,4 at 630°–660 °C over the nickel gauze [20], or at 430 °C in nickel, iron, or copper tubes (84%–87% yield) [21, 22]; by dehydro-fluorination of undecafluorocyclohexane [23, 24] or of a mixture of isomeric hepta-fluorocyclohexadienes at 460°–530 °C over the iron gauze (15%–20%) [24], or of 1-H,2-H,4-H-nonafluorocyclohexane [10, 11], 3-H,4-H- and 4-H,5-H-octafluoro-cyclohexenes-1 [11] with an aqueous solution of KOH (6%–13% yield). Hexafluoro-benzene obtained by each of these methods is isolated from the reaction mixture by the preparative GLC. Hexafluorobenzene is also prepared by fluorination of hexa-chlorobenzene with elemental fluorine in 1,1,2-trichloro-1,2,2,-trifluoroethane at room temperature to yield a mixture of chlorofluorocyclohexanes, with its subsequent dehalogenation at 330 °C over the iron gauze (22% yield; the by-products are chloropentafluorobenzene and other chloropolyfluorobenzenes) [25, 26]; by fluorination of hexachlorobenzene with ClF_3 and subsequent dehalogenation of the resulting mixture of polychlorofluorocyclohexenes or -hexanes at 260°–315 °C over the iron gauze (11%–21% yield; the by-products are chloropentafluorobenzene and other compounds) [27, 28]. Using CoF_3 as a fluorinating agent increases the yield of hexafluorobenzene to 60% at the stage of aromatization [28, 29]. Other methods for the preparation of hexafluorobenze are: pyrolysis of tribromofluoromethane at 630°–640 °C in the platinum tube (45%–48% yield) [9, 16, 30] or in the flow of nitrogen at 540 °C/4.5 atm (55% yield) [31, 32]; co-pyrolysis of tribromofluoromethane, bromoform or 1,1,1-tribromo-2,2,2-trifluoroethane at 540 °C and the pressure of 10–18 atm of nitrogen flow [33, 34]. Preparation of hexafluorobenzene by pyrolysis of a number of freons has also been reported [35, 37].

Chloropentafluorobenzene is generally prepared from hexachlorobenzene by the same methods as hexafluorobenzene.

The following method for the preparation of hexafluorobenzene together with chloropentafluorobenzene is a modification of the procedure described in [12, 18].

Procedure

$$C_6Cl_6 \xrightarrow{KF} C_6F_6 + C_6F_5Cl + C_6F_XCl_{6-X}$$

$$X = 1 - 4$$

A 1-l stainless steel autoclave equipped with a manometer, a gate valve, and a thermocouple in a well, is charged with 350 g of hexachlorobenzene and 1000 g of freshly calcinated, hot KF (Notes 1, 2). The hermetically closed autoclave is placed in a furnace, heated to 400 °C and allowed to stand at that temperature for 10 h; then at 510 °C, for 20 h (Note 3). Towards the end of that period the pressure in the autoclave has reached 80–100 atm. The autoclave is connected to a Liebig condenser and the product is distilled off to a cooled receiver (Note 4). The product (220–235 g) dried over $MgSO_4$ (Notes 5), is distilled through a fractionating column with a glass or Nichrome packing (18–20 theoretical plates, the reflux ratio 10), affording the following fractions: (a) hexafluorobenzene, bp 79°–80 °C, the yield is 80–84 g (35%–37%); (b) chloropentafluorobenzene, bp 117°–118 °C, the yield is 90–100 g (36%–40%); (c) dichlorotetrafluorobenzenes, bp 152°–154 °C, the yield is 26–30 g (10%–12%).

The still-residue containing low-fluorinated polychlorobenzenes is refluorinated (Note 6).

Notes

1. Potassium fluoride is dehydrated by heating it at 300°–350 °C on a stainless steel tray. After that it is granulated to ~0.5 mm grains and stored in stoppered bottles. Before the reaction the KF sample is calcinated in a muffle furnace at 450°–500 °C during 5–6 h, with constant crushing of the lumps. It is loaded into the autoclave while still hot.
2. Loading should be carried out as quickly as possible so as to avoid the loss of hexachlorobenzene, which sublimes in contact with hot KF and also when the highly

hydroscopic KF absorbs air moisture. Loading of wet KF will result in resinification of the reaction mixture.

3. When a larger autoclave is used, the operating temperature and the time of keeping should be changed. Thus, for the 8-l autoclave, the conditions are: at 410 °C it is kept for 24 h; at 450 °C for 12 h; and, finally, at 510 °C for 24 h.
4. It is possible to distil off the product with steam from the reaction mixture taken out of the autoclave cooled to room temperature.
5. As shown by GLC, the reaction mixture contains 40%–45% of C_6F_6 and 40% to 44% of C_6F_5Cl.
6. Refluorination of the still-residue is carried out under the same conditions. It should be borne in mind that the pressure is 100–120 atm. For each 800 g of the still residue, 1000 g of KF is used.

Octafluoronaphthalene

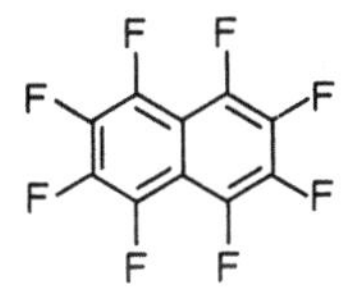

$C_{10}F_8$ MW 272

Octafluoronaphthalene is a white crystalline substance, soluble in common organic solvents. Bp 207°–209 °C/760 mm Hg [38], 80 °C/15 mm Hg [39], 92°–93 °C/10 mm Hg [19]; mp 83.5°–85.5 °C [40], 85°–86 °C [41], 86°–87 °C [19, 23, 38, 42], 87°–88 °C [43].

Method of Preparation

Octafluoronaphthalene is prepared by defluorination of perfluorodecaline at 460° to 500 °C over the iron gauze in a 47% yield [23, 42, 43], or in a 30% to 40% yield [41]; by fluorination of octachloronaphthalene with BrF_3 and SbF_5, followed by dehalogenation with zinc [39]; or by treatment of octachloronaphthalene with KF in sulpholane (52% yield) [19], or in the absence of a solvent (24% and 60% yields) [38, 44].

The following procedure is a modification of the one described in [38].

Procedure

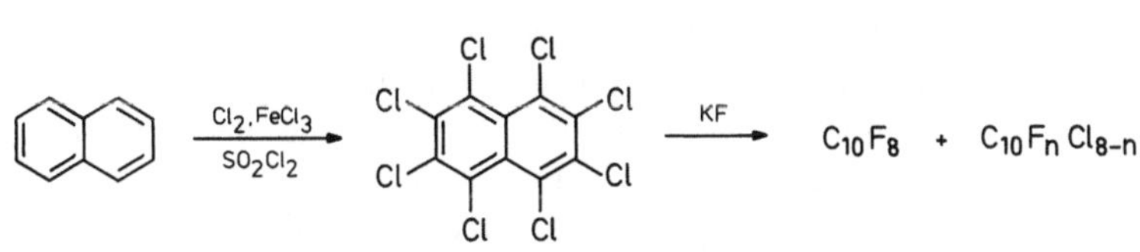

Octachloronaphthalene. In a 3.5-l four-necked, round-bottom flask fitted with a stirrer, a thermometer, a reflux condenser with a calcium chloride tube and a chlorine inlet tube, are placed 2 l of freshly distilled sulphuryl chloride, 200 g (1.6 mol) of ground naphthalene (mp ≧ 79.6 °C), 2 g of powdered iron or ferric chloride. The mixture is stirred to complete dissolution of naphthalene, whereupon a flow of chlorine is let in at a reasonable rate (Note 1). The temperature of the reaction mixture is maintained at 55°–60 °C by regulating the flow rate of chlorine; towards the end of the reaction, by warming the reaction mixture. After 5–6 h, octachloronaphthalene is precipitated and chlorination is continued for another 15–20 min (Note 2). The mixture is cooled, the precipitate is filtered off on a glass filter, washed on the filter with two or three portions (200 ml) of cooled sulphuryl chloride and dried in air till the odour of sulphuryl chloride has vanished. This gives the product with mp 167° to 183 °C. After two recrystallizations from CCl_4 the product is dried in air or in a desiccator at 70°–90 °C till it has gained constant weight. The yield of octachloronaphthalene is 320–340 g (50%–55%); mp 190°–196 °C (Note 3).

Octafluoronaphthalene. In a 1-l stainless steel autoclave (a rotating one is preferable) fitted with a thermometer, a manometer, and a gate valve, are charged in layers 400 g (1 mol) of octachloronaphthalene and 900 g (15.5 mol) of freshly calcinated KF (Note 4). The autoclave is heated to 380 °C for 4–5 h, then at 380 °C for 20 h and at 410 °C for 20 h, whereupon it is cooled to room temperature. The reaction product is transferred to a 2–3 l steam distillation flask, 1–1.5 l of water is added to it and the product is steam-distilled during ~30–40 h. This procedure gives 100–120 g of the mixture of chloropolyfluoronaphthalenes containing 30% of octafluoronaphthalene, as shown by GLC (Note 5).

To a mixture of chloropolyfluoronaphthalenes obtained from several runs (300 to 360 g all in all) is added 180 ml of benzene and the residual water azeotrope is distilled off. The mixture is then distilled on a fractionating column with a glass or nickel packing (25–30 theoretical plates) to afford 90–100 g of octafluoronaphthalene boiling at 207°–210 °C and 90–105 g of a mixture of α- and β-chloroheptafluoronaphthalenes boiling at 240°–245 °C.

The still-residue obtained from several runs is refluorinated as described above for pure octachloronaphthalene to give a mixture of chloropolyfluoronaphthalenes containing up to 80% of octafluoronaphthalene. The total yield of octafluoronaphthalene is 60–70 g (22%–26%), the yield of the mixture of α- and β-chloroheptafluoronaphthalenes is 70–80 g (24%–28%), based on 400 g of octachloronaphthalene.

Notes

1. Chlorine should first be dried by passing it through the concentrated H_2SO_4.
2. By the time octachloronaphthalene precipitate has been formed, the amount of chlorine passed through the mixture should be nearly theoretical; chlorination should be stopped when the density of the reaction mixture has reached 1.685 g/cm^3. Futher chlorination would cause dissolution of the precipitate and result in the formation of decachlorodihydronaphthalene.
3. If required, octachloronaphthalene is recrystallized from benzene.
4. KF is dehydrated as described above.
5. The product may be distilled off from the hot autoclave through a condenser into a cooled receiver.

Pentafluoropyridine

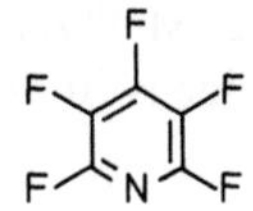

C_5F_5N MW 169

Pentafluoropyridine is a colourless liquid with the odour of an aromatic halogeno-derivative. It is very volatile and stable against light and air moisture. Bp 83.5 °C [45, 46], 84 °C [47–49]; mp —42 °C [45]; n_D^{20} 1.3818 [48], 1.3834 [47, 49], 1.3856 [45].

Method of Preparation

Pentafluoropyridine is obtained by fluorination of pyrydine with CoF_3, followed by defluorination of the initially formed perfluoropiperidine (a small yield) [45, 47, 49]; or by interaction of pentachloropyridine with KF (60%–70% yield) [46, 48].

The following procedure is based on the preparations described in [46, 48].

Procedure

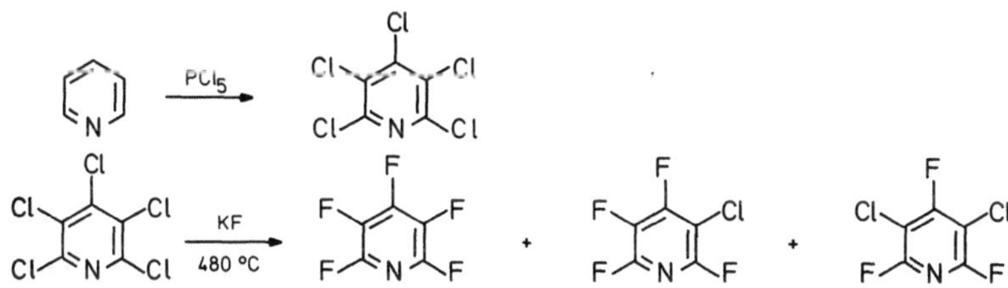

Pentachloropyridine. In a 2-l rotating steel autoclave placed in a furnace and fitted with a manometer, a valve and a thermocouple in a well, is placed 1700 g of PCl_5; then 60 ml of pyridine is quickly added (Note 1). The autoclave is heated to 370 °C and kept at that temperature for 20 h. Towards the end of that period the pressure has reached 190 atm. After 20 h, the autoclave is cooled to room temperature and hydrogen chloride formed in the reaction is vented off into a hood or collected in an alkaline trap. The autoclave is cautiously opened (a gas mask and gloves should be worn) and the reaction mixture is poured onto 3 kg of ice placed in a stainless steel vessel. Pentachloropyridine paste is washed with water till it is neutral to litmus, and distilled with overheated (150 °C) steam into a separatory funnel provided with a cooling jacket. The product is dried for a week over $CaCl_2$ in a desiccator. The yield of dry pentachloropyridine is 120–150 g (58%–72%); mp 124°–125 °C.

Pentafluoropyridine. A 1-l rotating steel autoclave fitted with a manometer, a valve and a thermocouple in a well, is charged with 240 g (0.95 mol) of pentachloropyridine and 900 g of freshly calcinated anhydrous KF (Note 2). The autoclave is heated to 380 °C and kept at that temperature for 20 h, and at 470°–480 °C for an additional 20 h. Then it is cooled to 120 °C and rotation is stopped, whereupon it is cooled to room temperature and further cooled with snow or ice-water. The product is distilled with steam into a separatory funnel, under the aqueous layer. This gives 100–120 g of polyfluorinated products. The mixture is dried over $MgSO_4$ and distilled through a fractionating column. The yield of pentafluoropyridine is 70–90 g (44% to 56%); bp 118.5°–119.5 °C; n_D^{20} 1.4307 (Note 4).

Notes

1. Pyridine has been previously distilled and dried over KOH. The autoclave should be thoroughly dried.
2. The temperature of KF should not exceed 200 °C.
3. Pentachloropyridine is stored in a hermetically sealed container.
4. If 3-chlorotetrafluoropyridine is desired as the major product, the reaction conditions should be as follows: pentachloropyridine 240 g, KF 600 g, the temperature in the autoclave 410 °C, reaction time 24 h. The yield of the mixture of products is 90–100 g.

Decafluorodiphenyl

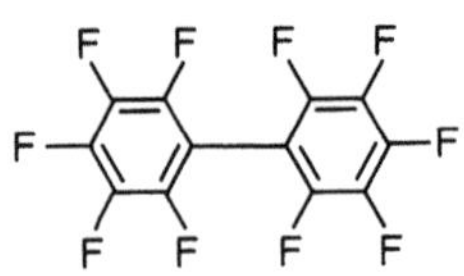

$C_{12}F_{10}$ MW 334

Decafluorodiphenyl is a white crystalline substance with a faint odour, volatile and soluble in common organic solvents. Mp 66.5 °C [50], 67 °C [51], 67°–68 °C [38, 52–64], 67°–69 °C [42], 67.5°–68 °C [52, 55, 56], 67.5°–69 °C [57], 68 °C [58], 68°–69 °C [59–61], 69 °C [26]. Subl.p. 50 °C at 1 mm Hg [52, 56], 85 °C at 10 mm Hg [53].

Method of Preparation

Decafluorodiphenyl is prepared by the Ullmann reaction from chloropentafluorobenzene [26, 55], bromopentafluorobenzene [52, 56, 60] and iodopentafluorobenzene [50] (yields are 69%–73%, 42%–87%, and 72%, respectively); with a small yield, by the reaction of decachlorodiphenyl with KF in sulpholane [59] or in the absence of a solvent [38]; by fluorination of decachlorodiphenyl with BrF_3 and SbF_5, followed by dehalogenation with zinc [39] or defluorination of perfluorodicyclohexyl [23, 42] (yields are up to 15%); by the reaction of pentafluorophenylmagnesium bromide with $CoCl_2$ [52, 56] or of pentafluorophenyldiazonium fluoride with cuprous bronze [3] (yields are 15%–23%). Decafluorodiphenyl may be formed as a by-product in the reaction of bromopentafluorobenzene with CuCN [54], 1-bromo-2,3,4,5-tetrafluorobenzene [53] and 1-chloro-1,2-difluoroethane [62].

The following procedure is a modification of the preparations described in [26, 55].

Procedure

$$C_6F_5Cl \xrightarrow{Cu,\ 350\ °C} C_{12}F_{10}$$

Copper powder. One kg of $CuSO_4$ is dissolved in 2 l of hot water; the solution is filtered and cooled to room temperature. To the solution is added, with vigorous stirring, 265 g of zinc dust in small portions (the addition requires ~2 h; the solution should not boil). Stirring is continued for another 4 h. The resulting copper powder is filtered off on a Büchner funnel, washed with four or five portions of water, then with two or three 150-ml portions of 10% HCl, with careful stirring of the filter residue, and washed with water till it is neutral to litmus. Residual zinc is removed as $ZnCl_2$. In order to remove residual $CuSO_4$, the copper powder obtained in the reaction is treated on the filter with 30% aqueous solution of NH_3, with careful stirring, and washed with water till it is neutral to litmus. The residue is washed with acetone, then with alcohol, and dried at 70 °C during 12 h. This gives 200 g of copper powder.

Decafluorodiphenyl. In a 1-l rotating steel autoclave provided with a valve, a manometer, and a thermocouple in a well, are placed 500 g (~2.5 mol) of chloropentafluorobenzene, 400 g of dry copper powder (to ensure good mixing, two or three copper rods may be inserted). The autoclave is heated at 350 °C for 24 h. By the end of that period the pressure has reached 60 atm. Decafluorodiphenyl is distilled off from the autoclave through a Liebig condenser into a receiver cooled in the ice-salt bath (Note 1). This yields 330–340 g of the mixture containing 90% of decafluorodiphenyl (71%–73% yield). The mixture is distilled with a dephlegmator. The fraction boiling at 207°–207.5 °C is collected. The yield of decafluorodiphenyl is 270–280 g (65%–67% yield); mp 67°–68 °C.

Note. The products may be isolated by steam distillation of the reaction mixture, which has been transferred into a flask from the autoclave cooled to room temperature.

Perfluorobenzocyclobutene

C_8F_8 MW 248

Perfluorobenzocyclobutene is a colourless liquid, soluble in common organic solvents. 117 °C [63, 64]; n_D^{21} 1.3868 [64].

Method of Preparation

Perfluorobenzocyclobutene is prepared by co-pyrolysis of tetrafluorophthalic anhydride with tetrafluoroethylene or octafluorocyclobutane at 650 °C (35%–40% yield) [63, 64]; or by the photolysis of octafluorostyrene [65].

Procedure

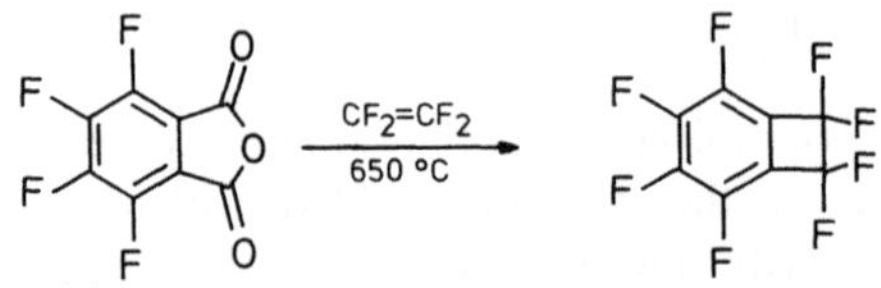

Through a quartz tube (Notes 1, 2) are simultaneously passed, during 1.5 h, 35.4 g (0.16 mol) of tetrafluorophthalic anhydride and tetrafluoroethylene (at the rate of 15 l/h). Pyrolysate is collected in a 100-ml two-necked flask fitted with an effective reflux condenser with a gas outlet tube, then distilled with steam. The organic layer is separated and dried over $MgSO_4$. This gives 29–30 g of the mixture containing 75% of perfluorobenzocyclobutene, as shown by GLC. Fractional distillation of 158 g of that mixture yields 78.5 g of perfluorobenzocyclobutene; bp 117 °C; n_D^{19} 1.3868.

Notes

1. The quartz tube (650 × 30 mm) is provided with a thermocouple in a well, a pressure-equalizing dropping funnel, an electric heater, a gas outlet tube, and an adapter connecting it to a receiver.
2. The adapter should be replaced every 5 h with a new one, owing to the corrosive action of the thermolysis products.

Perfluoroindane

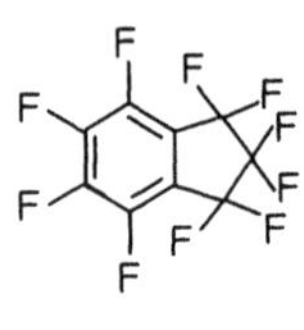

C_9F_{10} MW 298

Perfluoroindane is a colourless liquid with a specific odour, soluble in common organic solvents and slightly soluble in water. Bp 140 °C [66], 141°–142 °C [67], 136 °C [68]; n_D^{23} 1.3810 [66, 67], n_D^{25} 1.3775 [68].

Method of Preparation

Perfluoroindane is prepared by co-pyrolysis of pentafluorobenzene derivatives containing electron-donating substituents — OH, OCH_3, SH, NH_2 — with tetrafluoroethylene at 620°–720 °C (30%–60% yield) [66, 67]; by defluorination of perfluoroperhydroindane at 490 °C on the nickel catalyst (33% yield) [68]; by passing hexafluorocyclopentadiene over alkaline or alkaline earth metal fluorides at 400° to 700 °C [69] or by passing perfluoro (bicyclo [4,3,0]nonatriene-2,4,7) over the iron catalyst (28% yield) [70].

Procedure

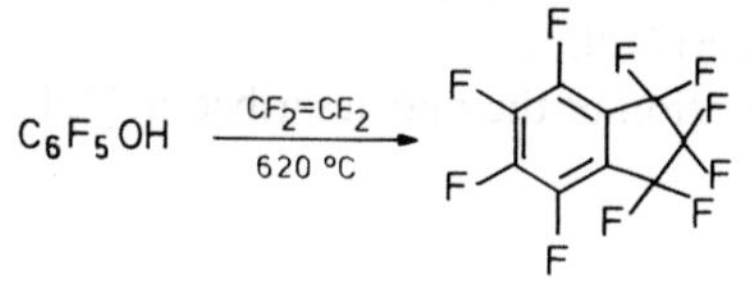

Through a quartz tube are simultaneously passed 1700 g (17 mol) of tetrafluoroethylene (15 l/h) and 637 g (3.45 mol) of pentafluorophenol (25 g/h). Pyrolysate is collected in a 1-l two-necked flask fitted with an inlet tube and an effective reflux condenser with a gas outlet tube. Then water (200 ml) is added and the mixture is distilled with steam. The organic layer is separated and dried over $MgSO_4$. This yields 750 g of the mixture containing 60% of perfluoroindane, as shown by GLC. The product is distilled through a fractionating column with a glass packing (35 to 40 theoretical plates, reflux ratio 10). A fraction boiling at 141°–142 °C is collected. The yield of perfluoroindane is 320 g (31%); n_D^{23} 1.3810.

Octafluorotoluene

(Pentafluorobenzotrifluoride)

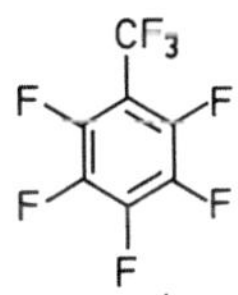

C_7F_8 MW 236

Octafluorotoluene is a volatile colourless liquid, soluble in common organic solvents. Bp 102°–103 °C [42], 103.1°–103.4 °C [71], 103.5 °C [17], 106.5°–108 °C [9]; mp < −70 °C [17]; n_D^{19} 1.3680 [42], n_D^{20} 1.3663 [71], 1.3664 [17]; d_4^{25} 1.66 [17]; bp 103.5° to 104.5 °C [72]; n_D^{20} 1.3664 [72]; d_4^{20} 1.668 [72].

Method of Preparation

Octafluorotoluene is prepared, with a small yield, by treatment of pentachlorobenzotrifluoride with KF in sulpholane [10]. In the absence of a solvent, the yield of octafluorotoluene is considerably increased [73]. The reaction of hexachlorobenzene with KF at 450°–550 °C yields octafluorotoluene in a small quantity [74]. Using hexafluorobenzene instead of hexachlorobenzene will increase the yield up to 20% [71]. Octafluorotoluene may be prepared also by the reaction of hexafluorobenzene with Teflon at 550 °C (35%–40% yield) [75]; by passing hexafluorobenzene vapour with iodotrifluoromethane or perfluoroolefins through a quartz tube heated to 700° to 850 °C (10%–20% yield) [76]; by pyrolytic defluorination of perfluoromethylcyclohexane (~25% yield) [23, 42]; by the pyrolysis of tribromofluoromethane [9, 33, 34], perfluorodicyclohexylmethane [77]; or by dehalogenation of polyfluorohalomethylcyclohexanes with zinc (yields up to 8%) [17, 39, 78, 79]; by co-pyrolysis of hexafluorobenzene with tetrafluoroethylene in a tube reactor of stainless or carbon steel (90%–96% yield; 62%–65% conversion) [72].

The following procedure is a modification of the one described in [75].

Procedure

$$C_6F_6 \xrightarrow{CF_2=CF_2} C_6F_5CF_3$$

Hexafluorobenzene (93 g; 0.5 mol) and tetrafluoroethylene are simultaneously passed through a stainless steel tube (Note) at 720 °C during 1.5 h at the rates of 60 g/h and 10 l/h, respectively. Pyrolysate is condensed into a water-cooled condenser and collected in a receiver placed in a cooling dry-ice acetone bath to afford 110–115 g of a mixture. Fractional distillation of the mixture on a fractionating column (~20 theoretical plates) yields 33–35 g of hexafluorobenzene and 66–73 g of octafluorotoluene. The yield of octafluorotoluene is 62%–65%, bp 103.5°–104.5 °C; n_D^{20} 1.3664; d_4^{20} 1.6680.

Note. The tube (200 × 20 mm) is fitted with a thermocouple in a well, a pressure-equalizing dropping funnel, a gas inlet tube, and an adapter, connecting it to a receiver.

Bis(pentafluorophenyl)difluoromethane

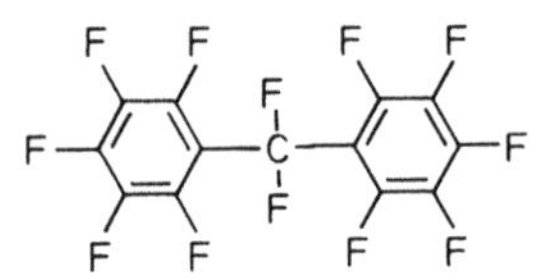

$C_{13}F_{12}$ MW 384

Bis(pentafluorophenyl)difluoromethane is a white crystalline substance, stable against light and air moisture, and soluble in common organic solvents. Mp 46°–48 °C [81], 49°–50 °C [82].

Method of Preparation

Bis(pentafluorophenyl)difluoromethane is prepared by the reaction of octafluorotoluene with pentafluorobenzene in the presence of SbF_5 and anhydrous HF (68% yield) [81], or by treatment of decafluorobenzophenone with SF_4 (88% yield) [82].

Procedure

$$C_6F_5CF_3 \xrightarrow[HF]{C_6F_5H,\, SbF_5} C_6F_5CF_2C_6F_5$$

In a 100-ml three-necked flask fitted with a stirrer, a pressure-equalizing dropping funnel, a calcium chloride tube, and a thermometer, is placed 50 g of SbF_5. Octafluorotoluene (14.2 g; 0.06 mol) is dropped in with vigorous stirring and the mixture is kept at 20 °C for 0.5 h. Then pentafluorobenzene (10.1 g; 0.06 mol) is added at the same temperature, and the mixture is stirred for 1 h.

Freshly distilled anhydrous HF (200 ml) is poured into a 0.5-l polyethylene bottle with a tight cover and the bottle is cooled to —10 °C. The reaction mixture contained in the flask is added in portions to HF so that the temperature in the bottle should not exceed 10 °C (Note 1). The mixture is kept in the polyethylene bottle at 20 °C for 5 h and then poured onto 1.5 kg of ice (Note 2). The precipitate is filtered off on a Bücher funnel, washed with water (250 ml) and dried, first in air, then over P_2O_5. The yield of bis(pentafluorophenyl)difluoromethane is 15.7–17 g (68%–74%), mp 47°–49 °C (from petroleum ether, bp 40°–60 °C).

Notes

1. Hydrogen fluoride reacts with water with considerable evolution of heat, therefore the addition should be carried out in dry atmosphere. Dissolution of SbF_5 in HF proceeds with evolution of heat and if it is carried out quickly, ejection of the reaction mass may take place.
2. HF and SbF_5 react with water with considerable evolution of heat, so this procedure should be carried out with caution, under an effective hood. The bottle should preferably be cooled with ice-water.

Heptafluorobenzyl bromidc

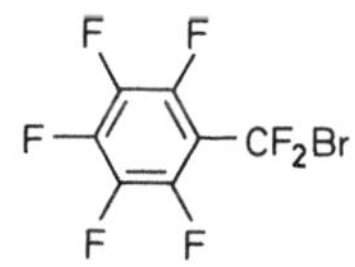

C_7BrF_7 MW 297

Heptafluorobenzyl bromide is a mobile colourless liquid with a specific odour, soluble in common organic solvents. Bp 151°–151.5 °C [83].

Method of Preparation

Heptafluorobenzyl bromide is prepared by co-pyrolysis of pentafluorophenol with tetrafluoroethylene with addition of bromine, or with tetrafluoro-1,2-dibromoethane at 570°–600 °C (23% yield) [83]; by the pyrolysis of tetrafluoro-2-bromethylpenta-fluorophenyl ether [83]; or by bromination of pentafluoro (difluoromethyl)benzene with bromine at 380°–390 °C (45% yield) [84].

The following method is based on the data reported in [83].

Procedure

$$C_6F_5OH \xrightarrow{BrCF_2-CF_2Br} C_6F_5CF_2Br$$

Pentafluorophenol (209.5 g; 1 mol) and 1,2-dibromotetrafluoroethane (397 g; 1.53 mol) are simultaneously passed at 600 °C during 7 h through a quartz tube (see Notes 1, 2 on p. 118). Pyrolysate is collected in a condensing system consisting of a 1-l water-cooled two-necked flask, fitted with a reflux condenser and a dry-ice acetone trap. The pyrolysis products are distilled with steam, the organic layer is separated and dried over $MgSO_4$. The product (395 g) (Note) is distilled on a fractionating column (~20 theoretical plates). A fraction boiling at 151°–151.5 °C is collected. The yield of heptafluorobenzyl bromide is 78 g (23 %).

Note. As shown by GLC, the reaction mixture contains 27 % of heptafluorobenzyl bromide and 42 % of perfluoro(benzylphenyl) ether.

Perfluoromesitylene

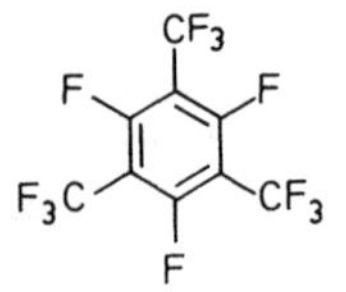

C_9F_{12} MW 336

Perfluoromesitylene is a volatile colourless liquid, soluble in common organic solvents and insoluble in water Bp 139°–139.5 °C [85].

Method of Preparation

Perfluoromesitylene is prepared by the reaction of hexafluorobenzene or octafluorotoluene and perfluoroxylene with excess Teflon in an autoclave at 450°–600 °C (42 % yield) [85]; by pyrolytic defluorination of perfluoro-(1,3,5-trimethylcyclohexane) at 470 °C on the iron catalyst (35 % yield) [86–87] or perfluoro (1,2-dimethylcyclohexane) at 600 °C over coal [88]. Perfluoromesitylene can be also prepared by the pyrolysis of Teflon (at 530°–560 °C; 100 kg/cm²) in the presence of a catalyst (Ni, Fe, Co) [89], or by the reaction of hexafluorobenze with a haloform and KF at 400° to 450 °C, in an autoclave [90].

The following procedure is based on the one described in [85].

Procedure

$$C_6F_6 \xrightarrow{\text{Teflon, 570 °C}} C_6F_3(CF_3)_3$$

In a 1-l rotating steel autoclave equipped with a manometer, a valve and a thermocouple in a well, are placed 225 g (1.21 mol) of hexafluorobenzene and 350 g of finely divided Teflon (Note 1). The autoclave is heated at 570 °C over a period of 8 h. After the autoclave has cooled, the pressure in it is brought down to that of the atmosphere (Note 2). The thermocouple well is replaced with a hollow steel tube and the reaction mixture is distilled with steam from the autoclave (Note 3). The organic layer is separated and dried over $MgSO_4$. This affords 340–360 g of a mixture containing 50 %–55 % of perfluoromesitylene, as shown by GLC. The mixture is distilled on a fractionating column (~40 theoretical plates). A fraction boiling at 139°–139.5 °C is collected. The yield of perfluoromesitylene is 150 g (37%) (Note 4).

Notes

1. Before the reaction the Teflon turnings are washed with acetone, then with ether, and dried in air, then in a desiccator at 150°–200 °C.
2. The reaction products may be isolated by steam distillation of the reaction mixture, which has been transferred from the autoclave, cooled to room temperature, into a flask.
3. The gaseous products of the reaction may include toxic perfluoroisobutylene, therefore care must be taken while reducing pressure in the autoclave and during distillation; work should be carried out under an efficient hood.
4. Perfluoromesitylene thus obtained contains 1%–3% of perfluoroindane, which is difficult to isolate.

Perfluoro-3,5-lutidine

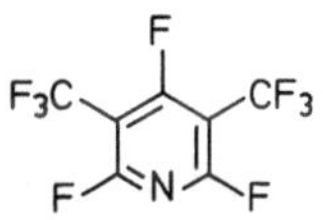

C_7F_9N MW 269

Perfluoro-3,5-lutidine is a volatile colourless liquid, soluble in most organic solvents. Bp 117.2°–117.4 °C [91].

Method of Preparation

Perfluoro-3,5-lutidine is prepared by the reaction of perfluoropyridine with KF or Teflon at 560°–570 °C (48% yield) [91, 92].

The following procedure is based upon the one described in [91].

Procedure

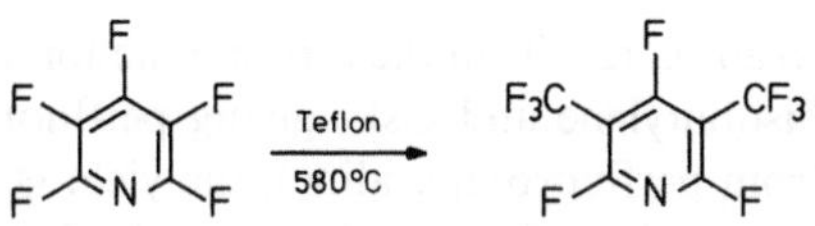

A 100-ml rotating steel autoclave equipped with a valve and a thermocouple in a well, is charged with 30 g (0.178 mol) of pentafluoropyridine and 26.6 g of Teflon (see Note 1 on p. 123) and heated at 560 °C over a period of 7 h. Then it is cooled to −10 °C, the pressure is reduced to that of the atmosphere, and the reaction mixture transferred into a 1-l flask. The autoclave is washed with 300 ml of water, the washings are placed into the same flask and the mixture is distilled with steam. The organic layer is separated and dried over $MgSO_4$. This procedure gives 42–43 g of a mixture containing 55% of perfluoro-3,5-lutidine, as shown by GLC. The mixture (130 g obtained by three runs) is distilled on a fractionating column with a glass packing (~40 theoretical plates). A fraction boiling at 117.2°–117.4 °C is collected. The yield of perfluoro-3,5-lutidine is 50–55 g (34%–37%).

Perfluoro-γ-*tert*-butylpyridine

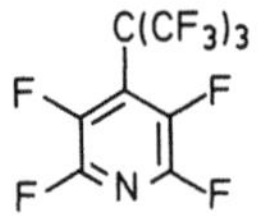

$C_9F_{13}N$ MW 369

Perfluoro-γ-*tert*-butylpyridine is a colourless liquid with a specific odour, soluble in common organic solvents. Bp 151°–153 °C, n_D^{20} 1.3670 [93].

Method of Preparation

Perfluoro-γ-*tert*-butylpyridine is prepared by the reaction of pentafluoropyridine with perfluoroisobutylene (Note) in the presence of CsF (92% yield) [93].

Procedure

$$C_5F_5N + (CF_3)_2C{=}CF_2 \xrightarrow{CsF} (CF_3)_3C{-}C_5F_4N$$

In a 100-ml three-necked flask provided with a stirrer, a gas inlet tube, and dry-ice acetone reflux condenser, are placed 5 g (0.03 mol) of pentafluoropyridine, 7.5 g (0.05 mol) of freshly calcinated, finely ground CsF, 30 ml of anhydrous sulpholane. Then 10 g (0.05 mol) of perfluoroisobutylene is added with vigorous stirring (Note). After the mixture has been stirred at room temperature for 5 h, it is filtered and the filtrate is distilled. The yields of perfluoro-γ-*tert*-butylpyridine is 10 g (92%), bp 151°–153 °C, n_D^{20} 1.3670.

Heating of the reaction mixture to 80 °C results in the further reaction of perfluoro-γ-*tert*-butylpyridine with perfluoroisobutylene and CsF, giving perfluoro(2,4,6-*tert*-butylpyridine); mp 165°–167 °C (from perfluoropiperidine), the yield is 40% [93].

Likewise, nitropentafluorobenzene and perfluoroisobutylene lead to perfluoro-*tert*-butylbenzene; bp 143°–145 °C, n_D^{20} 1.3960, 64,5% yield. This reaction continued at 50 °C for 2 h gives, in addition to perfluoro-*tert*-butylbenzene (21.4% yield), perfluoro-1,4-di-*tert*-butylbenzene, mp 149°–150 °C (from heptane), (23.4% yield) [93].

Note. Perfluoroisobutylene is extremely toxic. It should be handled only under a good hood.

Pentafluorobenzene

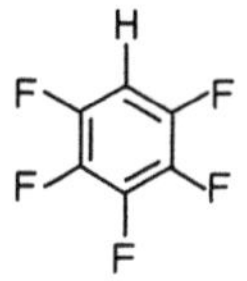

C_6HF_5 MW 168

Pentafluorobenzene is a volatile colourless liquid, soluble in common organic solvents. Bp 84.6 °C [14], 85 °C [58, 60, 94], 85°–86 °C [52], 89 °C [11], 84 °C at 760 mm Hg [95], 86 °C at 763 mm Hg [54]; n_D^{20} 1.3914 [94], 1.3920 [54].

Method of Preparation

Preparation methods of pentafluorobenzene may be classified in several groups.

1. Dehydrofluorination, with an aqueous solution of KOH, of isomeric octafluorocyclohexanes (45% yield) [60, 96] or of 1-H,4-H,5-H-heptafluorocyclohexene (20% yield) [11]; defluorination of undecafluorocyclohexane, 4-H-nonafluoro- and 4-H,5-H-octafluorocyclohexenes [23, 24] or of isomeric heptafluoro- and hexafluorocyclohexadienes [20, 21, 24] at 400°–600 °C over the iron gauze in a nickel reactor/tube; dehalogenation of the products of chlorofluorination of benzene with ClF_3 [97], or of dichlorodecafluoro- and trichlorononafluorocyclohexanes [98] at 300° to 430 °C over the iron gauze. The procedures described in [20–24, 97, 98] give, apart from pentafluorobenzene, hexafluorobenzene and a number of other compounds; pentafluorobenzene is isolated by the preparative GLC.

2. Pyrolysis of dibromofluoromethane (gives also hexafluorobenzene), or of the mixture of hexafluorobenzene with KOH, LiI, KI, KBr, or KCN at 500°–600 °C (20%–40% yield) [76].

3. Reduction of bromo- and chloropentafluorobenzenes with hydrogen over platinum, palladium, or nickel (gives a mixture of pentafluorobenzene with 1,2,3,4-tetrafluorobenzene) [99]. The best results have been obtained when these starting compounds were treated with zinc in glacial acetic acid [100], or with hydrogen over palladium black in methanol with an addition of sodium acetate or cuprous oxide mixed with acetic anhydride and dimethylformamide [54]. According to [94, 54, 100] the yield of pentafluorobenzene is 51%–84%.

Some other preparations of pentafluorobenzene have been reported: treatment of hexafluorobenzene with lithium aluminium hydride (61% yield) [40]; oxidation of pentafluorophenylhydrazine in the aqueous solution of $CuSO_4$ (77% yield) or sodium hypobromite (43% yield) [58]; hydrolysis of Grignard reagents obtained from bromo- or chloropentafluorobenzenes [25, 26, 101]; hydrolysis of pentafluorophenyllithium (49%–67% yield) [102]; decarboxylation of pentafluorobenzoic acid in an aqueous 30% solution of ammonia (79.4% yield) [95].

Pentafluorobenzene also formed in the haloform decomposition of pentafluoroacetophenone and its α-substituted derivatives with 63%–92% yield [103, 104].

Procedure

1. Hydrolysis of Pentafluorophenylmagnesium Chloride [40, 101]

$$C_6F_5Cl \xrightarrow{Mg} C_6F_5MgCl \xrightarrow{H_2O;\ HCl} C_6HF_5$$

Pentafluorophenylmagnesium chloride. In a 2-l three-necked flask fitted with a stirrer, a pressure-equalizing dropping funnel, and a long (8-bulb) reflux condenser with a calcium chloride tube, are charged three to five crystals (0.5–0.7 g) of iodine and 48.6 g (2 mol) of metallic magnesium (Note 1). The flask is heated to sublimation of iodine and cooled to room temperature. Then 900 ml of absolute ether is poured in. From a dropping funnel, 5–10 ml of a solution is added, prepared from 202.5 g (124 ml, 1 mol) of chloropentafluorobenzene, 187.8 g (86 ml, 1 mol) of 1,2-dibromoethane (Note 2), and 200 ml of absolute ether. The flask is heated in a water-bath till the beginning of the reaction (indicated by decolorization of the solution, its slight cloudiness, and spontaneous boiling), whereupon it is quickly immersed in a cooling bath (the temperature is —5° to —10 °C). The remaining part of the solution is dropped in, with vigorous stirring, over a period of 2–2.5 h, at such a rate as to maintain the reaction mixture in a condition near to boiling (Nöte 3). Then it is refluxed for 0.5–1 h.

Pentafluorobenzene. A solution of the Grignard reagent is poured into a mixture of 900 g of ice and 250 ml of concentrated HCl. The ether layer is separated and the aqueous one extracted with ether (6 × 300 ml). The ethereal solutions are combined and washed with 10% aqueous Na_2SO_3, 15% aqueous Na_2CO_3, and with water till they are neutral to litmus. The ether extract is dried over $MgSO_4$, the ether is distilled off in a water-bath and the residue distilled on a fractionating column (~20 theoretical plates). A fraction boiling at 86°–87 °C is collected. The yield of pentafluorobenzene is 110–120 g (65%–70%); $n_D^{19.5}$ 1.3913.

2. Reduction of Chloropentafluorobenzene in the Presence of Copper Powder

$$C_6F_5Cl \xrightarrow{H_2O;\ Cu} C_6HF_5$$

A 1-l rotating steel autoclave equipped with a manometer and a needle valve, is loaded with 500 g (~2.5 mol) of chloropentafluorobenzene, 400 g of dry copper powder, and 30 ml of water. The autoclave is heated to 300 °C and kept at that temperature for 24 h. The product is distilled from the autoclave through a laboratory-scale fractionating column (~20 theoretical plates). A fraction boiling at 86°–87 °C is collected. The yield of pentafluorobenzene is 365–375 g (87%–90%); $n_D^{19.5}$ 1.3913.

Notes

1. The magnesium turnings are thoroughly washed with ether and dried in air.
2. 1,2-Dibromoethane should be previously dried over $CaCl_2$.

Bromopentafluorobenzene

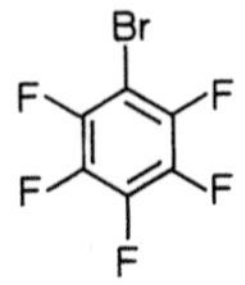

C_6BrF_5 MW 247

Bromopentafluorobenzene is a colourless liquid with a strong odour, soluble in common organic solvents. Bp 134°–135 °C [60], 134°–136 °C [102], 136.8 °C at 760 mm Hg [14], 138 °C at 761.8 mm Hg [101, 105]; n_D^{19} 1.4505 [60], 1.4548 [106], $n_D^{19.2}$ 1.4480 [105, 107].

Method of Preparation

Bromopentafluorobenzene is prepared by bromination of pentafluorobenzene in 65% oleum [52] or in 20% oleum with $AlBr_3$ addition (81% yield) [60]; by the reaction of pentafluorophenylmagnesium chloride [105, 107] or pentafluorophenyllithium [102] with bromine (60%–69% yield); or by the reaction of pentafluorobenzene with BrF_2SbF_6 salt [110]. Bromopentafluorobenzene is formed, together with hexafluorobenzene and other compounds, in co-pyrolysis of tribromofluoromethane with tetrabromomethane or bromoform at 540 °C [33]. Pyrolysis of the mixture of hexafluorobenzene with bromine at 650°–740 °C, or with tribromofluoromethane, dibromodifluoromethane, or 1,2-dibromotetrafluoroethane at 670°–680 °C, yields bromopentafluorobenzene (yields up to 30%) [76]. Bromopentafluorobenzene is also prepared by treatment of hexabromobenzene with KF at 380°–430 °C in the absence of solvents (~6% yield) (its isolation from the reaction mixture is carried out by the preparative GLC [106]); from pentafluoroaniline, by the Sandmeyer reaction (35%–63% yield) [3, 40]; by oxidation of pentafluorophenylhydrazine with bromine at −60 °C (24% yield) [58]; or by the reaction of iodopentafluorobenzene with bromine at 150 °C in a sealed tube [108].

The following method is a modification of the procedures described in [60, 109]; it involves bromination of pentafluorobenzene.

Procedure

$$C_6HF_5 \xrightarrow{Br_2;\,AlBr_3} C_6BrF_5$$

In a 1-l four-necked flask fitted with a stirrer, a reflux condenser with a gas inlet-outlet system, a thermometer, and a dropping funnel, is placed 480 g (3 mol) of bromine. Metallic aluminium (2.7 g; 0.1 mol) is added in portions with vigorous stirring (Note 1). After the addition is complete, the reaction mixture is cooled to 20 °C. Pentafluorobenzene (252 g; 1.5 mol) is added during 1 h; the temperature should not exceed 20 °C. The mixture is heated, with continuous stirring, at 50 °C for 4 h and at 70 °C for an additional 4 h. Then it is treated with an aqueous 10% solution of $NaHSO_3$ (Note 2) till the brown colour has disappeared. The organic layer is separated, washed with water (290 ml), dried over $MgSO_4$ and distilled. A fraction boiling at 134°–135 °C is collected. The yield of bromopentafluorobenzene is 435 g (85%); n_D^{20} 1.4492.

Notes

1. Metallic aluminium vigorously reacts with bromine, therefore it should be added in portions, as small tablets, and with continuous stirring of the mixture.
2. It is preferable to add to the reaction mixture 0.5 l of the concentrated solution of $NaHSO_3$ and then, while stirring, dry $NaHSO_3$ till the brown colour disappears.

Iodopentafluorobenzene

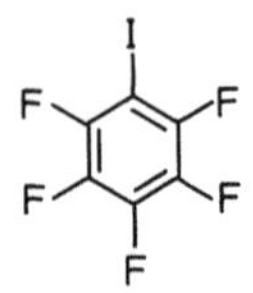

C_6F_5I MW 294

Iodopentafluorobenzene is a colourless or a slightly rosy liquid. Bp 159°–160 °C [101], 161°–163 °C [60, 103], 162.5°–163.5 °C at 760 mm Hg [3], 107 °C at 150 mm Hg [58], 73°–75 °C at 35 mm Hg [58], 77°–79 °C at 35 mm Hg [40], 60°–60.6 °C at 25 mm Hg [101]; n_D^{19} 1.4970 [60], $n_D^{19.5}$ 1.4965 [101], n_D^{20} 1.4990 [52, 56].

Method of Preparation

Iodopentafluorobenzene is prepared by iodination of pentafluorobenzene in 65% [52] or 20% oleum (71% yield) [60]; by the reaction of pentafluorophenylmagnesium chloride with iodine [101] or pentafluorophenylmagnesium bromide with iodotrifluoroethylene or 2-iodo-1-chloro-1,1,2-trifluoroethane [52, 56] (46%–50% yield); By the reaction of KI with the diazo solution formed by diazotization of pentafluoroaniline in anhydrous or 80% HF (41%–50% yield) [3, 40]; by oxidation of pentafluorophenylhydrazine with silver oxide in the presence of methyl iodide (60% yield) [58]; by the pyrolysis of silver pentafluorobenzoate in the presence of iodine (55% yield) [111]; or by pyrolsis of hexafluorobenzene in the presence of iodine, KI and iodotrifluoromethane [76].

The following method is a modification of the procedure described in [101].

Procedure

$$C_6F_5MgCl \xrightarrow{I_2} C_6F_5I$$

In a 1-l three-necked flask fitted with a stirrer, a reflux condenser with a calcium chloride tube and a pressure-equalizing dropping funnel are placed 39.7 g (0.153 mol) of powdered iodine and 375 ml of absolute ether. The Grignard reagent formed from 32 g of chloropentafluorobenzene (p. 126) is added, with careful stirring at room temperature from the dropping funnel, at such a rate as to avoid vigorous boiling. The mixture is allowed to stand overnight, whereupon it is poured into a mixture of 300 g of ice and 20 ml of concentrated HCl. The ether layer is separated and the aqueous layer is extracted with ether (4 × 75 ml). The ether extract is washed with 15% Na_2CO_3 solution, with a saturated solution of $Na_2S_2O_5$ and with water till it is neutral to litmus, and dried over $MgSO_4$. The ether is distilled off in a water-bath and the residue distilled in vacuum. A fraction boiling at 64°–65 °C/23 mm Hg is collected. The yield of iodopentafluorobenzene is 36–38 g (77%–81%); n_D^{23} 1.4950.

2,3,4,5,6-Pentafluorotoluene

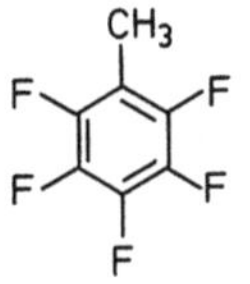

$C_7H_3F_5$ MW 182

2,3,4,5,6-Pentafluorotoluene is a volatile colourless liquid, soluble in common organic solvents and practically insoluble in water. Bp 115°–118 °C [103], 117.49 °C [112], 117°–118 °C at 760 mm Hg [113]; n_D^{20} 1.4023 [113], n_D^{21} 1.4021 [116].

Method of Preparation

2,3,4,5,6-Pentafluorotoluene is prepared by the reaction of hexafluorobenzene with methyllithium (70%–75% yield) in ether [40, 103, 114] or tetrahydrofuran [113], or with methylmagnesium iodide (2%–3% yield) [115, 116]; by the reaction of the Grignard reagent obtained from bromopentafluorobenzene or from iodopentafluorobenzene, with methyl iodide (27% yield) [116] or dimethylsulphate (56% yield) [103].

2,3,4,5,6-Pentafluorotoluene may be also obtained by pyrolytic defluorination of nonafluoro-2-methylcyclohexene (~38% yield), as suggested in [117].

The following procedure is a modification of the preparations described in [40, 103].

Procedure

$$CH_3I \xrightarrow{Li} CH_3Li$$

$$C_6F_6 \xrightarrow{CH_3Li} C_6F_5CH_3$$

Methyllithium. In a 1-l four-necked flask fitted with a stirrer, a reflux condenser with a calcium chloride tube, a pressure-equalizing dropping funnel, and an argon inlet tube, are placed 400 ml of absolute ether and 10 g (1.4 mol) of finely divided metallic lithium (Note 1). In the funnel are placed 100 g (44 ml, 0.7 mol) of methyl iodide and 100 ml of absolute ether. The solution of methyl iodide is dropped in, with vigorous stirring, before the reaction has started (Note 2). Then the reaction flask is cooled to −10° ÷ −15 °C and methyl iodide is added at such a rate as to keep up bubbling. After all the methyl iodide has been added, stirring is continued for an additional 30–40 min. A dark-grey liquid is transferred to the 500 ml pressure-equalizing funnel placed in a flow of dry argon (Note 3).

2,3,4,5,6-Pentafluorotoluene. In a 2-l four-necked flask provided with a stirrer, a reflux condenser with a calcium chloride tube, a 500-ml pressure-equalizing dropping funnel, and a gas inlet tube, is placed 130 g (82 ml, 0.7 mol) of hexafluorobenzene in 400 ml of absolute ether. The system is flushed with dry argon over a period of 5 min. A solution of methyllithium is added dropwise, in a slow flow of dry argon, with vigorous stirring, at such a rate as to keep the liquid in the flask slightly boiling. After the reaction mixture has been stirred for 0.5 h, it is poured into 700 ml of distilled water acidified with 20 ml of concentrated HCl. The resulting mixture is thoroughly stirred and the ether layer is separated. The aqueous layer is extracted with ether (3 × 300 ml); ether extracts are combined, washed with a saturated solution of $Na_2S_2O_5$ (Note 4), then with 15% Na_2CO_3 solution and with water till neutral to litmus. The ether solution is dried over $MgSO_4$, the ether is distilled off in a water-bath and the residue distilled on a fractionating column (15 theoretical plates). A fraction boiling at 117°–117.8 °C is collected (Note 5). The yield of 2,3,4,5,6-pentafluorotoluene is 77–90 g (60%–70%).

Notes

1. If it is impossible to prepare lithium wire, the lithium sample is crushed with a hammer to lumps 1–1.5 mm thick, which are cut with scissors to bars of the size 2 × 10–15 (mm).
2. The beginning of the reaction is indicated by evolution of gas bubbles and spontaneous boiling of the reaction mixture.
3. If there are any lumps of unchanged lithium, the solution must be filtered in a flow of dry argon. The pressure-equalizing dropping funnel is first flushed with dry argon. Then a funnel containing a thin layer of cotton is inserted, which keeps away lithium lumps but does not hinder the flow of the mixture. Dry argon is supplied through the pressure-equalizing tube; then the solution is quickly filtered.
4. The ethereal solution is washed with a saturated solution of $Na_2S_2O_5$ to remove free iodide formed as a by-product.
5. Tetrafluoroxylene is formed as a by-product (bp 143°–144 °C).

2,3,4,5,6-Pentafluorostyrene

(2,3,4,5,6-pentafluorophenylethylene)

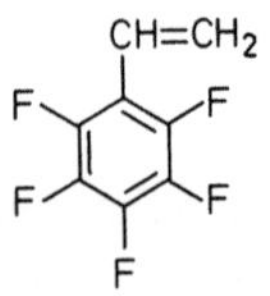

$C_8H_3H_5$ MW 194

2,3,4,5,6-Pentafluorophenylethylene is a colourless liquid with a specific strong odour, soluble in common organic solvents. Bp 139°–140 °C [60], 140°–141 °C at

760 mm Hg [52, 56], 34 °C at 25 mm Hg [40], 62°–63 °C at 50 mm Hg [101]; n_D^{20} 1.4414 [52, 56], n_D^{22} 1.4439 [101], 1.4444 [60].

Method of Preparation

2,3,4,5,6-Pentafluorophenylethylene is prepared by dehydration of α-(2,3,4,5,6-pentafluorophenyl)ethyl alcohol over P_2O_5 [60, 101] or Al_2O_3 [52, 56] (33%–62.5% yields); by the reaction of hexafluorobenzene with vinyllithium in ether (20% yield) [40]; or by dechlorination of 1,1,2-trichloro- or 1,2-dichloro-2-(pentafluorophenyl)-ethylene with zinc in ethanol (with a small yield; together with other products) [104].

The following procedure is a modification of the preparations described in [62, 101].

Procedure

$$C_6F_5MgCl \xrightarrow{CH_3CHO} C_6F_5{-}\underset{\displaystyle OH}{\underset{|}{C}H}{-}CH_3 \xrightarrow[-H_2O]{P_2O_5} C_6F_5{-}CH{=}CH_2$$

1-(Pentafluorophenyl)ethyl alcohol [101]. A Grignard reagent is prepared from 202.5 g (1 mol) of chloropentafluorobenzene (p. 126) in a 2-l four-necked flask fitted with a stirrer, a reflux condenser with a calcium chloride tube, a pressure-equalizing dropping funnel, and a thermometer. After it was cooled to —5° ÷ —10 °C, 160 g (3.6 mol) of acetaldehyde is added dropwise with stirring, at such a rate that the temperature of the reaction mixture should not rise above 0 °C. The reaction mixture is stirred at 0 °C for 2 h, then at room temperature for 1.5 h and allowed to stand overnight, whereupon it is poured into a mixture of 900 g of ice and 300 ml of concentrated HCl. The ether layer is separated and aqueous layer extracted with ether (4 × 350 ml). The ether solutions are combined, washed with 10% Na_2CO_3 solution (~250 ml for 1.7 l of the extract), washed with water till neutral to litmus and dried over $MgSO_4$. The ether is distilled off in a water-bath and the residue distilled in vacuum. A fraction boiling at 88°–91 °C/17 mm Hg is collected. The yield of 1-(pentafluorophenyl)ethyl alcohol is 165–170 g (78%–80%); n_D^{20} 1.4426.

2,3,4,5,6-Pentafluorophenylethylene. In a 0.5-l three-necked round-bottom flask fitted with a stirrer, a reflux condenser with a calcium chloride tube and a dropping funnel, are placed 103 g (0.72 mol) of P_2O_5 and 6–7 g of hydroquinone. Then 240 g (1.13 mol) of α-(pentafluorophenyl)ethyl alcohol is slowly added. The reaction mixture is vigorously stirred, heated in an oil-bath at 150 °C over a period of 2 h and cooled. The crude product is distilled from the flask at 50–60 mm Hg. At the end of the distillation the temperature of the bath is raised to 170 °C. The product (with an addition of 1–2 g of hydroquinone) is redistilled under a short dephlegmator and a fraction boiling at 140°–141 °C is collected. The yield of 2,3,4,5,6-pentafluorophenylethylene is 100–110 g (45%–50%); n_D^{21} 1.4460; it is stored with a small amount of hydroquinone.

Perfluorostyrene

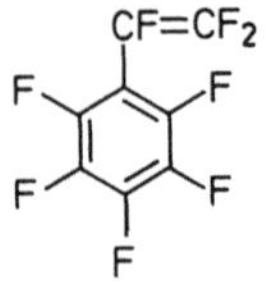

C_8F_8 MW 248

Perfluorostyrene is a colourless liquid, soluble in common organic solvents. Bp 120°–121 °C [118], 124°–125 °C [119], 121 °C [120].

Method of Preparation

Perfluorostyrene is prepared by dehalogenation of α- or β-chlorononafluoroethylbenzene or of a mixture of these compounds at 570°–620 °C over iron turnings (30% yield) [118]; by the reaction of pentafluorophenylcopper with trifluoroiodoethylene (55% yield) [119, 121, 122]; by defluorination of perfluoro(ethylbenzene) at 600 °C over iron or nickel (15% yield) [120, 123]; by the pyrolysis of octafluoro-β-chloro-α-hydroethylbenzene at 450°–650 °C (40%–45% yield) or its dehydrofluorination with an alkali (46% yield) [124]; by the pyrolysis of hexafluorocyclopentadiene dimer (0.5%–16% yield) [69, 125]; by thermolysis of nonafluoro-α-hydroethylbenzene or octafluoro-β-chloro-α-hydroethylbenzene at 500°–900 °C [126]; by dehydrochlorination of octafluoro-β-chloro-α-hydroethylbenzene with KOH [126]; by the reaction of nonafluoro-α-hydroethylbenzene with methyllithium with subsequent heating of the reaction product (85%–90% yield) [127]; by the reaction of pentafluorophenyllithium with excess tetrafluoroethylene in ether at −20 °C (10% yield) [127]; by dehalogenation of nonafluorobromethylbenzene or octafluoro-β-chloro-α-bromoethylbenzene at 600°–650 °C (95% yield) [119, 126, 127]; by debromination of octafluoro-α,β-dibromoethylbenzene with zinc in alcohol (89% yield) [128].

The following procedure is based on the one described in [118].

Procedure

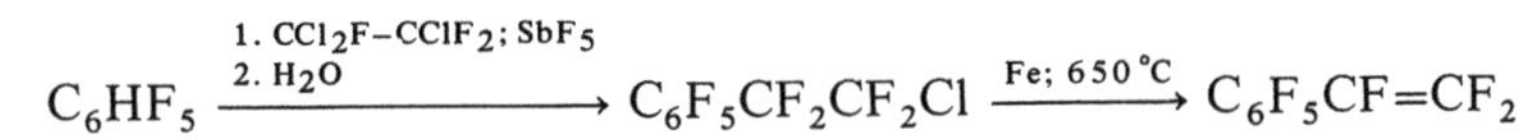

Nonafluoro-β-chloroethylbenzene. In a 0.5-l stainless steel reactor equipped with a stirrer, a thermometer, a pressure-equalizing dropping funnel, and a calcium chloride tube, is placed 434 g of SbF_5 (Note 1). A mixture of 109 g (0.65 mol) of pentafluorobenzene and 122 g (0.65 mol) of 1,1,2-trichloro-1,2,2-trifluoroethane is added at −1° to −5 °C with vigorous stirring over a period of 6 h. The temperature of the reaction mixture is raised to 20 °C with continuous stirring for 2 h. The mixture is poured onto 3 kg of ice (Note 2), the organic layer is separated and aqueous layer extracted with methylene chloride. The extract is combined with the main product, washed with water till it is neutral to litmus and dried over $MgSO_4$. The solvent is distilled off in a water-bath, using a short Vigreux column. This gives 130 g of a mixture containing 74% of nonafluoro-β-chlorobenzene, as shown by GLC. The reaction mixture is distilled on a fractionating column with a glass packing (~20 theoretical plates). A fraction boiling at 151°–152 °C is collected. The yield of nonafluoro-β-chloroethylbenzene is 91–93 g (50%–51%); bp 151°–152 °C [129].

Perfluorostyrene. Nonafluorochloroethylbenzene (10 g) is passed during 1 h in a flow of nitrogen (flow rate 3 l/h) through a quartz tube (400 × 17) (mm) filled with iron turnings (Note 3) and placed in a tube furnace at 650 °C. Pyrolysate is collected in dry-ice acetone traps. This yields 4.2 g of the reaction mixture containing 75% of perfluorostyrene, as shown by GLC. The product is distilled and a fraction boiling at 120°–121 °C is collected. The yield of perfluorostyrene is 2.6 g (30%).

Notes

1. As SbF_5 is rather hygroscopic, loading it in the reactor should preferably be carried out in a dry box. It is advisable to conduct the reaction in the atmosphere of an inert gas and to use freshly distilled SbF_5.
2. For the safety measures to be taken when handling SbF_5 and aqueous solutions containing hydrogen fluoride, see the Note on p. 121.
   The skin stained with SbF_5 should be quickly wiped with a dry woollen cloth and only after that washed with a large amount of water. Otherwise, SbF_5 will be hydrolysed with water, forming HF with considerable evolution of heat and thus producing chemical and thermal burns; heat also causes pore extension, facilitating penetration of HF deep into the skin.
3. The process largely depends on the turnings — castiron ones should preferably be used.

Heptafluoro-α-chlorostyrene

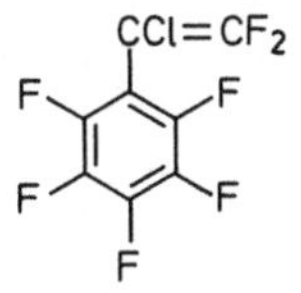

C_8ClF_7 MW 264.5

Heptafluoro-α-chlorostyrene is a colourless liquid with a specific strong odour, soluble in common organic solvents. Bp 144.5 °C; n_D^{20} 1.4232 [130].

Method of Preparation

Heptafluoro-α-chlorostyrene is prepared by co-pyrolysis of pentafluorobenzotrichloride with tetrafluoroethylene at 620 °C in a quartz tube (38% yield) [130].

Procedure

$$C_6F_5CF_3 \xrightarrow{AlCl_3;\ CH_3COCl} C_6F_5CCl_3 \xrightarrow[620\ ^\circ C]{CF_2=CF_2} C_6F_5CCl{=}CF_2$$

2,3,4,5,6-Pentafluorobenzotrichloride. In a 350-ml four-necked flask provided with a stirrer, a reflux condenser with a calcium chloride tube and a thermometer, are placed 66 g (0.28 mol) of octafluorotoluene and 165 ml of acetyl chloride. The mixture is cooled to —5 °C with stirring. Then anhydrous $AlCl_3$ is added in small portions over a period of 40–45 min, so as to maintain the temperature in the range of —5 °C to —1 °C. The cooling bath is removed and the reaction mixture allowed to warm gradually to room temperature. Then it is heated in a water-bath at 30 °C for 1 h and at 45 °C for 2 h, cooled, allowed to stand overnight, and poured onto 400–500 g of crushed ice. The organic layer is separated and the aqueous layer extracted with ether (4 × 150 ml). The ether extract is combined with the organic layer, washed with water (2 × 80 ml), with saturated aqueous $NaHCO_3$ and with water till it is neutral. The ether extract is dried over $MgSO_4$, the ether is distilled off in a water-bath and the residue distilled in vacuum under a dephlegmator. A fraction boiling at 79°–81 °C/ 11 mm Hg is collected. The yield of pentafluorobenzotrichloride is 38–40 g (48% to 51%); n_D^{20} 1.4810 (bp 109°–116 °C/48 mm Hg [103].

Heptafluoro-α-chlorostyrene. Pentafluorobenzotrichloride (350 g, 1.22 mol) and tetrafluoroethylene (940 g, 9.4 mol) are simultaneously passed at 620 °C (flow rate 15 l/h) through a quartz tube (400 × 20 mm), fitted with a dropping funnel, a thermocouple in a well and a gas inlet tube, and placed in an electric furnace. Pyrolysate is collected in a condensing system involving a 1-l two-necked flask with a reflux condenser and a gas outlet tube (Note). After thermolysis, 250 ml of water is added in the flask and the product is steam-distilled. The organic layer is separated and dried over $MgSO_4$. This gives 463 g of a mixture, containing 38% of heptafluoro-α-chlorostyrene, as shown by GLC. The product is distilled on a fractionating column with a glass packing (~40 theoretical plates). A fraction boiling at 144°–145 °C is collected. The yield of heptafluoro-α-chlorostyrene is 106 g (33%); n_D^{23} 1.4232.

Notes. The gaseous products of thermolysis may involve highly toxic perfluoroolefins and other compounds, therefore the procedure should be carried out with care and the gases should be vented off into an efficient hood.

Pentafluorophenylacetylene

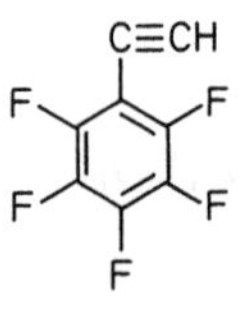

C_8HF_5 MW 192

Pentafluorophenylacetylene is a colourless liquid with a specific unpleasant odour, soluble in common organic solvents, slightly soluble in water; quickly darkens upon standing; highly volatile. Bp 130°–131 °C [104], 50°–52 °C at 37 mm Hg [131], 64°–65 °C at 75 mm Hg [132]; n_D^{20} 1.4418; d_4^{20} 1.4019 [132].

Method of Preparation

Pentafluorophenylacetylene is prepared by dehalogenation of 1,1,2-trichloro-2-(pentafluorophenyl)ethylene with magnesium (80% yield) [104]; by the reaction of (pentafluorophenyl)ethyltriethylsilane with aqueous NaOH (77% yield) [131]; by oxidation of 3-(pentafluorophenyl)propyn-2-ol-1 with nickel peroxide in the aqueous-dioxan solution of NaOH (70% yield) [133].

The following procedure is based on the one described in [104].

Procedure

$$C_6F_5Cl \xrightarrow[2.\ CCl_3CHO]{1.\ Mg;\ ether} C_6F_5CH(OH)CCl_3 \xrightarrow{PCl_5}$$

$$\rightarrow C_6F_5CHClCCl_3 \xrightarrow{AlCl_3} C_6F_5CCl{=}CH_2 \xrightarrow{Mg;\ THF} C_6F_5C{\equiv}CH$$

Pentafluorophenyltrichloromethyl carbinol. In a 2-l three-necked flask fitted with a stirrer, a reflux condenser, and a pressure-equalizing dropping funnel, are placed the Grignard reagent, obtained from 48.6 g (2 mol) of magnesium, 202.5 g (1 mol) of chloropentafluorobenzene and 187.5 g (1 mol) of 1,2-dibromoethane in 900 ml absolute ether. The mixture is cooled to 0 °C and 163 g (1.1 mol) of trichloroacetaldehyde (Note 1) in 200 ml of absolute ether is added. The reaction mixture is kept at 20 °C for 10 h, then refluxed for 1 h, whereupon 350 ml of 4 N H_2SO_4, then 300 ml of water, are slowly dropped in, during 1 h, with stirring and cooling of the mixture in an ice-bath. The ether layer is separated and the filtrate extracted with ether (3 × 200 ml). The ether extracts are combined and dried over $MgSO_4$. The ether is removed on a rotary vacuum evaporator, the residue is distilled and a fraction boiling at 95°–97 °C/1–2 mm Hg is collected. The yield of pentafluorophenyltrichloromethyl carbinol is 186.7 g (60%); mp 65°–67 °C (mp 67°–70 °C, bp 124°–126 °C/20 mm Hg [104]).

1,1,1,2-Tetrachloro-2-(pentafluorophenyl)ethane. In a 300-ml two-necked flask with a reflux condenser, connected by means of a flexible rubber tube to a 200-ml pear-shaped flask, is placed 189 g (0.6 mol) of pentafluorophenyltrichloromethyl carbinol. Finely divided PCl_5 (125 g, 0.6 mol) is added in portions from the pear-shaped flask over a period of 1.5 h, so that the mixture gently boils (Note 2). Then the reaction mixture is heated at 100 °C for 1 h, poured onto 0.5 kg of ice and extracted with ether (3 × 200 ml). The ether extract is dried over $MgSO_4$. After the ether has been distilled off in a water-bath, the residue is distilled in vacuum and a fraction boiling at 113°–114 °C/16 mm Hg is collected. The yield of 1,1,1,2-tetrachloro-2-(pentafluorophenyl)ethane is 145–150 g (72%–75%) (bp 110 °C/16 mm Hg [104]).

1,1,2-Trichloro-2-(pentafluorophenyl)ethylene. In a 250-ml round-bottom flask fitted with a reflux condenser are placed 167 g (0.5 mol) of 1,1,1,2-tetrachloro-2-(pentafluorophenyl)-ethane and 2.7 g (0.02 mol) of anhydrous $AlCl_3$, and the flask is heated at 120 °C for 2 h. The liberated HCl is trapped in aqueous Na_2CO_3. The solution is poured into 300 ml of water and extracted with ether (3 × 250 ml). The ether extract is dried over $MgSO_4$. After the ether has been distilled off in a water-bath, the residue is distilled in vacuum and a fraction boiling at 84–86 °C/15 mm Hg is collected. The yield of 1,1,2-trichloro-2-(pentafluorophenyl)ethylene is 137–140 g (92%–94%) (bp 84 °C/16 mm Hg [104]).

Pentafluorophenylacetylene. In a 300-ml three-necked flask fitted with a stirrer, a reflux condenser, and a dropping funnel, are placed 140 ml of absolute tetrahydrofuran (Note 3) and 12.1 g (0.5 mol) of magnesium powder. Then 1,1,2-trichloro-2-(pentafluorophenyl)ethylene (47.6 g, 0.16 mol) is added so that the mixture should gently boil (Note 4), whereupon it is heated at 80 °C for 1.5 h and distilled with steam (Note 5). The distillate is extracted with ether (3 × 150 ml) and the ether extract dried over $MgSO_4$. After the ether has been distilled off in a water-bath, the residue is distilled in vacuum and a fraction boiling at 30°–45 °C/30 mm Hg is collected. The yield of pentafluorophenylacetylene is 14.1–15.3 g (46%–50%).

Notes

1. Trichloroacetaldehyde (bp 97.5 °C) is prepared by simple distillation of a mixture of chloral hydrate and concentrated H_2SO_4 (one mol to mol). Traces of HCl in trichloroacetaldehyde are removed by redistillation over $CaCO_3$.
2. The reaction is exothermic; cooling with ice-water is necessary.
3. The yield of the product depends on the content of moisture in tetrahydrofuran. It is advisable to use freshly distilled tetrahydrofuran dried over sodium (for no less than 24 h).
4. At first not more than 5 g of 1,1,2-trichloro-2-(pentafluorophenyl)ethylene is added and the mixture is vigorously stirred till the reaction starts with violent boiling; then it is cooled in an ice-bath so as to keep it gently boiling.
5. As the reaction mixture foams during steam distillation, it should be transferred to a steam distillation flask of at least 1-l capacity.

trans-Decafluorostilbene

$C_6F_5CH{=}CH{-}C_6F_5$

$C_{14}H_2F_{10}$ MW 360

trans-Decafluorostilbene is a white crystalline solid, soluble in common organic solvents. Mp 101.4°–103.5 °C [134], 101 °C [135], 96.5°–97.5 °C [136].

Method of Preparation

trans-Decafluorostilbene is prepared by the reaction of pentafluorobenzyltripenylphosphonium bromide with pentafluorobenzaldehyde (32% yield) [134]; by the reaction of pentafluorobenzyltriethylphosphonium bromide with pentafluorobenzaldehyde (63% yield) [135]; or by the reaction of pentafluorophenylmagnesium bromide with *trans*-diiodoethylene in the presence of $CoCl_2$ (25% yield) [136].

The following procedure is based on the one described in [134].

Procedure

$$C_6F_5CH_2Br \xrightarrow[\text{2. } C_6F_5CHO\text{; heating}]{\text{1. } (C_6H_5)_3P\text{; } C_2H_5ONa} C_6F_5CH{=}CHC_6F_5$$

In a 150-ml three-necked flask equipped with a thermometer, a reflux condenser with a calcium chloride tube, a pressure-equalizing dropping funnel, and a gas inlet tube, are placed 5 g (~0.01 mol) of 2,3,4,5,6-pentafluorobenzyltriphenylphosphonium bromide (Note 1) and 0.5 g (0.0092 mol) of sodium ethoxide in 50 ml of absolute tetrahydrofuran. The mixture is kept over a period of 50 h in a flow of argon at 20 °C. Pentafluorobenzaldehyde (1.96 g, 0.01 mol) in 20 ml of absolute tetrahydrofuran is then added and the resulting mixture is allowed to stand at 20 °C for 24 h, whereupon it is refluxed during 6 h. The reflux condenser is replaced with a Wurtz column and the solvent is distilled off. The residue is treated with 100 ml of water and extracted with petroleum ether (2 × 50 ml), then with ether (2 × 30 ml). The ether extract is evaporated to one-half of its volume, the precipitate is filtered off (Note 2) and the residual ether combined with the petroleum ether extract. The solvents are evaporated on a rotary vacuum-evaporator. The residue is sublimet at 110 °C/3 mm Hg and recrystallized from alcohol. The yield of *trans*-decafluorostilbene is 1.1 g (32%); mp 101.5°–103.5 °C.

Notes

1. An example of the preparation of 2,3,4,5,6-pentafluorobenzyltriphenylphosphonium bromide is as follows. In a 0.3-l two-necked flask fitted with a thermometer and a reflux condenser are placed 37 g (0.14 mol) of triphenylphosphine and 18.5 g (0.07 mol) of 2,3,4,5,6-pentafluorobenzyl bromide in 200 ml of absolute benzene. The mixture is refluxed for 5 h, then cooled. The precipitate is filtered off on a Büchner funnel and recrystallized from the ethanol-ethylacetate mixture (1:1). The yield of 2,3,4,5,6-pentafluorobenzyltriphenylphosphonium bromide is 35.7 g (97%, based on 2,3,4,5,6-pentafluorobenzyl bromide); mp 247°–249 °C.
2. The precipitate (1.6 g) is triphenylphosphine oxide, mp 152°–156 °C.

Bis(pentafluorophenyl)acetylene

(Perfluorotolane)

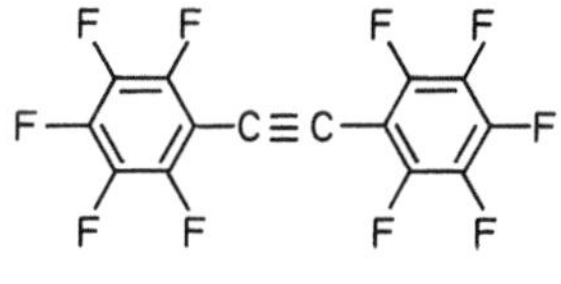

$C_{14}F_{10}$ MW 358

Bis(pentafluorophenyl)acetylene is a white crystalline substance, soluble in dipolar aprotic solvents, moderately soluble in alcohol, petroleum ether, acetone. Mp 122° to 123 °C [137], 123°–123.5 °C [138], 120°–121 °C [139]; sublimes at 122°–123 °C [137].

Method of Preparation

Bis(pentafluorophenyl)acetylene is prepared by thermolysis of pentafluoro(pentafluorobenzoyl)benzylidentriphenylphosphorane formed by the reaction of pentafluorobenzyltriphenylphosphonium bromide with n-butyllithium and subsequent treatment with pentafluorobenzoyl chloride (95% yield) [137]; by the reaction of pentafluorophenylmagnesium bromide with diiodoacetylene in the presence of $CoCl_2$ (56% yield) [138]; by the reaction of pentafluorophenyllithium with trifluorobromoethylene (14.5% yield) or with trifluoroethylene (9% yield) [139]; by the reaction of pentafluorophenylcopper with tribromoethylene (43% yield) [140]; by heating a mixture of *cis*- and *trans*-decafluoro-α-,β-dichlorostilbene with copper at 300 °C (50% yield) [141]; by the pyrolysis of pentafluorobenzotribromide over copper (60% yield) [141].

The following procedure is based on the one described in [141].

Procedure

$$C_6F_5CCl_3 \xrightarrow{450\,°C} C_6F_5CCl{=}CClC_6F_5 \xrightarrow[300\,°C]{Cu} C_6F_5C{\equiv}CC_6F_5$$

cis- and trans-Decafluoro-α,β-dichlorostilbenes. Pentafluorobenzotrichloride (8 g, 0.028 mol) and dry nitrogen are simultaneously passed at 450 °C (flow rate 10 l/h), over a period of 1 h, through a quartz tube (250 × 15 mm) fitted with a thermocouple in a well, a dropping funnel. and a gas inlet tube and placed in an electrical furnace. Pyrolysate is collected in a receiver provided with a reflux condenser and dry-ice acetone trap, then it is steam-distilled. The organic layer is separated and dried over $MgSO_4$. Sublimation at 50 °C/3 mm Hg gives the pure product. The yield of *cis*- and *trans*-decafluoro-α,β-dichlorostilbenes is 5–5.3 g (80%–83%); mp 51°–53 °C.

Bis(pentafluorophenyl)acetylene. A sealed tube with a mixture of *cis*- and *trans*-decafluoro-α,β-dichlorostilbenes (4.29 g, 0.01 mol) and copper powder (75 g) is heated at 300 °C during 30 h. Then the tube is carefully opened and the product is sublimed in vacuum at 120°–130 °C. This gives 1.8 g (50%) of bis(pentafluorophenyl)-acetylene, mp 122°–123 °C.

2,3,4,5,6-Pentafluorobenzyl Alcohol

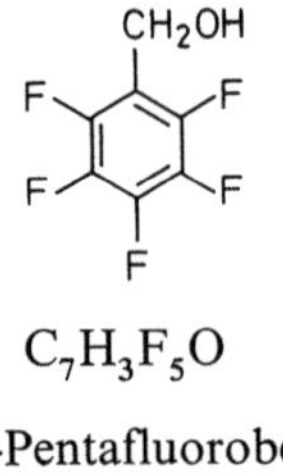

$C_7H_3F_5O$ MW 198

2,3,4,5,6-Pentafluorobenzyl alcohol is a colourless crystalline substance, soluble in common organic solvents, practically insoluble in water. Bp 114°–115 °C/60 mm Hg, 180°–182 °C/1 atm [103], 109°–110 °C/50 mm Hg [113], 113°–114.5 °C/57 mm Hg [143]; mp 30°–32 °C [103, 143], 33°–35 °C [113].

Method of Preparation

2,3,4,5,6-Pentafluorobenzyl alcohol is prepared by the reduction of pentafluorobenzaldehyde [103, 142] or pentafluorobenzoic acid [144] with $LiAlH_4$ (97% yields); by the hydrolysis of 2,3,4,5,6-pentafluorobenzyl chloride [103] or 2,3,4,5,6-pentafluorobenzyl acetate [113] (35%–50% yields); by the reaction of pentafluorophenylmagnesium chloride with formaldehyde (73% yield) [143]; or by treatment of 2,3,4,5,6-pentafluorobenzylamine with nitrous acid (28% yield) [113].

The following procedure is a modification of the one described in [145].

Procedure

$$C_6F_5MgCl \xrightarrow{CH_2O} C_6F_5CH_2OMgCl \xrightarrow{H_2O;\ HCl} C_6F_5CH_2OH$$

α-Polyoxymethylene. In a 1.5-l three-necked flask is placed 800 ml of 35%–40% formalin, 160–210 ml of which is distilled off at 45 °C/25 mm Hg. Then a stirrer is introduced into the flask and to a stirred solution, while it is still warm (~40 °C), is added dropwise a solution of 4 g KOH in 30 ml of water. The solution foams and starts to deposit a precipitate. Stirring is continued for an additional 1.5 h and the solution is allowed to stand overnight. The precipitate is filtered off, washed with ether on a filter and dried over P_2O_5 in a vacuum-desciccator. Dry α-polyoxymethylene melts at 190 °C.

Pentafluorobenzyl alcohol. In a 0.5-l two-necked round-bottom flask equipped with a nitrogen inlet tube and an outlet tube for nitrogen and formaldehyde vapours, is placed 80–90 g of dry α-polyoxymethylene. The flask is heated on a Wood's metal bath at 180°–190 °C with simultaneous passing of dry nitrogen. The formaldehyde vapour is bubbled (Note 1) in nitrogen flow through a cooled to −5 °C solution of pentafluorophenylmagnesium chloride in ether, prepared from 202.5 g (1 mol) of chloropentafluorobenzene (p. 126). Passing of formaldehyde is stopped after the reaction mixture has separated into layers (one-fifth of the starting amount of α-polyoxymethylene is recovered). The mixture is boiled for 30 min, cooled to room temperature and poured into a mixture of 900 g of crushed ice and 250 ml of concentrated HCl. The ether layer is separated and the aqueous layer extracted with ether (5 × 300 ml). The ethereal solutions are combined, washed with two portions of 10% solution of $NaHSO_3$, 15% solution of Na_2CO_3, with water to pH ~ 7 and dried over $MgSO_4$. The ether is distilled off, and the residue is distilled in vacuum. A fraction boiling at 86°–87 °C/15 mm Hg is collected. The yield of pentafluorobenzyl alcohol is 130–140 g (66%–70%); $n_D^{28.5}$ 1.4360 (Note 2).

Notes

1. The bubbler may get clogged; it should be provided with a glass rod inserted through one of two holes by means of a Teflon or rubber gland to dislodge the material which clogs it.
2. The product gets crystallized upon standing.

Pentafluorophenylacetonitrile

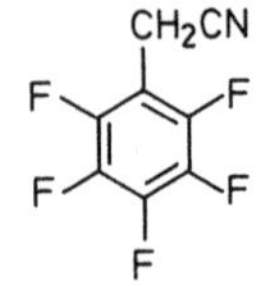

$C_8H_2F_5N$ MW 207

Pentafluorophenylacetonitrile is a colourless liquid, soluble in common organic solvents. Bp 105°–107 °C/8 mm Hg (146], 102°–106 °C/12 mm Hg [147]; n_D^{20} 1.4412 [147].

Method of Preparation

Pentafluorophenylacetonitrile is prepared by the hydrolysis of ethyl pentafluorophenylcyanoacetate (70% yield) [146] or by the reaction of pentafluorobenzyl chloride with NaCN (80% yield) [147].

The following procedure is based on the data reported in [146].

Procedure

$$C_6F_6 \xrightarrow[K_2CO_3;\ DMF]{NCCH_2COOC_2H_5} C_6F_5CH(CN)COOC_2H_5 \xrightarrow[H_2SO_4]{CH_3COOH} C_6F_5CH_2CN$$

In a 1-l four-necked flask provided with a stirrer, a thermometer, a reflux condenser with a calcium chloride tube and a dropping funnel, are placed 140 g (1 mol) of anhydrous K_2CO_3 and 650 ml of absolute dimethylformamide. The mixture is heated to 152 °C with stirring. Heating is stopped and 113 g (1 mol) of ethyl cyanoacetate is added dropwise to the stirred mixture. The temperature of the mixture (of bright orange colour) should be kept at about 110°–120 °C. Hexafluorobenzene (186 g, 1 mol) is dropped in at the above temperature and the dark-brown mixture is stirred at 110°–120 °C for 3 h. Then it is cooled, poured into 3 l of ice-water, acidified with 20% H_2SO_4 till K_2CO_3 has dissolved and allowed to stand for 2 h. The upper layer is decanted and the organic layer dissolved in 700 ml of ether. The ether extract is washed with water till it is slightly acid, then neutralized with 5% $NaHCO_3$ solution (to pH ~ 5–7). The ether is removed on a rotary vacuum-evaporator, the residue (223 g) is transferred to a 1-l flask with a reflux condenser, and 520 ml of 50% acetic acid is added together with 20 ml of concentrated H_2SO_4. The mixture is refluxed for 12 h, then cooled to 20 °C, and diluted to 1 i with water; a dark organic layer is deposited at the bottom of the flask. The mixture is cooled in an ice-bath and transferred to a separatory funnel. The organic layer is separated and dissolved in 800 ml of ether. The ether extract is washed with water (to pH ~ 4–5), then neutralized with 5% aqueous solution of $NaHCO_3$ and dried over $MgSO_4$. The ether is removed on a rotary vacuum-evaporator and the residue distilled in vacuum. A fraction boiling at 105°–107 °C/8 mm Hg is collected. The yield of pentafluorophenylacetonitrile is 115–120 g (70%–73%); n_D^{20} 1.4412.

2-(Pentafluorophenyl)ethyl Alcohol

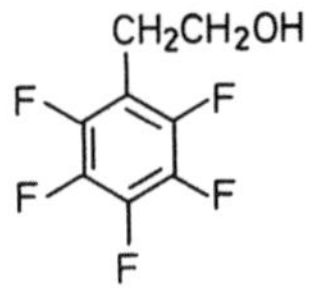

$C_8H_5F_5O$ MW 212

2-(Pentafluorophenyl)ethyl alcohol is a colourless liquid, soluble in common organic solvents, insoluble in water. Bp 88°–89 °C/13 mm Hg [148], 98 °C/13 mm Hg [149], 104 °C/16 mm Hg [103]; n_D^{20} 1.4428 [149], n_D^{24} 1.4428 [148].

Method of Preparation

2-(Pentafluorophenyl)ethyl alcohol is prepared, with a high yield, by the reaction of ethylene oxide with pentafluorophenylmagnesium chloride [149] or bromide [148], or by the reduction of ethyl pentafluorophenylacetate with $LiAlH_4$ (94% yield) [103].
The following procedure is a modification of the one described in [148].

Procedure

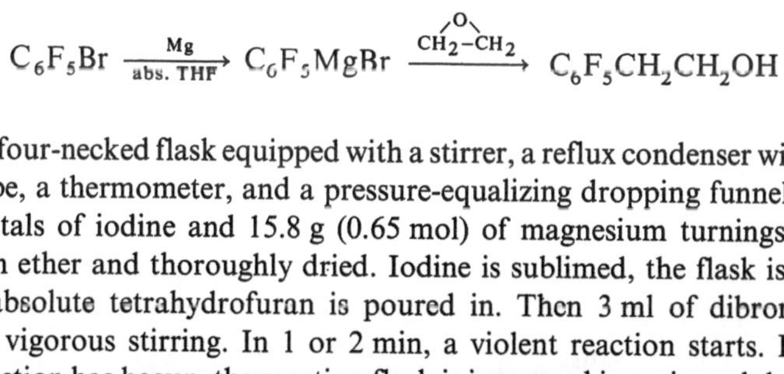

In a 350-ml four-necked flask equipped with a stirrer, a reflux condenser with a calcium chloride tube, a thermometer, and a pressure-equalizing dropping funnel, are placed several crystals of iodine and 15.8 g (0.65 mol) of magnesium turnings, previously washed with ether and thoroughly dried. Iodine is sublimed, the flask is cooled and 350 ml of absolute tetrahydrofuran is poured in. Then 3 ml of dibromoethane is added with vigorous stirring. In 1 or 2 min, a violent reaction starts. Immediately after the reaction has begun, the reaction flask is immersed in an ice-salt bath. Bromopentafluorobenzene (75 g, 0.3 mol) is slowly added from the dropping funnel to the cooled mixture maintained at −3° to 0 °C. After all bromopentafluorobenzene has been added, the mixture is allowed to stand at that temperature for 1 h, whereupon 33 ml of ethylene oxide is added during 15–20 min; the temperature should not rise above 0 °C. The temperature is then slowly raised to 20 °C and the mixture is allowed to stand overnight. After that it is poured into a mixture of 500 g of ice and 100 ml of concentrated HCl. The organic layer is separated and the aqueous layer extracted with ether (3 × 200 ml). The organic layer and the extracts are combined, washed with 10% Na_2CO_2 solution, then with water to pH ~ 7 and dried over $MgSO_4$. The solvents are distilled off through a Vigreux column, first in a water- and then in a oil-bath; the temperature of the bath is gradually raised to 100 °C. The residue is distilled in vacuum and a fraction boiling at 96°–98 °C/12 mm Hg is collected. The yield of 2-(pentafluorophenyl)ethyl alcohol is 43–45 g (67%–70%); n_D^{22} 1.4445.

Pentafluorobenzaldehyde

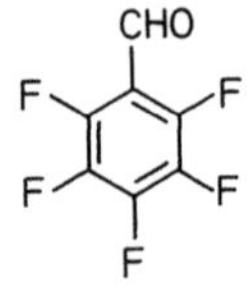

C_7HF_5O MW 196

Freshly distilled pentafluorobenzaldehyde is a colourless liquid with a strong odour; a lachrymator; soluble in common organic solvents and practically insoluble in water. Mp 20 °C [103]; bp 168°–170 °C [60, 103], 178°–180 °C/760 mm Hg [102], 67.5° to 70 °C/16 mm Hg [103, 150], 59°–59.5 °C/11 mm Hg [101], 50°–54 °C/10 mm Hg [151]; n_D^{17} 1.4505 [60], n_D^{25} 1.4520 [101].

Method of Preparation

Pentafluorobenzaldehyde is prepared by the reaction of pentafluorophenylmagnesium bromide [60, 103, 150], iodide [103, 150] or chloride [101], or of pentafluorophenyllithium [102] with ethyl-ortho-formiate [103] or 3,4-dihydro-6-methyl-3-(para-tolyl)-quinazoline methiodide [60]. Higher yields of pentafluorobenzaldehyde have been obtained by the reaction with N-methylformanilide [101–103, 150]. Pentafluorobenzaldehyde may be prepared also by the reaction of pentafluorobenzonitrile with anhydrous $SnCl_2$ in ethereal solution of HCl with subsequent hydrolsis of the intermediate (62% yield) [151].

The following procedure is a modification of the one described in [101].

Procedure

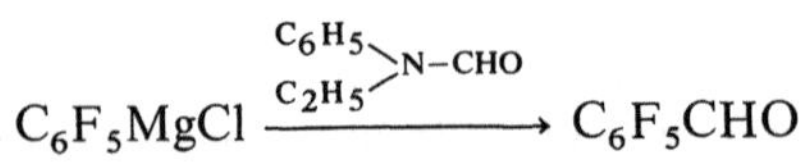

N-Ethylformanilide (197 ml, 1.5 mol) (Note 1) is added dropwise at 5°–8 °C to a stirred solution of pentafluorophenyl-magnesium chloride obtained from 202.5 g of chloropentafluorobenzene (see p. 126). The mixture is boiled during 4 h in a water-bath and allowed to stand overnight. Then it is poured, in small portions, into a stirred mixture of 900 g of ice and 450 ml of concentrated HCl. The ether layer is separated and the aqueous layer extracted several times with ether. The ethereal solutions are combined (the total volume is 5–6 l), neutralized or made mildly alkaline to litmus with 15% Na_2CO_3 solution and dried over $MgSO_4$. The ether is distilled off in a water-bath and the residue distilled in vacuum in an oil-bath. A fraction boiling at 51°–53 °C/7 mm Hg is collected. The yield of pentafluorobenzaldehyde is 135–145 g (69%–74%); n_D^{25} 1.4520 (Note 2).

Notes

1. N-Methylformanilide may be used as an alternative.
2. Pentafluorobenzaldehyde is oxidized when exposed to light and air. It should be kept in a tightly stopped dark container. Pentafluorobenzaldehyde is about ten times as toxic as benzaldehyde, therefore it should be handled with caution.

2,3,5,6-Tetrafluoro-4-hydroxybenzaldehyde

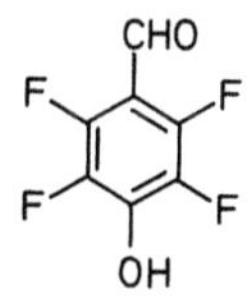

$C_7H_2F_4O_2$ MW 194

2,3,5,6-Tetrafluoro-4-hydroxybenzaldehyde is a white crystalline substance, soluble in common organic solvents. Mp 136°–137 °C [152].

Method of Preparation

2,3,5,6-Tetrafluoro-4-hydroxybenzaldehyde is prepared by the reaction of pentafluorobenzaldehyde diethylacetal with KOH in *tert*-butyl alcohol (68 % yield) [152].

The following procedure is based on the data reported in [152].

Procedure

$$C_6F_5CHO \xrightarrow[C_2H_5OH]{HCOOC_2H_5} C_6F_5CH(OC_2H_5)_2 \xrightarrow[tert\text{-}BuOH]{KOH} 4\text{-}OHC_6F_4CHO$$

Pentafluorobenzaldehyde diethylacetal. In a 0.5-l round-bottom flask equipped with a reflux condenser with a calcium chloride tube, are placed 78.4 g (0.4 mol) of pentafluorobenzaldehyde, 59.2 g (0.8 mol) of ortho-formate, 70 ml of anhydrous ethanol and 0.3 ml of concentrated HCl. The mixture is refluxed for 0.5 h, cooled and neutralized with 3 g of K_2CO_3. Ethanol and ethyl formate are removed on a rotary vacuum-evaporator, the residue is distilled with 250 ml of water and extracted with ether (3 × 150 ml). The ether extract is dried over $MgSO_4$. The solvent is distilled off in a water-bath and the residue distilled in vacuum. A fraction boiling at 65°–66 °C/ 2–3 mm Hg is collected. The yield of pentafluorobenzaldehyde diethylacetal is 97.2 g (90 %); n_D^{20} 1.4217.

2,3,5,6-Tetrafluoro-4-hydroxybenzaldehyde. In a 1-l round-bottom flask equipped with a reflux condenser with a calcium chloride tube, are placed 81 g (0.3 mol) of pentafluorobenzaldehyde diethylacetal, 42 g (0.75 mol) of KOH and 400 ml of *tert*-butyl alcohol. The reaction mixture is refluxed for 1.5 h and cooled. Then 400 ml of water is added to it and the aqueous solution of *tert*-butyl alcohol (about 550 ml) is distilled off. The residue is acidified with 250 ml of 6 N HCl and extracted with ether (4 × 150 ml). The ether extract is dried over $MgSO_4$. The solvent is removed on a rotary vacuum-evaporator and the residue recrystallized from CCl_4. The yield of 2,3,5,6-tetrafluoro-4-hydroxybenzaldehyde is 46.6 g (80%); mp 136°–137 °C.

2,3,4,5,6-Pentafluoroacetophenone

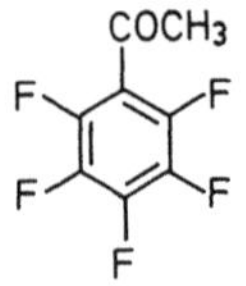

$C_8H_3F_5O$ MW 210

2,3,4,5,6-Pentafluoroacetophenone is a colourless liquid with a specific odour, soluble in common organic solvents. Bp 69 °C/22 mm Hg [104], 54°–56 °C/5 mm Hg [151], 48°–50 °C/5 mm Hg [101], 51°–53 °C/2.5 mm Hg [151]; $n_D^{17.5}$ 1.4330.

Method of Preparation

2,3,4,5,6-Pentafluoroacetophenone is prepared by the reaction of dimethylcadmium or methylmagnesium iodide with pentafluorobenzoyl chloride or pentafluorobenzonitrile [151], or by the reaction of pentafluorophenylmagnesium halide with acetyl chloride [103, 153] or acetic oxide [101], with 30%–61% yields. Pentafluoroacetophenone is also prepared by oxidation of 1-(pentafluorophenyl)ethanol with chromic anhydride in acetic acid (71% yield) [104], or by the reduction of α-bromo-2,3,4,5,6-pentafluoroacetophenone with potassium iodide (44.5% yield) [154]. It is formed, with a small yield, by the hydrolysis of pentafluorobenzoyl acetate with subsequent decarboxylation [155], or by hydration of pentafluorophenylacetylene [104].

The following procedure is a modification of the one described in [101].

Procedure

$$C_6F_5MgCl \xrightarrow{(CH_3CO)_2O} CH_3COO{-}\underset{\displaystyle OMgCl}{\overset{\displaystyle C_6F_5}{C}}{-}CH_3 \xrightarrow{H_2O} C_6F_5COCH_3$$

In a 2-l three-necked flask provided with a stirrer, a reflux condenser with a calcium chloride tube and a pressure-equalizing dropping funnel, is placed a mixture of 190 ml (194 g, 1.9 mol) of acetic oxide and 540 ml of absolute ether. The flask is cooled to −10 °C in an ice-salt bath. A cooled solution of a Grignard reagent obtained from 202.5 g (1 mol) of chloropentafluorobenzene (see p. 126) is added dropwise at −10 °C with vigorous stirring during 1–1.5 h. The resulting light-beige mixture is stirred at the above temperature for 1 h and poured into a mixture of 900 g of ice and 240 ml of concentrated HCl. The ether layer is separated and the aqueous layer extracted with ether (6 × 300 ml). The ethereal solutions are combined, washed with 15% solution of

Na_2CO_3, then with water till neutral to litmus and dried over $MgSO_4$. The ether is distilled off in a water-bath and the residue distilled in an oil-bath under vacuum. A fraction boiling at 51°–51.5 °C/6–7 mm Hg is collected. The yield of 2,3,4,5,6-pentafluoroacetophenone is 117–125 g (58%–60%); $n_D^{17.5}$ 1.4330.

Decafluorobenzophenone

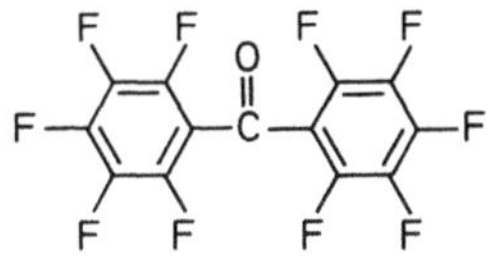

$C_{13}F_{10}O$ MW 362

Decafluorobenzophenone is a white crystalline substance, soluble in ether, benzene and chloroform. Mp 89°–90 °C [101], 91°–92 °C [103, 153].

Method of Preparation

Decafluorobenzophenone is prepared by oxidation of decafluorobenzhydrol with chromic anhydride in acetic acid (60% yield) [103, 153]; by the reaction of dimethyl carbinol with pentafluorophenyllithium (70% yield) [156], or of chlorocarbonate with pentafluorophenylmagnesium chloride in ether (33% yield) [101]; by acylation of pentafluorobenzene with pentafluorobenzoyl chloride in the presence of $AlCl_3$ [157, 158]; or by the reaction of perfluorobenzyl cation with pentafluorobenzene in SbF_5 with subsequent hydrolysis of bis(pentafluorophenyl)fluoromethyl cation [159].

Procedure

1. Oxidation of Decafluorobenzhydrol

$$(C_6F_5)_2CHOH \xrightarrow{CrO_3;\ CH_3COOH} C_6F_5COC_6F_5$$

Decafluorobenzhydrol. A solution of 88 ml (1.09 mol) of ethyl formate in 480 ml of absolute ether is gradually poured from a pressure-equalizing dropping funnel, with vigorous stirring and cooling in an ice-salt bath, into a Grignard reagent obtained from 202.5 g (1 mol) of chloropentafluorobenzene (see p. 126). The reaction mixture is stirred at room temperature during 1 h, then boiled for 3 h in a water-bath (the temperature of the bath is 45 °C) and allowed to stand overnight. The cooled reaction mixture is poured in portions into a stirred mixture of 900 g of ice and 450 ml of concentrated HCl. The ether layer is separated and the aqueous layer extracted with ether (5 × 350 ml). The ethereal solutions are combined, washed with 15% solution of Na_2CO_3 (250 ml for 1800 ml of the ethereal solution) and with water till neutral to litmus. The ether extract is dried over $MgSO_4$. The most part of the ether is distilled off in a water-bath and the residue distilled to dryness on a rotary vacuum-evaporator.

The resulting solid is thoroughly ground in a porcelain mortar, washed with cold petroleum ether (bp 70°–100 °C) then filtered and dried in air. The yield of decafluorobenzhydrol is 90–96 g (50%–53%); mp 76 °C.

Decafluorobenzophenone. In a 2-l three-necked round-bottom flask provided with a stirrer, a reflux condenser, and a thermometer, are placed 60 g (0.165 mol) of decafluorobenzhydrol, 60 g (0.6 mol) of CrO_3, and 1200 ml of glacial acetic acid. The mixture is heated at 105 °C in an oil-bath for 15–20 min (the temperature of the bath 120°–125 °C). Then it is cooled to room temperature with stirring and poured into 3 l of water. The precipitate is filtered off on a Büchner funnel, washed with water, then with 6% solution of Na_2CO_3 and again with water till it is neutral to litmus. The product is dried, first in air, then over P_2O_5 in a vacuum-desiccator. The yield of decafluorobenzophenone is 45–48 g (77%–80%); mp 91°–92 °C.

2. Reaction of Perfluorobenzyl Cation with Pentafluorobenzene in SbF_5 [159]

$$C_6F_5CF_3 + C_6HF_5 \xrightarrow[2.\ H_2O]{1.\ SbF_5} C_6F_5COC_6F_5$$

In a 100-ml three-necked flask fitted with a stirrer, a thermometer, and a pressure-equalizing dropping funnel with a calcium chloride tube, is placed 90 g of SbF_5. Then 21.8 g (0.1 mol) of octafluorotoluene is added at 20 °C with stirring. The mixture is allowed to stand for 0.5 h, whereupon 15 g (0.1 mol) of pentafluorobenzene is added. After 1 h, the mixture is poured onto 0.5 kg of ice. The precipitate is filtered off on a Büchner funnel, washed with 250 ml of water, then with 200 ml of 6% Na_2CO_3 solution and again with water till it is neutral to litmus. The product is dried, first in air, then over P_2O_5 in a vacuum-desiccator. The yield of decafluorobenzophenone is 26–27 g (80%–83%); mp 89°–90 °C.

Pentafluorobenzoyl Acetate

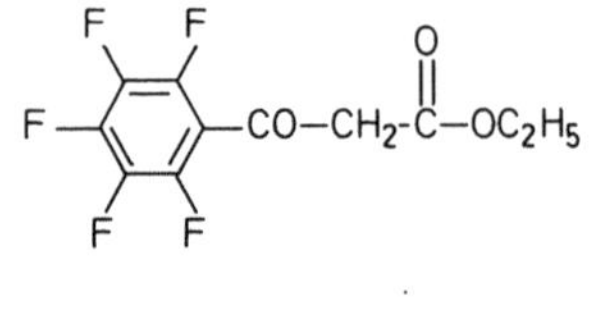

$C_{11}H_7F_5O_3$ MW 282

Pentafluorobenzoyl acetate is a water-immiscivle colourless liquid, soluble in common organic solvents. Bp 103.5 °C/4 mm Hg [160]; n_D^{21} 1.4604 [160].

Method of Preparation

Pentafluorobenzoyl acetate is prepared by partial hydrolysis of pentafluorobenzoylmalonic ester; by condensation of ethyl pentafluorobenzoate with ethyl acetate in the presence of diisopropylaminomagnesium bromide (38%–48% yield) [160–161]; or by oxidation of ethyl β-hydroxy-β-pentafluorophenylpropionate (17% yield) [161].

The following procedure is a modification of the one described in [160].

Procedure

$$C_6F_5COOH \xrightarrow{PCl_5} C_6F_5COCl\,;$$

$$CH_2(COOC_2H_5)_2 \xrightarrow{Mg,\ C_2H_5OH} C_2H_5OMgCH(COOC_2H_5)_2\,;$$

$$C_2H_5OMgCH(COOC_2H_5)_2 + C_6F_5COCl \rightarrow C_6F_5COCH(COOC_2H_5)_2\,;$$

$$C_6F_5COCH(COOC_2H_5)_2 \xrightarrow[H_2SO_4]{H_2O} C_6F_5COCH_2COOC_2H_5$$

Pentafluorobenzoyl chloride. In a 1-l round-bottom flask are quickly placed 450 g (2.1 mol) of pentafluorobenzoic acid and 500 g (2.3 mol) of PCl_5. After hydrogen chloride has ceased to evolve, the mixture is refluxed for 1 h and allowed to stand overnight. Then it is filtered through a glass filter. The filtrate is distilled and a fraction boiling at 51°–53 °C/14 mm Hg is collected. The yield of pentafluorobenzoyl chloride is 380–420 g (78%–87%); n_D^{20} 1.4550.

Magnesium ethoxymalonate. In a 2-l three-necked flask provided with a stirrer, a pressure-equalizing dropping funnel, and a reflux condenser with a calcium chloride tube, is placed 20.8 g (0.85 mol) of dry magnesium turnings. Magnesium is activated with iodine in a conventional way. Then 89 ml of absolute alcohol and 2 ml of dry carbon tetrachloride are added. The mixture is vigorously stirred for 5–10 min till heat starts to evolve, whereupon 118 g (0.74 mol) of malonic ester is dropped in, with stirring, at such a rate as to keep the reaction mixture boiling (Note 1). Then the mixture is cooled with constant stirring till it becomes viscous, 200 ml of dry ether (Note 2) is added dropwise, with great caution, and another 250 ml portion of ether is quickly poured in. To ensure complete dissolution of magnesium, the reaction mixture is heated for 3 h in a boiling water-bath.

Pentafluorobenzoyl acetate. Freshly distilled pentafluorobenzoyl chloride (182.5 g, 0.79 mol) in 200 ml of absolute ether is added dropwise to the cooled solution of magnesium ethoxymalonate, with constant cooling of the flask in an ice-water bath. The mixture is boiled for 10 min and cooled, whereupon first 200 ml of water, then 100 ml of 20% H_2SO_4 are added dropwise. The ether is distilled off from the reaction mixture, which is then steam-distilled. The distillate (4–4.5 l) is saturated with sodium chloride and extracted five or six times with ether (300–350 ml of ether for each 1.5 l of the distillate). The combined ether extract is washed with $NaHCO_3$, then with water till it is neutral to litmus and dried over $MgSO_4$. The ether is removed in a water-bath, and the residue distilled. A fraction boiling at 96°–98 °C/3 mm Hg is collected. The yield of pentafluorobenzoyl acetate is 78–85 g (35%–38%); n_D^{20} 1.4625.

Notes

1. If the malonic ester is added too slowly, the reaction mixture solidifies, whereas too fast an addition may cause ejection of the mixture from the flask, therefore the rate of the reaction should be controlled by regulating the rate of addition of the malonic ester. The flask should be immersed in a cooling bath.
2. Too fast an addition of the ether may lead to the ejection of the reaction mixture from the flask.

Perfluoro(butyrophenone)

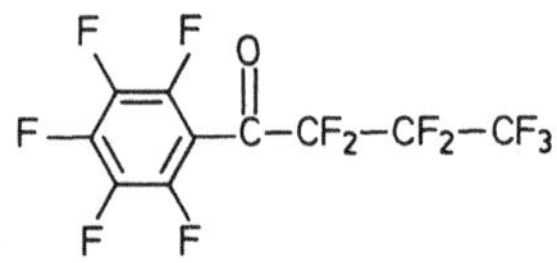

$C_{10}F_{12}O$ MW 364

Perfluoro(butyrophenone) is a colourless liquid with a strong odour, soluble in most organic solvents, partially soluble in water. Bp 158°–160 °C [162], 62°–64 °C/19 mm Hg [163–165]; n_D^{23} 1.3657 [162].

Method of Preparation

Perfluoro(butyrophenone) is prepared by the reaction of pentafluorophenylmagnesium bromide with perfluorobutyryl chloride (66% yield) [169]; by oxidation of (heptafluoro-n-propyl)-pentafluorophenyl carbinol with chromic anhydride in acetic acid (76%–92% yield) [163–165]; by the reaction of pentafluorobenzene with perfluorobutyryl chloride or fluoride (43%–58% yield) or with perfluorobutyric acid anhydride (81% yield) in the presence of SbF_5 [162].

The following method is based on the data given in [162].

Procedure

$$C_6HF_5 \xrightarrow{(n-C_3F_7CO)_2O,\ SbF_5} C_6F_5COC_3F_7$$

Perfluorobutyric acid adhydride (68.5 g, 0.167 mol) and 250 g of SbF_5 are placed in a 150-ml three-necked flask provided with a stirrer, a thermometer, and a pressure-equalizing dropping funnel with a calcium chloride tube. Pentafluorobenzene (28 g, 0.167 mol) is added at 20 °C with vigorous stirring during 0.5 h. The reaction mixture is heated at 50°–60 °C over a period of 2 h, then poured onto 1.5 kg of ice and extracted with methylene chloride. The extract is washed with 250 ml of water and dried over $MgSO_4$. Methylene chloride is removed by distillation in a water-bath, the residue is distilled and a fraction boiling at 158°–160 °C is collected. The yield of perfluoro-(butyrophenone) is 46–49 g (76%–81%); n_D^{23} 1.3657.

3,4,5,6-Tetrafluorobenzoquinone-1,2

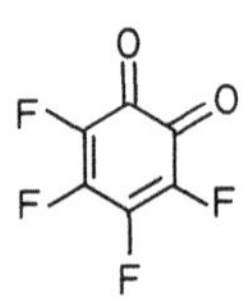

$C_6F_4O_2$ MW 180

3,4,5,6-Tetrafluorobenzoquinone-1,2 is a crystalline substance of yellow-orange colour, soluble in nitromethane and benzene. Mp 55°–62 °C [166], 125°–150 °C [167].

Method of Preparation

3,4,5,6-Tetrafluorobenzoquinone-1,2 is prepared by oxidation of pyrocatechol, o-aminophenol or o-phenylenediamine with hydrogen peroxide or potassium permanganate in 1% hydrofluoric acid, with a small yield [167]; or by oxidation of tetrafluoropyrocatechol with silver oxide [168] or with nitric acid (77% yield) [166].

The following method is a modification of the procedure described in [166].

Procedure

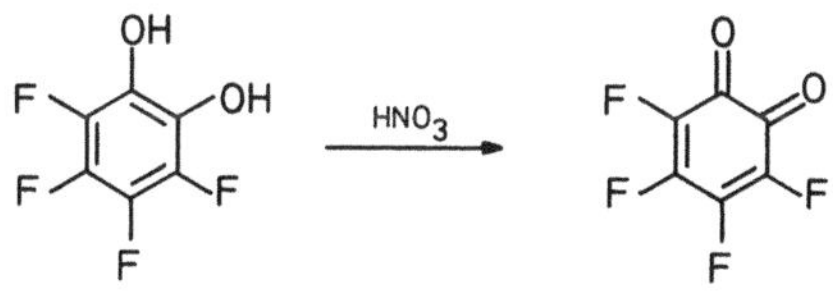

A solution of 1.5 g (0.008 mol) of tetrafluoropyrocatechol in 6 ml of nitromethane is placed in a 50-ml four-necked flask equipped with a stirrer, a thermometer, a dropping funnel, and an air condenser with a calcium chloride tube. The solution is cooled to —25° ÷ —30 °C, and 0.8 ml of HNO_3 (d 1.40) is added dropwise with stirring; the temperature is maintained at —25 °C. The mixture is stirred at that temperature for further 20 min, then poured into 250 ml of water and quickly extracted with benzene. The extract is dried over $MgSO_4$ and the solvent is removed on a rotary vacuum-evaporator at a temperature not higher than 30 °C. The yield of tetrafluorobenzoquinone-1,2 is 1–1.2 g (73–80%); bright-orange crystals; mp 55°–70 °C (Note).

Note. On standing at room temperature, 3,4,5,6-tetrafluorobenzoquinone-1,2 undergoes polymerization (turns red) and its melting point increases. It should be used for further synthesis immediately after its preparation.

2,3,5,6-Tetrafluorobenzoquinone-1,4

(Fluoranil)

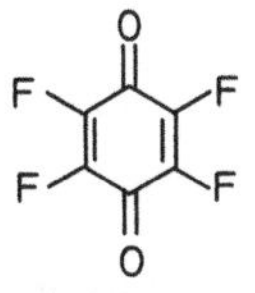

$C_6F_4O_2$ MW 180

Fluoroanil is a light-yellow crystalline substance with a strong odour. It is a low-subliming compound, soluble in common organic solvents, partially soluble in water. Mp 165°–170 °C [169], 177°–178 °C [170], 178°–179 °C [3], 179 °C [171–173].

Method of Preparation

Fluoranil is prepared by oxidation of hexafluorobenzene (32% yield) [172], pentafluorophenol (50% yield) [170], tetrafluoro-para-aminophenol (53% yield) [51],

2,3,5,6-tetrafluoro-N,N'-diphenyl-1,4-phenylenediamine (16% yield) [169] with nitric acid; by oxidation of tetrafluorohydroquinone with lead dioxide in ether (64% yield) [171] or with potassium permanganate in 2% HF solution, with a small yield [167]; or by the reaction of anhydrous KF with chloranil (25% yield) [171].

The following procedure is a modification of the one described in [172].

Procedure

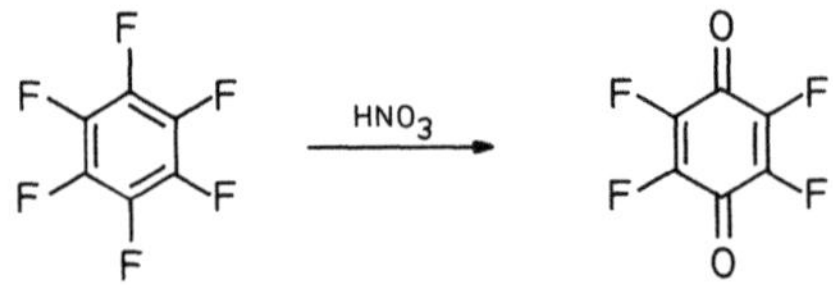

Hexafluorobenzene (186 g, 1 mol) and HNO_3 (1 l) (d 1.52) are placed in a 2-l two-necked flask provided with a stirrer and a reflux condenser. The mixture is reflux for 6 h, then cooled and poured onto 1200 g of ice. The fluoranil precipitate (32–33 g) is filtered off on a glass filter, washed with water, thoroughly pressed and dried in air. The filtrate is extracted with benzene (3–4 × 500 ml) and the extract is dried over $CaCl_2$. Distillation to remove benzene gives further 12–13 g of fluoranil. The total yield of fluoranil is 44–46 g (24%–26%); mp 177°–179 °C.

Octafluoroanthraquinone

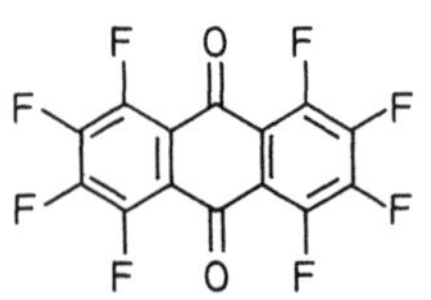

$C_{14}F_8O_2$ MW 352

Octafluoroanthraquinone is a yellow crystalline substance, slightly soluble in organic solvents. Mp 342°–343 °C, 341°–342 °C [174, 175]; subl. p. 280 °C/20 mm Hg [174, 175].

Method of Preparation

Octafluoroanthraquinone is prepared by the reaction of KF with tetrafluorophthalic anhydride (72% yield) or tetrachlorophthalic anhydride in the absence of solvents (43% yield) [174, 175]; or in dimethylformamide (~2% yield) [176].

The following procedure is a modification of the one described in [175].

Procedure

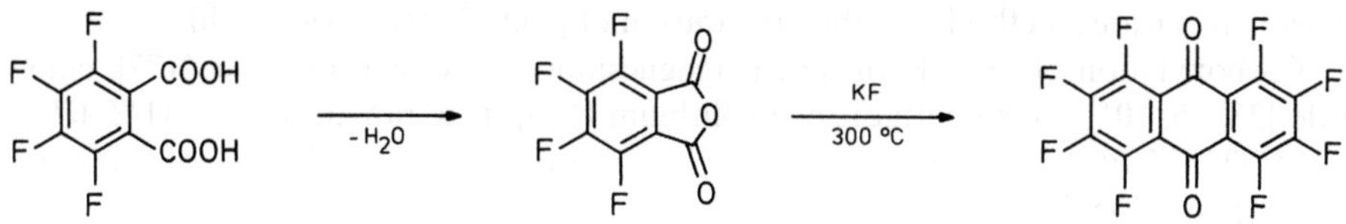

Tetrafluorophthalic anhydride. In a 150-ml two-necked flask are placed 40 g (0.17 mol) of tetrafluorophthalic acid, 100 ml of benzene and 1 ml of concentrated H_2SO_4. The mixture is boiled for 24 h, during which time the water-benzene azeotrope is removed. Benzene (~500 ml) is continuously introduced into the flask at such a rate as to keep the amount of the reaction mixture constant. The hot benzene solution is separated from H_2SO_4 and cooled. The precipitate is filtered and sublimed at 150 °C /20 mm Hg. The yield of tetrafluorophthalic anhydride is 30 g (~85%); mp 90° to 91.5 °C.

In contact with the air tetrafluorophthalic anhydride is gradually hydrolysed to tetrafluorophthalic acid, therefore it should be kept over P_2O_5 in a desiccator.

Octafluoroanthraquinone. In a 10-ml steel reactor are placed about 2 g of freshly calcinated KF, then 3 g (0.014 mol) of tetrafluorophthalic anhydride and again KF. The total amount of KF introduced into the reactor is 6 g (0.1 mol). The reactor is quickly closed, immersed into a bath heated to 320°–330 °C (the temperature in the bath after the reactor has been immersed in it should be 300 °C) and kept at 300 °C for 30 min. The reaction mixture is sublimed at 280 °C/20 mm Hg.

First a small amount of tetrafluorophthalyl difluoride formed in the reaction is sublimed. It condenses on a condenser and should be removed before octafluoroanthraquinone starts to sublime. It may be removed by washing it off with acetone.

Sublimed octafluoroanthraquinone is quite pure. It may be further purified by washing it with a small amount of ether. The yield is 1.6–1.7 g (65%–70%). Crystallization from nitrobenzene affords analytically pure octafluoroanthraquinone, mp 342° to 343 °C (in a sealed capillary tube).

Pentafluorobenzoic Acid

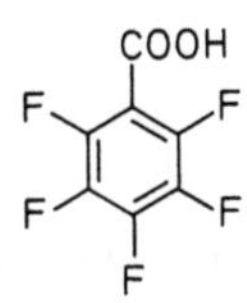

$C_7HF_5O_2$ MW 212

Pentafluorobenzoic acid is a white crystalline substance, soluble in ether, alcohol, acetone and slightly soluble in carbon tetrachloride and water. Mp 101°–103 °C [52, 56], 102 °C [60], 103°–104 °C [66], 104 °C [178], 104°–105 °C [42], 105°–105.5 °C [101], 106 °C [116], 106°–107 °C [144], 106.5°–107.5 °C [116].

Method of Preparation

There are various methods for the preparation of pentafluorobenzoic acid:
1. Carboxylation of pentafluorophenylmagnesium iodide [60], bromide [177], chloride [25, 26, 101], or pentafluorophenyllithium [116] in tetrahydrofuran (41%–67% yield) [25, 26, 101, 177] (in ether carboxylation proceeds with difficulty or does not proceed at all [56]);
2. Oxidation of the side chain of pentafluorobenzaldehyde [103], pentafluorophenylethylene [60], pentafluorophenylpropylene [111, 179], perfluorostyrene [178], 1-chloro-1,2-difluoro-2-(pentafluorophenyl)ethylene [62] with sodium bichromate in H_2SO_4 [103], with chromic anhydride in acetic acid [113], with nitric acid [62] or with potassium permanganate in acetone [60, 111, 178, 179] (60%–80% yields);
3. Hydrolysis, with concentrated sulphuric acid, of pentafluorobenzotrichloride at 100°–110 °C (97% yield) [103], or pentafluorobenzotrifluoride at 110 ° or 160 °C (25%–62.5% yield) [42, 144].
4. The reaction of pentafluorobenzene with phosgene in the presence of SbF_5 with subsequent treatment with water (60% yield) [162], or hydrolysis of octafluorotoluene in SbF_5 (66% yield) [112].

The following procedures are based on the data reported in [101, 112, 162].

Procedure

1. Reaction of Pentafluorobenzene with Phosgene in SbF_5 [162]

$$C_6HF_5 \xrightarrow[2.\ H_2O]{1.\ COCl_2;\ SbF_5} C_6F_5COOH$$

In a 50-ml stainless steel autoclave provided with a valve and a thermocouple in a well, are loaded 10 g (0.06 mol) of pentafluorobenzene and 50 g of SbF_5. The autoclave is cooled to −20° ÷ −30 °C and 6.4 g (0.065 mol) of phosgene is added. The mixture is allowed to stand at 20 °C over a period of 40 h. The pressure in the autoclave is then reduced to that of the atmosphere, its contents are poured onto 0.3 kg of ice and extracted with ether (4 × 150 ml). The ether is removed on a rotary vacuum-evaporator and the residue recrystallized from a 1:1 toluene-petroleum ether (bp 70°–100 °C) mixture. The yield of pentafluorobenzoic acid is 7.6 g (60%); mp 104° to 105 °C.

2. Hydrolysis of Octafluorotoluene in SbF_5 [112]

$$C_6F_5CF_3 \xrightarrow[2.\ H_2O]{1.\ SbF_5} C_6F_5COOH$$

In a 150-ml two-necked flask fitted with a stirrer and a dropping funnel is placed 200 g of SbF_5. Octafluorotoluene (47.6 g, 0.2 mol) is added dropwise, with mild stirring and cooling of the flask in ice-water over a period of 0.5 h. The mixture is maintained at 20 °C for 0.5 h, poured onto 1 kg of ice and extracted with ether. The ether extract is dried over $MgSO_4$. The ether is removed on a rotary vacuum-evaporator and the residue recrystallized from the 1:1 toluene-petroleum ether (bp 70°–100 °C) mixture. The yield of pentafluorobenzoic acid is 30–33 g (71%–78%); m.p. 104 to 105 °C.

3. Carboxylation of Pentafluorophenylmagnesium Chloride [101]

$$C_6F_5MgCl \xrightarrow{CO_2} C_6F_5COOMgCl \xrightarrow{H_2O} C_6F_5COOH$$

Most part of the ether is distilled from the Grignard reagent obtained from 20 g (0.1 mol) of chloropentafluorobenzene (p. 126), with constant stirring and heating (the temperature of the bath is up to 60 °C). Residual ether is removed in vacuum of a water-jet pump with continuous stirring (Note 1). The reaction mixture is then cooled to 0 °C and 100 ml of absolute tetrahydrofuran is added to it (Note 2), whereupon a flow of dry carbon dioxide is passed at a reasonable rate during 12–15 h, with cooling of the flask (Note 3). The contents of the flask are poured into a mixture of 90 g of ice and 35 ml of concentrated HCl, and steam-distilled till all the acid has been removed (the total amount of the distillate is ~500 ml). The forerun (~110 ml) contains mainly tetrahydrofuran and pentafluorobenzoic acid, the next fraction includes water, tetrahydrofuran traces and the acid. The forerun is evaporated to dryness to give 1.5–2 g of pentafluorobenzoic acid. The second fraction is subjected to salting-out with NaCl and extracted five or six times with ether. The ether extracts (~200 ml) are dried over $MgSO_4$. Distillation to remove the solvent affords a solid, which is first dried in air, then ground and dried further in vacuum at 50°–60 °C/15–20 mm Hg. This procedure gives an additional 9.5–10 g of the acid. The total yield of pentafluorobenzoic acid is 11–12 g (51%–56%); mp 101°–102 °C (Note 4). Recrystallization from a mixture of equal volumes of toluene and petroleum ether (bp 70°–100 °C) yields 8–9 g (38%–42%) of the acid, mp 105°–105.5 °C.

Notes

1. Partial removal of the ether results in a sharply decreased yield of the acid. Further work with the dry Grignard reagent should be conducted quickly, since upon standing the dry product may explode.
2. Tetrahydrofuran is added from the dropping funnel, at first dropwise, with hand-stirring. After the first portion has dissolved, the remaining tetrahydrofuran is added in one portion, with stirring and cooling.
3. Instead of passing CO_2, solid carbon dioxide (200–250 g) may be placed into a flask and allowed to stand overnight.
4. Stable yields of pentafluorobenzoic acid are obtained when treating 20–100 g (0.1–0.5 mol) of chloropentafluorobenzene. Treatment of greater amounts (more than 1 mol of chloropentafluorobenzene) does not result in stable yields of the acid.

Bis(pentafluorobenzoyl)peroxide

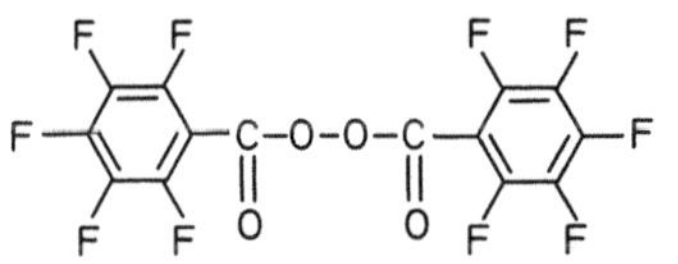

$C_{14}F_{10}O_4$ MW 422

Bis(pentafluorobenzoyl)peroxide is a white crystalline substance, soluble in common organic solvents, slightly soluble in water. Mp 76°–78 °C [180], 72 °C [181].

Method of Preparation

Bis(pentafluorobenzoyl)peroxide is prepared by the reaction of pentafluorobenzoyl chloride with hydrogen peroxide in an alkaline medium (76% yield) [180, 181].

The following procedure is based on the data reported in [180].

Procedure

$$C_6F_5COCl \xrightarrow{H_2O_2;\ NaOH} C\ F_5\text{—}COO\text{—}OOC\text{—}C_6F_5$$

In a 0.5-l three-necked flask fitted with a stirrer, a thermometer, and a pressure-equalizing dropping funnel with a calcium chloride tube, is placed 120 ml of 15% aqueous solution of NaOH. The solution is cooled to 0 °C, then 90 ml of 30% H_2O_2 is slowly added to it during 0.5 h, so that the temperature of the mixture should not rise above 0 °C. The solution is cooled to —10 °C and 60 g (0.26 mol) of pentafluorobenzoyl chloride is slowly added to it with vigorous stirring during about 1 h. Stirring is continued for further 0.5 h at —8° to —10 °C. The precipitate is removed by filtration on a Büchner funnel, washed with 0.5 l of ice-water, then with 1.5 l of water at 20 °C and dried in air. This gives 44 g of the product which is further recrystallized from petroleum ether (bp 40°–60 °C). The yield of bis(pentafluorobenzoyl)peroxide is 38–39 g (69%–71%); mp 76°–78 °C.

Pentafluorobenzonitrile

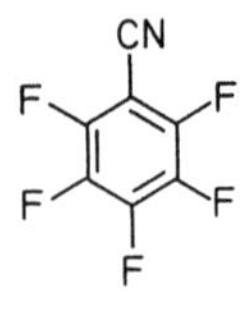

C_7F_5N MW 193

Pentafluorobenzonitrile is a colourless liquid with a specific odour, soluble in common organic solvents. Bp 160°–161 °C [182], 161°–162 °C [181], 162.5 °C [180], 185° to 190 °C [109, 183, 184], 99°–105 °C/98 mm Hg [185], 69°–71 °C/12 mm Hg [151]; n_D^{20} 1.4429 [180], 1.4400 [181], $n_D^{23.8}$ 1.4764 [109, 183, 184]; d_4^{25} 1.56 [181].

Method of Preparation

Pentafluorobenzonitrile is prepared by the reaction of bromopentafluorobenzene with CuCN in dimethylformamide (70%–82% yield) [180, 181], (31% yield) [182]; by the reaction of iodopentafluorobenzene with CuCN in pyridine (93% yield) [109, 183, 184]; by the reaction of pentachlorobenzonitrile with KF in an autoclave at 300°–480 °C (70% yield) [185]; by the reaction of P_2O_5 with pentafluorobenzamide

(54% yield) [151]; by the pyrolysis of hexafluorobenzene at 570 °C over KCN (5% conversion) [186].

The following procedure is based on the data reported in [151].

Procedure

$$C_6F_5COCl \xrightarrow{NH_3;\,ether} C_6F_5CONH_2 \xrightarrow{P_2O_5} C_6F_5CN$$

Pentafluorobenzamide. In a 1-l two-necked flask bearing a reflux condenser with a calcium chloride tube and a wide gas inlet tube, are placed 75 g (0.325 mol) of pentafluorobenzoyl chloride and 800 ml of absolute ether. Gaseous NH_3 is passed during 3 h into the flask cooled in ice-water. The mixture is then transferred to a 2-l separatory funnel, washed with 1 l of water, the ether layer is separated and dried over $MgSO_4$. The ether is removed on a rotary vacuum-evaporator, and the residue recrystallized from benzene. The yield of pentafluorobenzamide is 65.2 g (95%); mp 150 °C.

Pentafluorobenzonitrile. In a 250-ml two-necked flask provided with a reflux condenser with a calcium chloride tube and a thermometer, are placed 65 g (0.3 mol) of pentafluorobenzamide and 85.2 g (0.6 mol) of P_2O_5. The mixture is heated at 200 °C for 1 h. The reflux condenser is then replaced with a Wurtz column and the mixture is distilled. A fraction boiling at 161°–162 °C is collected. The yield of pentafluorobenzonitrile is 47–48 g (79%–81%); n_D^{20} 1.4430.

Tetrafluorophthalic Acid

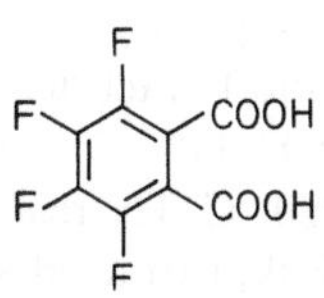

$C_8H_2F_4O_4$ MW 238

Tetrafluorophthalic acid is a white crystalline substance. Mp 151°–152 °C [116], 152°–153 °C [187], 152°–154 °C [188], 153°–154 °C [189–191], 154°–156 °C [54], 155°–156 °C [192].

Method of Preparation

Tetrafluorophthalic acid is prepared by the hydrolysis of perfluoro-*ortho*-xylene [189–191], dimethyl tetrafluorophthalate, tetrafluorophthalyl difluoride or dinitrile [54]; by oxidation of octafluoronaphthalene with nitric acid [188, 192] (75%–91% yields); or by metallation of 1,2,3,4-tetrafluorobenzene with n-butyllithium with subsequent carboxylation of the dilithium derivative (9% yield) [116].

Procedure

1. Conversion of Tetrachlorophthalic Anhydride [187]

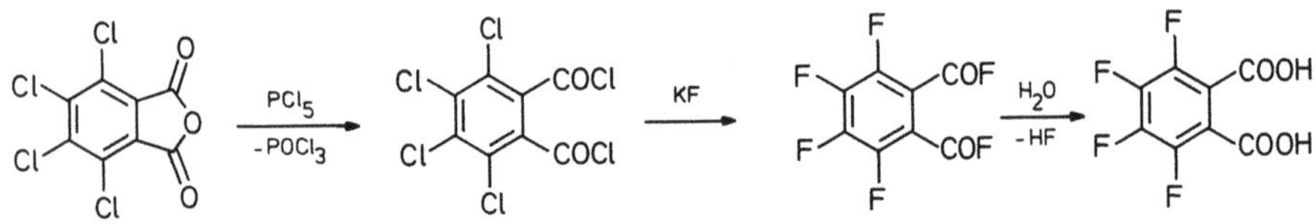

Tetrachlorophthalyl dichloride. In a 2-l round-bottom flask provided with a reflux condenser with a calcium chloride tube, are placed in layers 740 g of ground PCl_5 and 700 g of dry tetrachlorophthalic anhydride (mp no less than 255 °C). The reaction mixture is heated in an oil-bath to the temperature of 150°–160 °C in the bath, and maintained at that temperature over a period of 10–15 h. The reaction mixture should be transparent. The reflux condenser is replaced with a straight one and $POCl_3$ is distilled off at 150 °C and atmospheric pressure; the residue is removed by distillation in a slight vacuum (12–15 mm Hg). The melted mixture is poured into another flask in such a way that PCl_5 sublimated on the walls of the first flask should remain there. Petroleum ether (1000 ml) (bp 70°–100 °C) is gradually poured with stirring into the mass cooled to 40°–50 °C. The mixture is refluxed for 20–30 min and filtered. Refluxing with petroleum ether is repeated and conducted to complete dissolution of the melted mass. After the filtrate has been cooled, the precipitate formed is filtered off and washed on the filter with petroleum ether (2 × 100 ml). The yield of tetrachlorophthalyl dichloride is 520 g (62%); mp 133°–134 °C (Note 1).

Tetrafluorophthalyl difluoride. A 1-l rotating steel autoclave provided with a manometer, a valve, and a thermocouple in a well, is charged (as quickly as possible) with 700 g of freshly calcinated anhydrous KF and 200 g (0.58 mol) of tetrachlorophthalyl dichloride, placed in layers. The autoclave is heated to 160°–170 °C and maintained for 12 h, then for 6 h at 290°–300 °C, and for 6 h at 310°–320 °C. The reaction mixture is quickly transferred from the cooled autoclave into a flask, extracted several times with petroleum ether (bp 40°–60 °C) (the total amount is 2–2.5 l) and filtered. The solvent is distilled off under a dephlegmator and the residue is distilled on a fractionating column (16–20 theoretical plates). A fraction boiling at 182°–183 °C is collected. This procedure gives 30–35 g of tetrafluorophthalyl difluoride.

The still-residue is refluorinated under the same conditions to yield further 10–15 g of tetrafluorophthalyl difluoride. The total yield is 40–50 g (28%–35%) (Note 2); n_D^{22} 1.4383.

Tetrafluorophthalic acid. Tetrafluorophthalyl difluoride (64 g, 0.26 mol) and water (150 ml) are refluxed during 1 h in a 250-ml flask. The mixture is then cooled and extracted with ether (5 × 500 ml), the extract is dried over $MgSO_4$. The ether is completely removed, first in a water-bath, then on a rotary vacuum-evaporator. The residue is dried during 1 h in a vacuum-desiccator at 100 °C. The yield of tetrafluorophthalic acid is 59 g (95%); mp 151.5°–153 °C.

Notes

1. If required, recrystallization from 2.5 l of petroleum ether (bp 70°–100 °C) is carried out.
2. It is not practical to fractionate large quantities of the mixture, since prolonged or repeated heating causes partial decomposition of the product.

2. Oxidation of Octafluoronaphthalene [192]

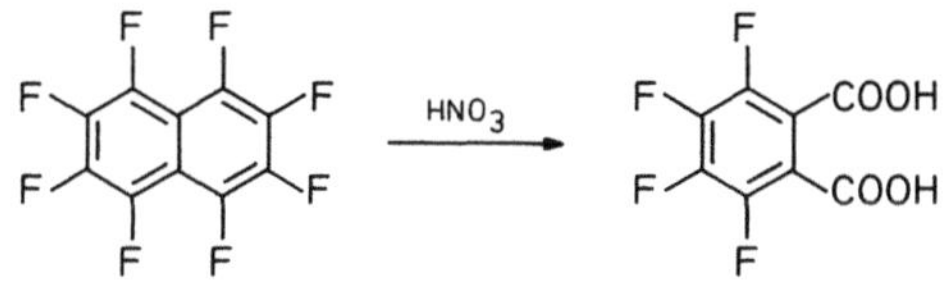

Octafluoronaphthalene (90 g, 0.33 mol) is added in portions to the concentrated nitric acid (350 ml) (d 1.52) boiled in a 0.5-l round-bottom flask, and the reaction mixture is refluxed during 20 h. The mixture is then cooled, poured onto 800 g of ice and the precipitate is filtered off. Recrystallization from benzene and sublimation at 150°–160 °C/0.1 mm Hg afford 55.9 g (71 %) of tetrafluorophthalic acid, mp 155° to 156 °C.

Tetrafluoroterephthalic Acid

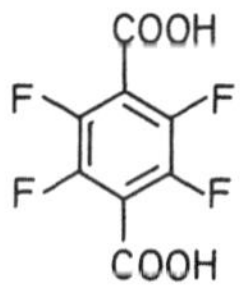

$C_8H_2F_4O_4$ MW 238

Tetrafluoroterephthalic acid is a white crystalline substance. Mp 281.5°–282.5 °C [116], 282°–283 °C [111], 283°–284 °C [103, 189, 190], 284°–284.5 °C [187, 193], 285°–287 °C [54].

Method of Preparation

Tetrafluoroterephthalic acid is prepared by the hydrolysis of perfluoro-*para*-xylene [189, 190], tetrafluoroterephthalyl difluoride [187, 193] or dinitrile [54]; by carboxylation of Grignard reagents or aryllithium reagents generated from 2,3,5,6-tetrafluorobenzene [116], or from 1,4-dibromo-2,3,5,6-tetrafluorobenzene [194]. The yields of tetrafluoroterephthalic acid prepared by these methods are 70 %–80 %. With lower yields (10 %–32 %) tetrafluoroterephthalic acid is formed in oxidation of bis-(*para*-propenyl)- [111] or bis-(*para*-bromomethyl)-tetrafluorobenzene [103].

The following procedure is a modification of the preparations reported in [187, 193].

Procedure

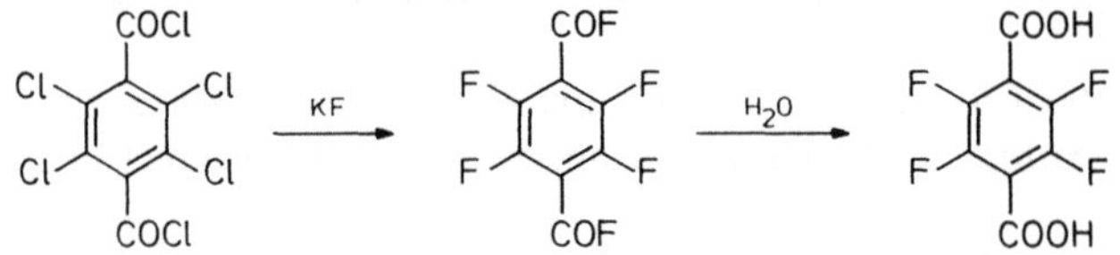

Tetrafluoroterephthalyl difluoride. In a 200-ml rotating steel autoclave equipped with a valve and a thermocouple in a well, are placed 75 g of tetrachloroterephthalyl dichloride (Note) and 170 g of freshly calcinated KF. The autoclave is heated at 160°–170 °C for 18 h, and at 280°–290 °C for 48 h. Sublimation of the reaction mixture at 175°–180 °C/20 mm Hg gives 34–36 g (66%–68%) of tetrafluoroterephthalyl difluoride, mp 94°–95 °C (from heptane).

Tetrafluoroterephthalic acid. Tetrafluoroterephthalyl difluoride (24 g) and 350 ml of water are refluxed for 5 h, then the mixture is cooled, filtered and crystallized from water. The yield of tetrafluoroterephthalic acid is 17–18 g (72%–76%); mp 284° to 284.5 °C.

Note. For the preparation of tetrachloroterephthalyl dichloride see [187, 195].

Tetrafluoroterephthalodinitrile

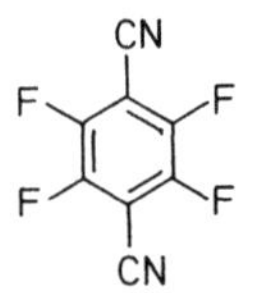

$C_8F_4N_2$ MW 200

Tetrafluoroterephthalodinitrile is a white crystalline substance, partially soluble in common organic solvents, slightly soluble in water. Mp 196°–198 °C [180, 196], 195°–196 °C [197], 197°–198 °C [185].

Method of Preparation

Tetrafluoroterephthalodinitrile is prepared by the reaction of tetrachloroterephthalodinitrile with KF in an autoclave (72%–75% yield) [196, 197]; by the reaction of 1,4-dibromo-2,3,5,6-tetrafluorobenzene with CuCN in dimethylformamide (61.5% yield) [180], (29.5% yield) [185].

The following procedure is based on the one described in [180].

Procedure

1,4-Dibromo-2,3,5,6-tetrafluorobenzene. In a 0.5-l three-necked flask fitted with a stirrer, a thermometer, and a reflux condenser with a calcium chloride tube, are placed 100 ml of 60% oleum, 160 g (54 ml) (1 mol) of bromine, 2 g of anhydrous $AlBr_3$ and 50 g (0.3 mol) of 1,2,4,5-tetrafluorobenzene. The reaction mixture is heated at 50 °C with stirring during 4 h and poured onto 1.5 kg of ice (Note). Then 10% aqueous $NaHSO_3$ is added to decolorization of the solution. The precipitate is filtered off, washed with water (2 × 150 ml) on a Büchner funnel and dried. The yield of 1,4-dibromo-2,3,5,6-tetrafluorobenzene is 72–78 g (70%–76%); mp 76°–77 °C (from aqueous ethanol) (mp 76°–77 °C [198, 199]).

Tetrafluoroterephthalodinitrile. 1,4-Dibromo-2,3,5,6-tetrafluorobenzene (308 g, 1 mol), freshly prepared anhydrous CuCN (225 g, 2.5 mol) and absolute dimethylformamide (200 ml) are placed in a 0.6-l round-bottom flask fitted with a stirrer and a reflux condenser with a calcium chloride tube. The reaction mixture is heated at 160 °C, with stirring for 3 h, then cooled and pored into 1 l of water, whereupon 600 ml of 20% aqueous $FeCl_3$ is added to it. The mixture is extracted with ether (5 × 250 ml) and the ether extract dried over $MgSO_4$. The ether is removed on a rotary vacuum-evaporator and the residue dried in a desiccator at 130°–150 °C. The yield of tetrafluoroterephthalodinitrile is 120–125 g (60%–62.5%); mp 196°–198 °C.

Note. The reaction mixture should better be transferred to a separatory funnel with a long narrow stem extending to the bottom of a flask for decomposition of the reaction mixture.

2,3,5,6-Tetrafluorohomoterephthalic Acid

$C_9H_4F_4O_4$ MW 252

2,3,5,6-Tetrafluorohomoterephthalic acid is a white crystalline substance, partially soluble in common organic solvents, readily soluble in diluted alkaline solutions. Mp 220°–221 °C [199].

Method of Preparation

2,3,5,6-Tetrafluorohomoterephthalic acid may be obtained by saponification, with aqueous KOH, of diethyl 2,3,5,6-tetrafluorohomoterephthalate formed in the reaction of ethyl pentafluorobenzoate with sodium acetoacetate [199].

The following method is based on the data reported in [199].

Procedure

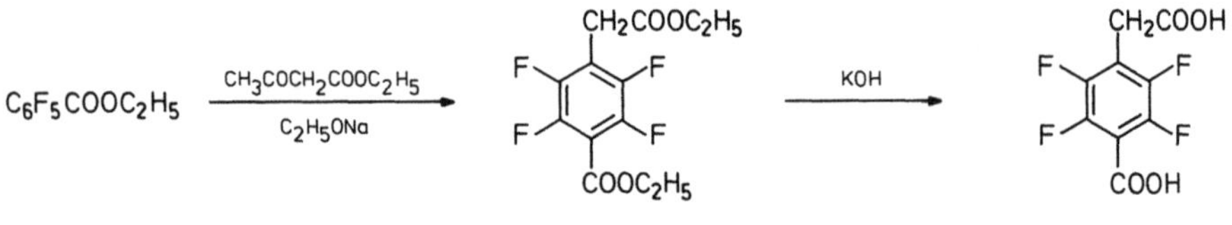

Diethyl 2,3,5,6-tetrafluorohomoterephthalate. In a 100-ml two-necked flask fitted with a reflux condenser and a pressure-equalizing dropping funnel with a calcium chloride tube, are placed 37 ml of absolute ethanol, then 2.9 g (0.126 mol) of sodium is added in portions. After the reaction is complete, 16.7 g (0.128 mol) of acetacetic ether is added. After 0.5 h, the reflux condenser is replaced with a Wurtz column and ethanol is distilled off, with gentle heating, in vacuum of a water-jet pump. Residual ethanol (traces) is distilled off in vacuum of an oil pump; the temperature of the bath is raised to 100 °C. To a cooled precipitate is added 30 g (0.125 mol) of ethyl pentafluorobenzoate. The Wurtz column is replaced with the reflux condenser and the reaction mixture is heated at 140°–145 °C for 3 h, then the temperature of the bath is raised to 180 °C over a period of 0.5 h. After that the mixture is cooled and 14 ml of water is added to it, whereupon it is acidified with a cooled solution of 4 ml of concentrated H_2SO_4 in 15 ml of water. The ether layer is separated and the aqueous layer extracted with ether. The ether extracts are combined, washed with 10 ml of 3% aqueous Na_2CO_3, with 10 ml of water, and dried over $MgSO_4$. The ether is distilled off in a water-bath and the residue distilled. A fraction boiling at 141°–145 °C/5 mm Hg is collected. The yield of diethyl 2,3,5,6-tetrafluorohomoterephthalate is 21.4 g (55.7%); n_D^{25} 1.4590.

2,3,5,6-Tetrafluorohomoterephthalic acid. Diethyl 2,3,5,6-tetrafluorohomoterephthalate (21 g, 0.068 mol) and KOH (30 g, 0.536 mol) in 800 ml of water are placed in a 1-l round-bottom flask with a reflux condenser. The mixture is reflux for 4 h, then acidified with concentrated HCl to pH ~2 and extracted several times with 100 ml portions of ether. The ether extract is dried over $MgSO_4$. The ether is distilled off on a rotary vacuum-evaporator and the residue recrystallized from water. The yield of 2,3,5,6-tetrafluorohomoterephthalic acid is 13.3 g (77%); mp 220°–221 °C.

2,3,4,5-Tetrafluorobenzoic Acid

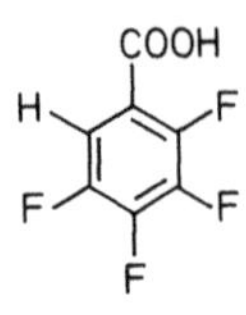

$C_7H_2F_4O_2$ MW 194

2,3,4,5-Tetrafluorobenzoic acid is a white crystalline compound, soluble in ether, acetone, and slightly soluble in petroleum ether. Mp 86°–86.5 °C [187], 92°–92.5 °C [116].

Method of Preparation

2,3,4,5-Tetrafluorobenzoic acid is prepared by metallation of 1,2,3,4-tetrafluorobenzene with n-butyllithium in tetrahydrofuran, with subsequent carboxylation (21% yield) [116], or by decarboxylation of tetrafluorophthalic acid in dimethylformamide (64% yield) [187].

The following method is a modification of the one described in [187].

Procedure

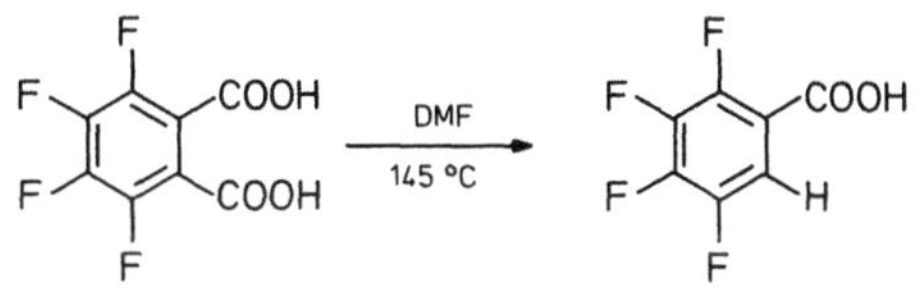

Tetrafluorophthalic acid (7.5 g, 0.031 mol) and anhydrous dimethylformamide (30 ml) are placed in a 100-ml three-necked flask fitted with a stirrer, a reflux condenser with a calcium chloride tube, and a thermometer. The reaction mixture is heated at 145 °C with stirring for 1.5 h, then cooled and poured into 40 ml of water. The oily layer is separated and 10 ml of concentrated HCl is added to it with stirring. The precipitate is filtered off and dried. The yield of the acid is 5.5 g; mp 86°–88 °C. Recrystallization from petroleum ether (bp 40°–60 °C) gives 3.8–4.1 g (63%–67%) of 2,3,4,5-tetrafluorobenzoic acid; mp 91°–92 °C.

3,4,5,6-Tetrafluorosalicylic Acid

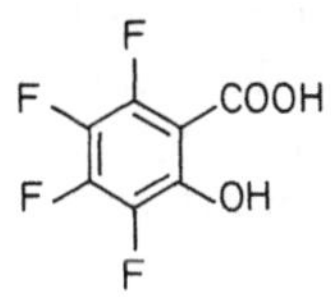

$C_7H_2F_4O_3$ MW 210

Tetrafluorosalicylic acid is a white crystalline substance, well soluble in ether and acetone, slightly soluble in water. Mp 165°–168 °C [200], 168.5°–170 °C [201], 169° to 170 °C [143], 169°–172 °C [202].

Method of Preparation

Tetrafluorosalicylic acid is prepared by oxidation of oxygen-containing fluorohetero-cyclic compounds with potassium permanganate in acetone (30%–64% yield) [143, 200–203]; by boiling 5,6,7,8-tetrafluoro-3-carbethoxy-1,2-benzoxazinone-4 with aqueous potassium carbonate (97% yield) [200].

The following procedure is a modification of the one described in [201].

Procedure

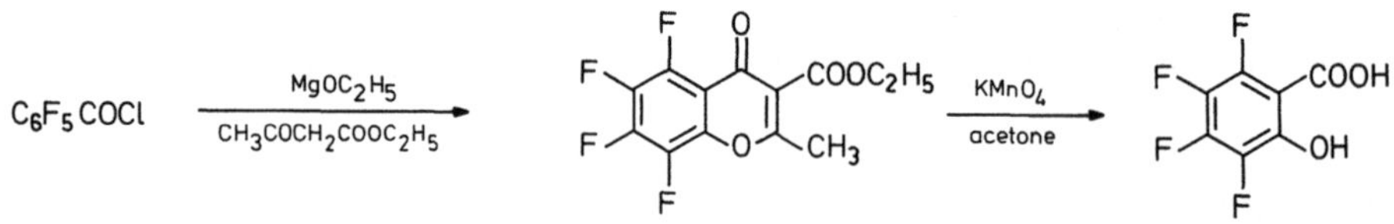

5,6,7,8-Tetrafluoro-2-methyl-3-carbethoxychromone. In a 500-ml three-necked flask provided with a stirrer, a reflux condenser with a calcium chloride tube, and a pressure-equalizing dropping funnel, are placed 6.7 g (0.28 mol) of magnesium turnings, previously washed with ether and dried in air, 88 ml of absolute ethanol and 0.8 ml of carbon tetrachloride. As soon as a vigorous reaction has started, a solution of 24 ml of absolute ethanol and 120 ml of absolute benzene are gradually added from the dropping funnel. After all the magnesium has been dissolved, 33.1 g (0.25 mol) of acetacetic ether is added dropwise and the mixture is stirred at room temperature for 1 h, whereupon a solution of 54.8 g (0.25 mol) of pentafluorobenzoyl chloride in 120 ml of absolute benzene is added. Stirring is continued for 1 h at room temperature and for further 15 min at 50 °C. The mixture then cooled to 5°–10 °C, a solution of 13.6 ml of concentrated H_2SO_4 in 80 ml of water is added to it and the benzene layer is separated. The aqueous layer is extracted with ether (4 × 30 ml). The combined ether-benzene extract is dried over $MgSO_4$. Distillation of the extract to remove the solvents and unchanged acetoacetic ether gives the solid product. It is washed with petroleum ether and dried over $CaCl_2$ in a vacuum-desiccator. The yield of 5,6,7,8-tetrafluoro-2-methyl-3-carbethoxychromone is 30–32 g (41%–44%); mp 88°–89.5 °C. Recrystallization from petroleum ether yields 27 g of the product, mp 90°–91 °C (mp 91°–91.5 °C [204]).

Tetrafluorosalicylic acid. In a 700-ml three-necked flask provided with a stirrer, a reflux condenser with a calcium chloride tube, and a thermometer, are placed solutions of $KMnO_4$ (10.4 g) in 360 ml of dry acetone and of 5,6,7,8-tetrafluoro-2--methyl-3-carbethoxychromone (10 g, 0.033 mol) in 40 ml of dry acetone. The reaction mixture is stirred at room temperature for 16 h, then evaporated on a rotary vacuum-evaporator to remove acetone; the residue is refluxed with 0.5–0.6 l of water during 1 h. The MnO_2 precipitate is filtered off and washed several times with boiling water on a filter. The filtrate is evaporated to one-half of its volume, acidified with HCl to pH ~ 3–4 and continuously extracted with ether during 20 h. Distillation to remove the ether gives 0.6 g of the crude product, which is then crystallized from water. The yield of tetrafluorosalicylic acid is 4.7–4.9 g (68%–71%); mp 169.5°–171 °C.

2,3,5,6-Tetrafluoro-4-hydroxybenzoic Acid

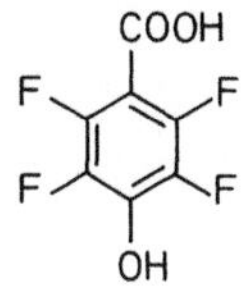

$C_7H_2F_4O_3$ MW 210

2,3,5,6-Tetrafluoro-4-hydroxybenzoic acid is a white crystalline substance, soluble in most organic solvents. Mp of the crystalline hydrate is 154 °C [205], 157 °C [206], 154°–156 °C [182, 207], 152°–154 °C [152]; pK_a 2.27 (COOH), 5.17 (OH) [208].

Method of Preparation

2,3,5,6-Tetrafluoro-4-hydroxybenzoic acid is obtained by the reaction of octafluorotoluene with KOH in water at 145°–150 °C/8–9 atm (80% yield) [205]; by the reaction of hydroiodic acid with 2,3,5,6-tetrafluoro-1-ethoxy-4-trifluoromethylbenzene (30% yield) [206]; by the reaction of n-butyllithium with 2,3,5,6-tetrafluorophenol in tetrahydrofuran with subsequent carbonization of the product (82.5% yield) [207]; by the reaction of aqueous KOH with pentafluorobenzoic acid (74% yield) [152] or with pentafluorobenzonitrile [182]; by oxidation of 2,3,5,6-tetrafluoro-4-hydroxybenzaldehyde with $K_2Cr_2O_7$ in H_2SO_4 (37% yield) [152].

The method suggested is based on the data reported in [205].

Procedure

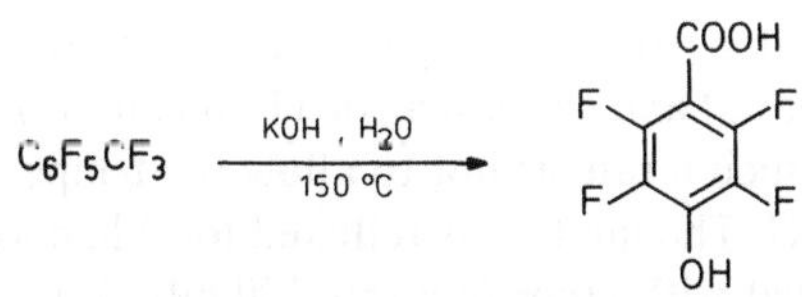

Octafluorotoluene (50 g, 0.212 mol) and a solution of KOH (70 g, 1.25 mol) in 350 ml of water are heated in a 0.5-l rotating steel autoclave at 145°–150 °C during 6 h. The autoclave is allowed to cool, the reaction mixture is transferred from the autoclave into a beaker (the remaining water is washed off the autoclave walls with water) and acidified with diluted H_2SO_4 to pH ~ 2. Then it is extracted with ether (3 × 250 ml), washed with 150 ml of water and dried over $MgSO_4$. The ether is removed on a rotary vacuum-evaporator. This gives 44 g of the crude product, which is then dried over P_2O_5 and recrystallized from benzene. The yield of 2,3,5,6-tetrafluoro-4-hydroxybenzoic acid monohydrate is 40 g (90%); mp 154 °C (Note).

Note. The ether extract should be dried thoroughly, since 2,3,5,6-tetrafluoro-4-hydroxybenzoic acid is hygroscopic.

3,4,5,6-Tetrafluoroanthranylic Acid

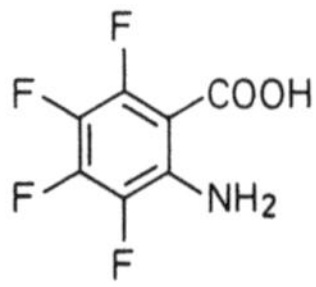

$C_7H_3F_4NO_2$ MW 209

Tetrafluoroanthranylic acid is a white crystalline substance. Mp 136.5°–137.5 °C [187], 141°–142 °C [189–191], 143°–144 °C [209].

Method of Preparation

Tetrafluoroanthranylic acid is prepared by the reaction of sodium hypobromite with tetrafluorophthalimide (31% yield) [189–191]; by the Schmidt reaction, from tetrafluorophthalic acid (67% yield) [187]; by the hydrolysis of 2-amino-3,4,5,6-tetrafluorobenzamide (92% yield); by metallation of 2,3,4,5-tetrafluoroaniline with butyllithium with subsequent carboxylation (19% yield) [208].

The procedure suggested here is a modification of the one described in [187].

Procedure

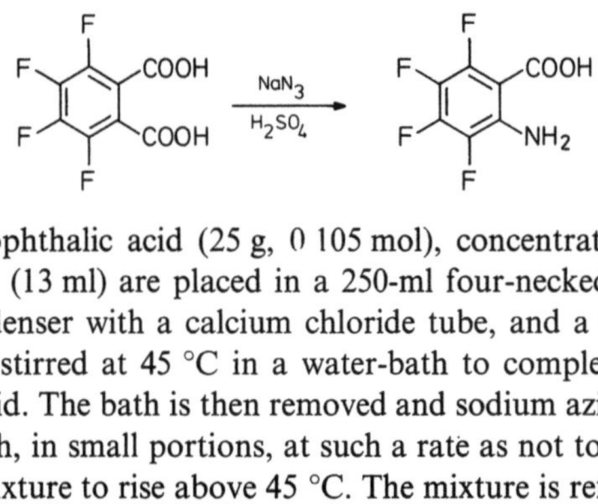

Tetrafluorophthalic acid (25 g, 0 105 mol), concentrated H_2SO_4 (97.5 ml) and dry chloroform (13 ml) are placed in a 250-ml four-necked flask fitted with a stirrer, a reflux condenser with a calcium chloride tube, and a thermometer. The mixture is vigorously stirred at 45 °C in a water-bath to complete dissolution of tetrafluorophthalic acid. The bath is then removed and sodium azide (13.6 g, 0.21 mol) is added during 1.5 h, in small portions, at such a rate as not to allow the temperature of the reaction mixture to rise above 45 °C. The mixture is refluxed for 6 h, cooled, poured onto 200 g crushed ice and extracted with ether (8 × 150–200 ml). The ether extract is dried over $MgSO_4$ and the ether is distilled off. The residue is made mildly alkaline by dissolving it in 10% $NaHCO_3$ (~300 ml) and filtered. The filtrate is acidified with concentrated HCl. The precipitate is filtered off and dried. The yield of tetrafluoroanthranylic acid is 16–17 g (73%–77%); mp 138.5°–139.5 °C.

3,4,5,6-Tetrafluoro-2-anilinobenzoic Acid

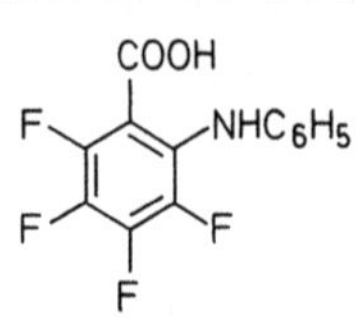

$C_{13}H_7F_4NO_2$ MW 285

3,4,5,6-Tetrafluoro-2-anilinobenzoic acid is a white crystalline substance, soluble in ether, alcohol, and diluted alkaline solutions. Mp 195.5°–197 °C [210].

Method of Preparation

3,4,5,6-Tetrafluoro-2-anilinobenzoic acid is prepared by the reaction of pentafluorobenzoic acid with a Grignard reagent obtained from aniline and ethylmagnesium bromide in ether (59% yield) [210].

The following method is based on the data reported in [210].

Procedure

$$C_6H_5NH_2 \xrightarrow[\text{ether}]{C_2H_5MgBr} C_6H_5NHMgBr$$

$$C_6F_5COOH \xrightarrow[\text{ether}]{C_6H_5NHMgBr} o\text{-}C_6H_5NH\text{—}C_6F_4COOH$$

In a 1-l three-necked flask equipped with a stirrer, a pressure-equalizing dropping funnel, and a reflux condenser with a calcium chloride tube, is placed 7.5 g (0.326 mol) of magnesium turnings. Magnesium is activated with iodine in a conventional way and 300 ml of ether is added to it. Then a solution of 43 g (0.314 mol) of ethyl bromide in 100 ml of ether is dropped in with vigorous stirring. After magnesium has dissolved, a solution of freshly distilled aniline (23 g, 0.25 mol) in 200 ml of ether is added at 20 °C during 0.5 h. The mixture is stirred for 15 min, whereupon pentafluorobenzoic acid (21.2 g, 0.1 mol) in 200 ml of ether is added to it during about 0.5 h at such a rate that the temperature of the mixture should not rise above 20 °C. The mixture is refluxed for 4 h, then cooled, poured onto a mixture of 500 g of ice and 250 ml of 10% HCl and extracted with ether (3 × 200 ml). The ether extract is washed with water and dried over $MgSO_4$. The ether is removed on a rotary vacuum-evaporator, the residue is dissolved while heating in 250 ml of 10% aqueous NH_3 and filtered. The filtrate is acidified with 10% HCl till it is neutral. The precipitate is filtered off on a Büchner funnel, washed with 250 ml of water and dried to give 19 g of the product, which is then recrystallized from 50% aqueous alcohol. The yield of 3,4,5,6-tetrafluoro-2-anilinobenzoic acid is 17–18 g (60%–64%); mp 195.5°–197 °C.

2,3,5,6-Tetrafluoro-4-aminobenzoic Acid

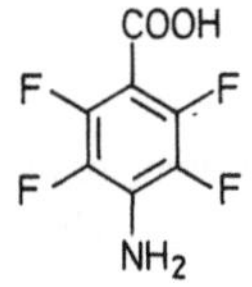

$C_7H_3F_4NO_2$ MW 209

2,3,5,6-Tetrafluoro-4-aminobenzoic acid is a white crystalline substance, partially soluble in most organic solvents, well soluble in ether and tetrahydrofuran. Mp 182°–184 °C [207].

Method of Preparation

2,3,5,6-Tetrafluoro-4-aminobenzoic acid is prepared by the reaction of n-butyllithium with 2,3,5,6-tetrafluoroaniline in tetrahydrofuran with subsequent carbonization of the product (36.9% yield) [207].

The following procedure is based on the data reported in [207].

Procedure

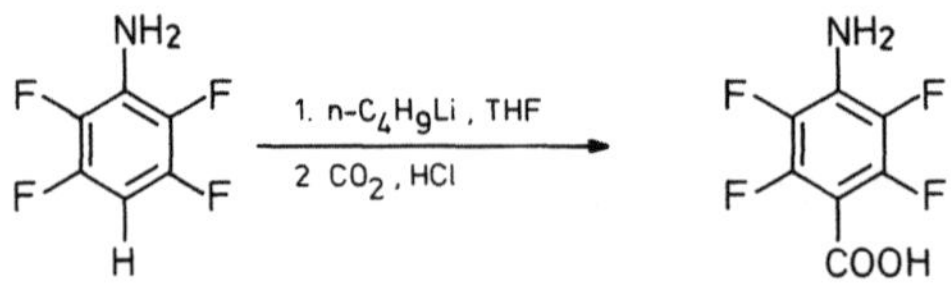

In a 1-l four-necked flask fitted with a stirrer, a thermometer, a bubbler, and a pressure-equalizing dropping funnel with a calcium chloride tube, are placed 2.1 g (0.3 mol) of finely divided lithium and 250 ml of absolute hexane. n-Butyl bromide (41.4 g, 0.3 mol) is added to the stirred mixture at 20 °C; the flask is flushed with argon. After 0.5 h, 300 ml of absolute tetrahydrofuran is added and the reaction mixture is cooled to −70 °C. 2,3,5,6-Tetrafluoroaniline (16.5 g, 0.1 mol) in 40 ml of anhydrous tetrahydrofuran is added slowly during 0.5 h and the mixture is stirred at that temperature for 3 h. A flow of CO_2 is bubbled through the mixture at the same temperature over a period of 4 h, whereupon the temperature of the mixture is slowly raised to 20 °C during 1 h, with continuous passing of CO_2. Then 250 ml of 3 N HCl is added, the organic layer is separated and the aqueous layer extracted with ether (3 × 100 ml). The extracts are dried over $MgSO_4$. The solvents are distilled off on a rotary vacuum-evaporator and the residue recrystallized from benzene. The yield of 2,3,5,6-tetrafluoro-4-aminobenzoic acid is 8.5–9 g (40.7%–43%); mp 182°–184 °C.

Pentafluorophenylacetic Acid

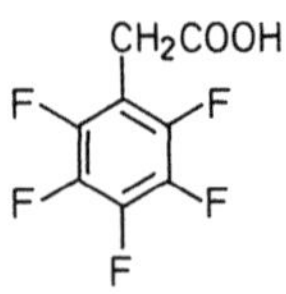

$C_8H_3F_5O_2$ MW 226

Pentafluorophenylacetic acid is a colourless crystalline substance, soluble in ether, alcohol, diluted alkaline solutions; pK_a 3.48 (water, 25 °C) [211]. Mp 103°–105 °C [212, 213], 108°–110 °C [113], 109 °C [103].

Method of Preparation

Pentafluorophenylacetic acid is prepared by carboxylation of a Grignard reagent obtained from pentafluorobenzyl bromide (14% yield) [113], or by the hydrolysis of pentafluorophenylacetonitrile (87% yield) [103], or ethyl pentafluorophenylcyanoacetate with subsequent decarboxylation [212, 213].

The following procedure is a modification of the one described in [103].

Procedure

$$C_6F_5CH_2OH \xrightarrow[CCl_4]{PCl_5} C_6F_5CH_2Cl \xrightarrow{NaCN} C_6F_5CH_2CN \xrightarrow[H_2SO_4,\ CH_3CO_2H]{H_2O:} C_6F_5CH_2COOH$$

2,3,4,5,6-Pentafluorobenzyl chloride. In a 2-l three-necked flask provided with a stirrer and a reflux condenser with a calcium chloride tube connected with an alkaline trap for HCl, are placed melted pentafluorobynzyl alcohol (159.5 g, 0.8 mol) and anhydrous carbon tetrachloride (500 ml); then ground PCl_5 (208 g, 1 mol) is added in portions through a delivery funnel, with vigorous stirring. The reaction proceeds with evolution of heat and HCl vapours. The reaction mixture is refluxed for 1 h, cooled to 60°–70 °C, and slowly poured onto 900 g of ice. Then a solution of Na_2CO_3 (400 g) in water (1800 ml) is added. It is stirred at room temperature for 4–5 h until the gas is no longer evolved. The lower organic layer is separated and the aqueous layer is extracted with carbon tetrachloride (4 × 250 ml). The extracts are combined, washed twice with water and dried over $MgSO_4$. Carbon tetrachloride is distilled off in an oil-bath (the temperature of the bath at the end of this procedure is 110° to 115 °C) and the residue is distilled in vacuum from the Favorsky flask (five or six sets of indentations). A fraction boiling at 53°–55 °C/12 mm Hg is collected. The yield of pentafluorobenzyl chloride is 112–115 g (65%–67%); n_D^{18} 1.4440 (bp 160° to 163 °C/760 mm Hg [214], 156°–157 °C/750 mm Hg [113], 80 °C/47 mm Hg [103]; n_D^{20} 1.4435 [103]).

Pentafluorophenylacetonitrile. In a 700-ml three-necked flask provided with a stirrer, a reflux condenser, and a thermometer, is placed 32.8 g (0.68 mol) of NaCN (Note 1) in 160 ml of water, whereupon ethanol (260 ml) and pentafluorobenzyl chloride (109.4 g, 0.51 mol) are sequentially added to it. The reaction mixture is thoroughly stirred to homogeneity. *Vigorous stirring to the end of the process is required.* The mixture is heated in a water-bath at 70 °C for 1 h (it should not boil), during which time it gradually darkens and acquires a dark-brown colour. Then it is poured onto 1.5 kg of ice and allowed to cool. The cooled solution is extracted with ether (4 × 250 ml). The ether extracts are combined, washed three or four times with water until neutral to the universal test paper, and dried over $CaCl_2$ *(not over $MgSO_4$)*. The ether is removed on a rotary vacuum-evaporator, the residue is distilled in vacuum and a fraction boiling at 100°–102 °C/12 mm Hg is collected. The yield of pentafluorophenylacetonitrile is 85–88 g (81%–84%); n_D^{22} 1.4440 (Note 2) (bp 102°–106 °C per 12 mm Hg, n_D^{20} 1.4412 [215]).

Pentafluorophenylacetic acid. In a 1-l three-necked flask fitted with a stirrer, a reflux condenser, and a thermometer, are placed 85 ml of concentrated H_2SO_4 and 85 ml of glacial acetic acid. Pentafluorophenylacetonitrile (85 g, 0.4 mol) is then added and the mixture is heated at 125°–127 °C for 1 h. Then the mixture is poured, while still hot, onto 400 g of ice. The precipitate is filtered off on a Büchner funnel, washed with ice-water, filtered off and dried in air. This procedure gives 80 g of the product. After recrystallization from petroleum ether (bp 70°–100 °C) the yield of pentafluorophenylacetic acid is 76–79 g (83%–85%); mp 108°–109 °C.

Notes

1. Potassium cyanide may be used as an alternative. The reaction, decomposition and extraction should be carried out under a good hood. The gloves, goggles, apron, washings and all the glassware that has been in contact with the reaction mixture, are washed with 10% solution of ferrous sulphate (1.5–2 l). All solutions neutralized with ferrous sulphate are mixed with 10% solution of ferric chloride and poured into a special bottle.
2. Pentafluorophenylacetonitrile darkens after having stood at room temperature for 1 day. It should be kept at 5°–10 °C in a tightly stoppered glass bottle.

Pentafluorophenylmalonic Ester

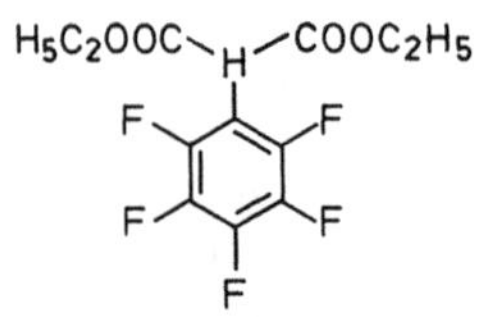

$C_{13}H_{11}F_5O_4$ MW 236

Pentafluorophenylmalonic ester is a colourless liquid with a specific odour. Bp 126 °C /0.5 mm Hg [212], 126°–127 °C/3.5 mm Hg; n_D^{20} 1.4372 [216].

Method of Preparation

Pentafluorophenylmalonic ester is prepared by the reaction of hexafluorobenzene with sodium-malonic ester (39% yield) [212, 216].

The following procedure is a modification of the one described in [212].

Procedure

$$C_6F_6 + CH_2(COOC_2H_5)_2 \xrightarrow{\text{NaH; DMF}} C_6F_5CH(COOC_2H_5)_2$$

In a 350-ml four-necked flask provided with a stirrer, a reflux condenser with a calcium chloride tube, a dropping funnel, and a thermometer, are placed sodium hydride (8.4 g, 0.348 mol) and absolute dimethylformamide (180 ml), then the distilled malonic ester (51.6 g, 0.324 mol) is dropped in with stirring. After hydrogen has ceased to evolve, hexafluorobenzene (60 g, 0.324 mol) is slowly added. The reaction mixture is heated at 130 °C for 6 h, then cooled, poured into 250 ml of 5% solution of HCl and extracted with ether (5 × 100–120 ml). The extract is washed with water (3 × 70 ml) and dried over $MgSO_4$. The ether is distilled off in a water-bath, then unchanged hexafluorbenzene is removed by distillation in an oil-bath. The residue is distilled in vacuum and a fraction boiling at 116°–118 °C/1.5–2 mm Hg is collected. The yield of pentafluorophenylmalonic ester is 43–46 g (40%–43%); based on the unconverted hexafluorobenzene (~26 g): 68%–71%; n_D^{24} 1.4362.

2,3,4,5,6-Pentafluorocinnamic Acid

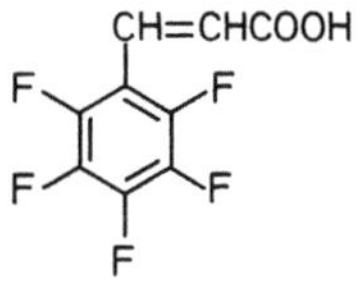

$C_9H_3F_5O_2$ MW 238

2,3,4,5,6-Pentafluorocinnamic acid is a white crystalline substance. Mp 152 °C, 152.5 °C [103, 217].

Method of Preparation

Pentafluorocinnamic acid is prepared by means of the Perkin reaction, from pentafluorobenzaldehyde, acetic oxide, and potassium acetate, or by means of Knevenagel reaction, from pentafluorobenzaldehyde and malonic acid [103, 217]. Yields range from 36% to 44%.

The following procedure is a modification of the one described in [103].

Procedure

$$C_6F_5CHO \xrightarrow[-H_2O]{CH_2(COOH)_2} C_6F_5CH{=}C(COOH)_2 \xrightarrow[-CO_2]{H_2SO_4}$$

$$C_6F_5CH{=}CHCOOH$$

In a 100–150 ml three-necked flask fitted with a stirrer, a reflux condenser with a calcium chloride tube and a thermometer, are placed pentafluorobenzaldehyde (22.5 g, 0.114 mol), malonic acid (36 g, 0.327 mol) and absolute pyridine (50 ml). The mixture is heated to 100 °C with stirring and allowed to stand for 1.5–2 h, then cooled, poured into a mixture of crushed ice (100 g) and concentrated HCl (50 ml) and extracted with ether (4 × 30 ml). The ether extract is dried over $MgSO_4$ and the ether is distilled off in a water-bath. Concentrated H_2SO_4 (50 ml) is added to the residue, the mixture is heated to 100 °C and maintained at that temperature for 15 min, then cooled and poured onto 200 g of ice. The white precipitate is filtered off and dissolved in 5% Na_2CO_3 solution. The solution is filtered and the filtrate is acidified with concentrated H_2SO_4. A white precipitate is filtered off, washed with water and dried in air. The yield of pentafluorocinnamic acid is 11–12 g (41%–44%); mp 149°–151 °C.

Dimethyl 2,2′,3,3′,5,5′,6,6′-octafluorodiphenyl-4,4′-dicarboxylic Ester

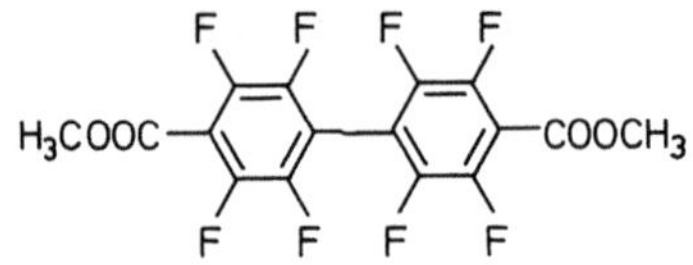

$C_{16}H_6F_8O_4$ MW 404

Dimethyl octafluorodiphenyl-4,4′-dicarboxylic ester is a colourless crystalline substance, soluble in common organic solvents, slightly soluble in water. Mp 98°–100 °C [218].

Method of Preparation

Dimethyl octafluorodiphenyl-4,4′-dicarboxylic ester is prepared by the reaction of methyl pentafluorobenzoate with hexaethyltriamidophosphite at 20 °C (60% yield) [218].

The following procedure is based on the data reported in [218].

Procedure

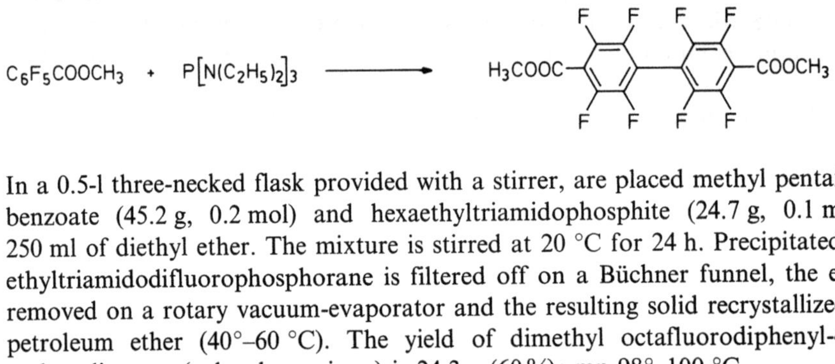

In a 0.5-l three-necked flask provided with a stirrer, are placed methyl pentafluorobenzoate (45.2 g, 0.2 mol) and hexaethyltriamidophosphite (24.7 g, 0.1 mol) in 250 ml of diethyl ether. The mixture is stirred at 20 °C for 24 h. Precipitated hexaethyltriamidodifluorophosphorane is filtered off on a Büchner funnel, the ether is removed on a rotary vacuum-evaporator and the resulting solid recrystallized from petroleum ether (40°–60 °C). The yield of dimethyl octafluorodiphenyl-4,4′-dicarboxylic ester (colourless prisms) is 24.3 g (60%); mp 98°–100 °C.

Pentafluorophenyol

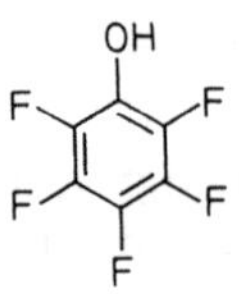

C_6HF_5O MW 184

Pentafluorophenol is a colourless crystalline substance with an odour resembling that of phenol; highly hygroscopic and soluble in water [16]; dissociation constant

$4.79 \cdot 10^{-6}$ at 17.5 °C [219]. Bp 72°–73 °C/48 mm Hg, 144°–146 °C [115], 142°–144 °C [219], 143 °C [16], 144°–145 °C [40, 220], 145–146 °C/760 mm Hg [112]. Mp 25° to 30 °C [16], 38.5°–39.5 °C [219]; n_D^{26} 1.4263 [115], n_D^{20} 1.4270 [16].

Method of Preparation

Pentafluorophenol is prepared by the reaction of hexafluorobenzene with potassium hydroxide in absolute pyridine (52% yield) [115], in *tert*-butyl alcohol (71% yield) [16], or in water (80%–85% yield) [40, 220]; by demethylation of pentafluoroanisole with HI [115] or HBr (yields up to 20%) [219], or with anhydrous aluminium chloride (58% yield) [219]; or by the pyrolysis of hexafluorobenzene in the presence of water [76].

The following procedure is a modification of the preparations described in [40, 220].

Procedure

$$C_6F_6 \xrightarrow{KOH;\ H_2O} C_6F_5OH$$

In a 1-l rotating steel autoclave are placed hexafluorobenzene (95 ml, 0.8 mol) and a solution of KOH (100 g) in water (600 ml), and the mixture is heated at 145°–150 °C during 8 h, then cooled, transferred into a beaker (residual mixture is washed off the walls of the autoclave with water) and acidified with concentrated HCl to pH ~ 2. The organic layer is separated and the aqueous layer extracted with ether (5 × 200 ml). The organic layer and the ether extracts are combined, washed with water and dried over $MgSO_4$ (Note 1). The ether is distilled off in a water-bath. The residue (140 to 150 g), which contains, as shown by GLC, 81%–86% of pentafluorophenol, is distilled on a fractionating column with a glass packing (12–15 theoretical plates). A fraction boiling at 143°–145 °C is collected. The yield of pentafluorophenol is 90–96 g (60% to 65%) (Note 2).

Notes

1. The ether extract should be thoroughly dried, as pentafluorophenol is hygroscopic.
2. Pentafluorophenol should be kept in the dark, in a tightly stoppered glass or polyethylene bottle with a paraffin-coated cap.

2,3,5,6-Tetrafluoro-4-trifluoromethylphenol

(Perfluoro-para-cresol).

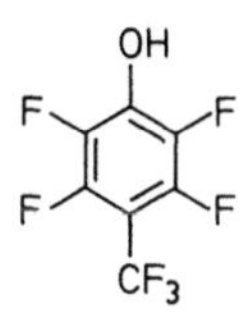

C_7HF_7O MW 234

2,3,5,6-Tetrafluoro-4-trifluoromethylphenol is a colourless liquid with a faint phenol-like odour; quickly darkens when exposed to air; soluble in common organic solvents, well soluble in diluted alkaline solutions, partially soluble in water; highly hygroscopic. Bp 142°–144 °C [221], 142 °C [222].

Method of Preparation

2,3,5,6-Tetrafluoro-4-trifluoromethylphenol is obtained by the reaction of octafluorotoluene with KOH in tert-butyl alcohol (57%–75% yield) [221–223]; by the reaction of octafluorotoluene with lithium *tert*-butoxide in *tert*-butyl alcohol (72% yield) [221]; or by treatment of 2,3,5,6-tetrafluoro-1-ethoxy-4-trifluoromethylbenzene with bases [206].

The following procedure is based on the data reported in [221].

Procedure

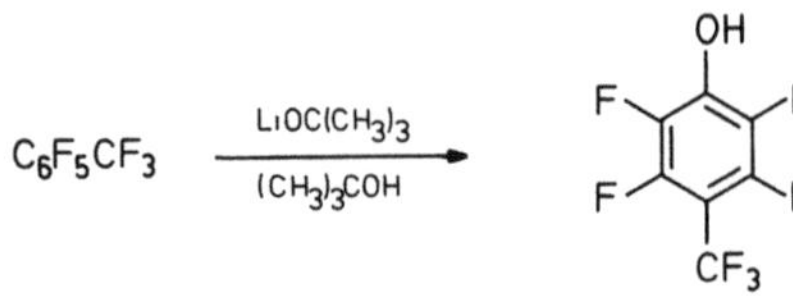

Absolute *tert*-butyl alcohol (600 ml) is placed in a 1-l three-necked flask provided with a stirrer, a reflux condenser with a calcium chloride tube, and a pressure-equalizing dropping funnel with a gas inlet tube. Finely divided metallic lithium (6.6 g, 0.95 mol) is added portionswise, with vigorous stirring and flushing of the flask with argon. After all lithium has been dissolved, octafluorotoluene (200 g, 0.85 mol) is added. The reaction mixture is refluxed for 15 h, then cooled and decanted to remove precipitated LiF. The stirrer is replaced with a Wurtz column, the reflux condenser is removed and *tert*-butyl alcohol is distilled off in an oil-bath. After 450 ml of alcohol has been distilled off, the temperature of the bath is raised to 180°–190 °C and 2,3,5,6-tetrafluoro-4-trifluoromethylphenol is distilled. A fraction boiling at 130°–165 °C is collected. Then the product is redistilled and a fraction boiling at 163°–164 °C is collected. The yield of 2,3,5,6-tetrafluoro-4-trifluoromethylphenol is 107 g (54%) (Note).

Note. 2,3,5,6-Tetrafluoro-4-trifluoromethylphenol darkens after having been exposed to air for a few days. It should be kept in a tightly stoppered dark bottle.

Octafluoro-4,4′-dihydroxybiphenyl

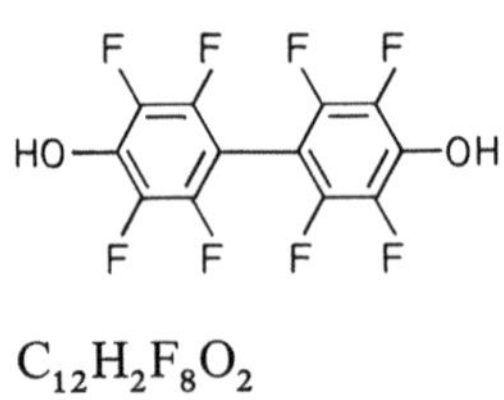

$C_{12}H_2F_8O_2$ MW 330

Octafluoro-4,4′-dihydroxybiphenyl is a white crystalline substance, soluble in common organic solvents and diluted alkaline solutions. Mp 198.5°–199.5 °C [55].

Method of Preparation

Octafluoro-4,4′-dihydroxybiphenyl is prepared by the reaction of decafluorobiphenyl with KOH in *tert*-butyl alcohol (68% yield) or in water (85% yield) [55].

The following method is based on the data reported in [55].

Procedure

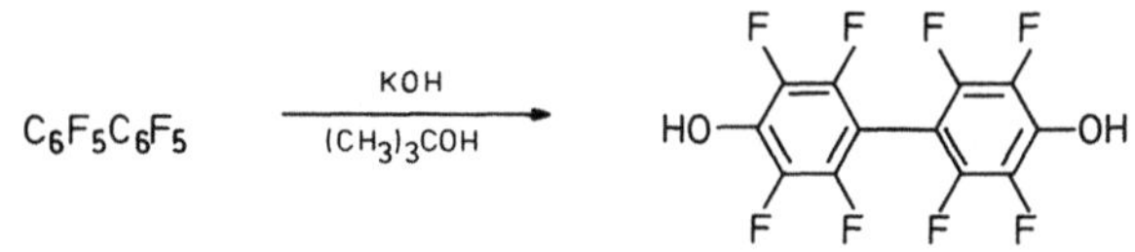

Decafluorobiphenyl (50 g, 0.15 mol), KOH (75 g, 1.34 mol) and *tert*-butyl alcohol (800 ml) are placed in a 1.5-l round-bottom flask equipped with a reflux condenser. The mixture is refluxed during 1 h, then 600 ml of water is added and an aqueous solution of *tert*-butyl alcohol is distilled off (Note 1). The reaction mixture is cooled, filtered and extracted with ether (2 × 200 ml) (Note 2). The aqueous layer is acidified with 6N HCl and extracted with ether (3 × 250 ml). The extract is dried over $MgSO_4$. Removal of the ether on a rotary vacuum-evaporator affords 40 g of the product which is then recrystallized from benzene. The yield of octafluoro-4,4′-dihydroxybiphenyl is 33–35 g (67%–71%); mp 198.5°–199.5 °C.

Notes

1. Distillation to remove aqueous *tert*-butyl alcohol is stopped when the temperature has increased up to 99°–100 °C.
2. The ether extraction is conducted to remove unchanged decafluorobiphenyl.

Bis(2,3,5,6-tetrafluoro-4-hydroxyphenyl)methane

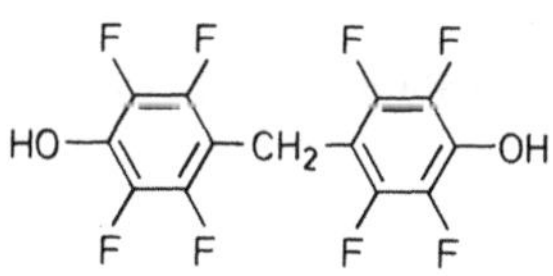

$C_{13}H_4F_8O_2$ MW 344

Bis(2,3,5,6-tetrafluoro-4-hydroxyphenyl)methane is a white crystalline substance, soluble in common organic solvents and diluted alkaline solutions. Mp 174°–176 °C [225].

Method of Preparation

Bis(2,3,5,6-tetrafluoro-4-hydroxyphenyl)methane is prepared by the reaction of equimolar amount of formaldehyde with 2,3,5,6-tetrafluorophenol in concentrated mineral acids (75%–90%) [225].

The following procedure is based on the data reported in [225].

Procedure

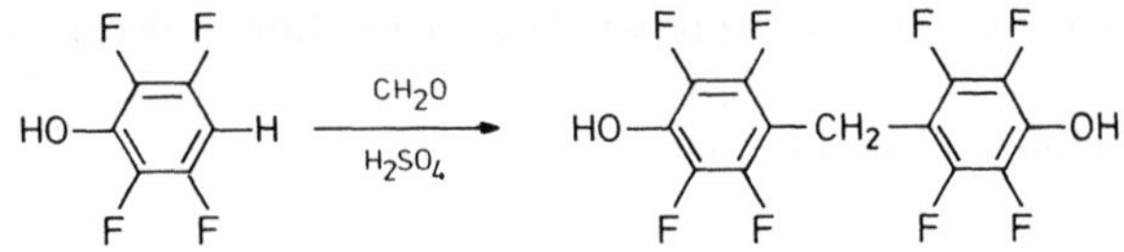

In a 0.5-l three-necked flask provided with a stirrer, a pressure-equalizing dropping funnel with a calcium chloride tube, and a thermometer, are placed 20 g (0.667 mol) of paraform in 120 ml of concentrated H_2SO_4 (d 1.84). The mixture is cooled to 0° to 5 °C and 100 g (0.603 mol) of 2,3,5,6-tetrafluorophenol in 160 ml of concentrated H_2SO_4 (d 1.84) is added in portions, with stirring, during 1 h. The reaction mixture is stirred at 5°–10 °C for 2 h and at 20 °C for 12 h, then poured into 700 ml of water. The precipitate is filtered off on a Büchner funnel, washed with water and dried to give a light-brown product which is dissolved in 1 l of benzene and, after an addition of 50 g of activated carbon, refluxed during 0.5 h. The solution is filtered and the filter residue is washed with hot benzene (100 ml). The combined benzene solutions are evaporated on a rotary vacuum-evaporator to give 98 g of the product which is then recrystallized from benzene. The yield of bis(2,3,5,6-tetrafluoro-4-hydroxyphenyl)methane is 90–92 g (88%–90%); mp 174°–176 °C.

3,4,5,6-Tetrafluoropyrocatechol

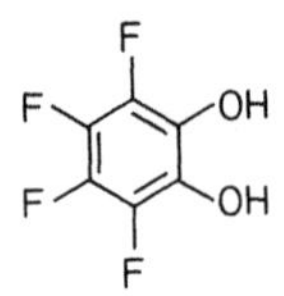

$C_6H_2F_4O_2$ MW 182

Tetrafluoropyrocatechol is a white crystalline substance; hygroscopic; sublimes at 60°–80 °C/15 mm Hg. Mp 67.5°–69 °C [168].

Method of Preparation

Tetrafluoropyrocatechol is obtained by dealkylation of 5,6,7,8-tetrafluoro-1,4-benzodioxan, which is heated with $AlCl_3$ in benzene (57% yield) [168].

Procedure [168]

2-Pentafluorophenoxyethyl alcohol. Hexafluorobenzene (157 g, 98 ml) (0.84 mol), KOH (46 g) and ethylene glycol (500 ml) are placed in a 1-l round-bottom flask fitted with a reflux condenser with a calcium chloride tube. The mixture is heated for 1 h at 80 °C, for 1 h at 90 °C, and for 1 h at 110 °C, then cooled, poured into 1.5 l of water and acidified with H_2SO_4. The white flaky precipitate is filtered off and the filtrate extracted with ether. The ether extract is dried over $MgSO_4$, the ether is distilled off in a water-bath and the residue distilled in vacuum. A fraction boiling at 110 °C per 15 mm Hg is collected. The yield of 2-pentafluorophenoxyethanol is 82–83 g (43%–44%); n_D^{23} 1.4411.

5,6,7,8-Tetrafluoro-1,4-benzodioxan. 2-Pentafluorophenoxyethanol (55 g, 0.24 mol) and anhydrous K_2CO_3 (60 g) in 360 ml of anhydrous dimethylformamide are refluxed during 20 h in a 0.6-l round-bottom flask provided with a reflux condenser with a calcium chloride tube. The mixture is then cooled, poured into 1.5 l of water and acidified with concentrated H_2SO_4. After 3 h the white crystalline precipitate is filtered off and dried. The yield of 5,6,7,8-tetrafluoro-1,4-benzodioxan is 22–23 g (45%–46%); mp 78°–79 °C (mp 78°–79 °C [168, 226]).

Tetrafluoropyrocatechol. 5,6,7,8-Tetrafluoro-1,4-benzodioxan (10 g) and anhydrous $AlCl_3$ (40 g) in 350 ml of absolute benzene are refluxed during 6 h in a 0.6-l round-bottom flask. The mixture is then cooled, poured into a beaker with ice and evaporated to remove benzene. The precipitate is filtered off, the filtrate is acidified with diluted H_2SO_4 and extracted with ether. The ether extract is dried over $MgSO_4$, the ether is removed on a rotary vacuum-evaporator and the crystallized residue sublimed at 100 °C/20 mm Hg. The yield of tetrafluoropyrocatechol is 5–5.7 g (57%–65%); mp 68°–69 °C (Note).

Note. Tetrafluoropyrocatechol darkens after having stood in air for some days. It should be kept in a tightly stoppered glass bottle.

2,4,5,6-Tetrafluoroesorcine

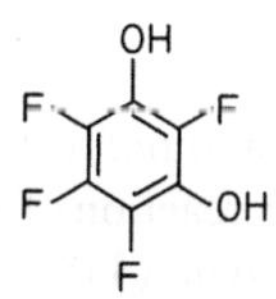

$C_6H_2F_4O_2$ MW 182

Tetrafluororesorcine is a white crystalline substance, soluble in water [220]. Bp 217° to 218 °C/760 mm Hg [220]; mp 95°–96 °C. On standing exposed to air it forms a hydrate, mp 72°–75 °C [220].

Method of Preparation

Tetrafluororesorcine is obtained by heating hexafluorobenzene with an aqueous solution of KOH at 150 °C and increased pressure (29% yield) [220].

The following method involves heating of pentafluorophenol with an aqueous solution of KOH under atmospheric pressure.

Procedure

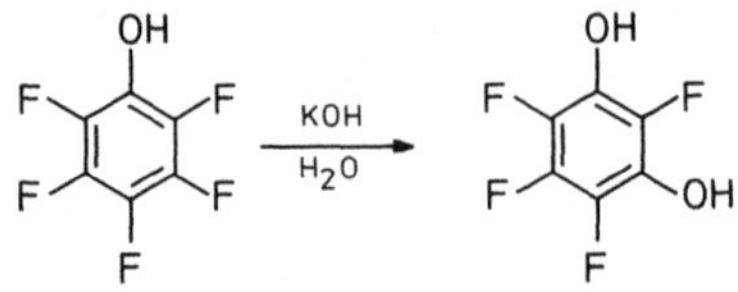

In a 250-ml three-necked flask fitted with a stirrer, a reflux condenser and a thermometer, are placed pentafluorophenol (27.6 g, 0.15 mol) and KOH (50.6 g, 0.9 mol) in 35 ml of water. The reaction mixture is heated in an oil-bath at 165°–170 °C with vigorous stirring for 6 h. After it has been cooled, it is poured onto crushed ice, acidified with concentrated HCl and extracted with ether (4 × 150–200 ml). The ether extract is dried over $MgSO_4$, the ether is distilled off in a water-bath and the residue is evaporated to dryness on a rotary vacuum-evaporator at 70 °C/10–15 mm Hg. The precipitate is washed with a small amount of cold carbon tetrachloride, then filtered and dried. The yield of tetrafluororesorcine is 11–13 g (43%–46%); mp 73° to 76 °C (the hydrate).

2,3,5,6-Tetrafluorohydroquinone

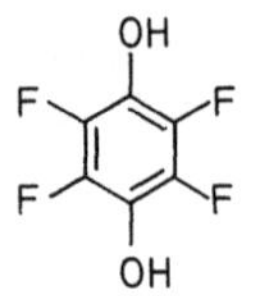

$C_6H_2F_4O_2$ MW 182

Tetrafluorohydroquinone is a white crystalline product, soluble in common organic organic solvents. Mp 161 °C [227], 166°–167 °C [171], 168°–169 °C [173].

Method of Preparation

Tetrafluorohydroquinone is prepared by the reaction of fluoranil with hydrogen over skeletal nickel [173] or over platinum dioxide [171]; by the reaction of tetrafluoro-1,4-di(β-hydroethoxy)benzene with anhydrous $AlCl_3$ in benzene (46% yield) [173], 74% yield [227]).

The following method involves reduction of fluoranil with zinc in hydrochloric acid.

Procedure

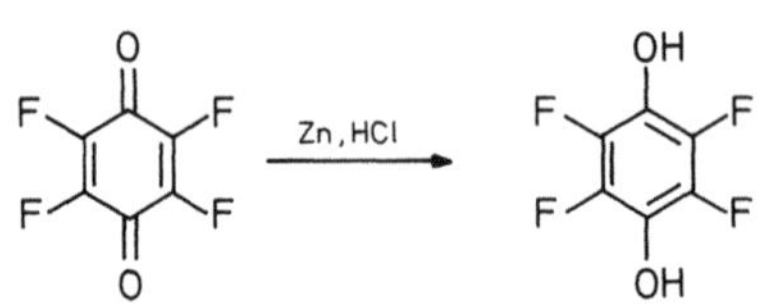

In a 350-ml four-necked flask provided with a stirrer, a reflux condenser with a calcium chloride tube, a thermometer, and a dropping funnel, are placed 30 g of zinc dust and 200 ml of ether. Concentrated HCl (10–15 ml) is then added from the dropping funnel with vigorous stirring; the temperature should not rise above 18 °C. When hydrogen begins to evolve, fluoranil (5 g, 0.028 mol) is added in small portions to the suspension, over a period of 1.5 h; simultaneously, further 150 ml of concentrated HCl is added dropwise. Stirring is continued for 4 to 6 h and the reaction mixture is allowed to stand overnight, whereupon it is filtered. The filtrate is extracted with benzene (3–4 × 20 ml), the extract is washed with water till it is neutral, and dried over $MgSO_4$. The main part of benzene is distilled off, the crystalline precipitate is filtered off and dried. The yield of tetrafluorohydroquinone is 4.6–4.8 g (92%–96%); mp 169°–170 °C.

3,4,5,6-Tetrafluoro-2-aminophenol

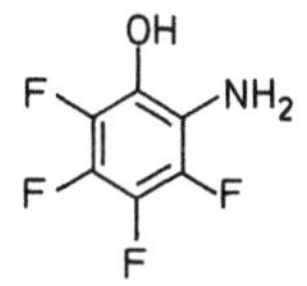

$C_6H_3F_4NO$ MW 181

3,4,5,6-Tetrafluoro-2-aminophenol is a colourless crystalline substance, soluble in common organic solvents. It is quickly oxidized and darkens when exposed to air. Mp 97.5°–99 °C [228], 91°–92 °C [229]; sublimes at 80 °C/0.1 mm Hg [228].

Method of Preparation

3,4,5,6-Tetrafluoro-2-aminophenol is prepared by the reduction of 3,4,5,6-tetrafluoro-2-nitrophenol with tin in HCl (92% yield) [228], or with hydrogen over palladium black (70% yield) [229].

Procedure

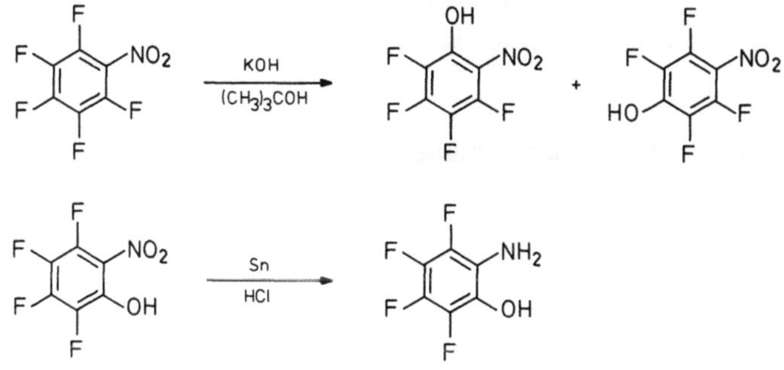

3,4,5,6-Tetrafluoro-2-nitrophenol. Pentafluoronitrobenzene (75 g, 0.35 mol) and dry *tert*-butyl alcohol (450 ml) are placed in a 1-l two-necked flask provided with a stirrer and a thermometer. Finely divided KOH (53 g, 0.95 mol) is added in portions at

20 °C, with continuous stirring during 1 h, at such a rate that the temperature of the reaction mixture should not rise above 20 °C (Note 1). The mixture is stirred at 20 °C for 18 h and at 50 °C for an additional 30 min. Then 300 ml of water is added into the flask, the stirrer is replaced with a Wurtz column and aqueous *tert*-butyl alcohol (about 500 ml) is distilled off. To the residue 250 ml of 6 M solution of H_2SO_4 is carefully added and the mixture is steam-distilled (the amount of the distillate is about 750 ml). The distillate is extracted with ether (4 × 250 ml) (Note 2) and the extract is dried over $MgSO_4$. After the ether has been removed in a water-bath, the residue is distilled in vacuum and a fraction boiling at 78°–79 °C/2–3 mm Hg is collected. The yield of 3,4,5,6-tetrafluoro-2-nitrophenol is 40–41 g (54%–55%); mp 53 °C.

3,4,5,6-Tetrafluoro-2-aminophenol. In a 350-ml three-necked flask provided with a stirrer, a reflux condenser, and a pressure-equalizing dropping funnel, are placed 3,4,5,6-tetrafluoro-2-nitrophenol (30 g, 0.142 mol), tin (60 g, 0.55 mol) and water (50 ml). The mixture is heated till 3,4,5,6-tetrafluoro-2-nitrophenol has melted (~50 °C), whereupon concentrated HCl (150 ml) is added in portions with vigorous stirring during 1 h (Note 3). After the mixture has started to boil violently, it is refluxed during 6 h in a water-bath, then cooled and neutralized with an aqueous 2 M $NaHCO_3$ solution to pH ~6–7 (Note 4), whereupon 200 ml of saturated solution of NaCl is added and the mixture is extracted with ether (5 × 250 ml). The extract is dried over $MgSO_4$, the ether is removed on a rotary vacuum-evaporator and the residue sublimed in vacuum at 80 °C/3 mm Hg. The yield of 3,4,5,6-tetrafluoro-2-aminophenol is 23–24 g (88%–93%); mp 97°–98 °C (Note 5).

Notes

1. The reaction of pentafluoronitrobenzene with the alkali is exothermic.
2. The still-residue contains 2,3,5,6-tetrafluoro-4-nitrophenol. It may be isolated by ether extraction of the still-residue.
3. When the reaction starts, violent heating and foaming of the reaction mixture occurs, so the mixture should be cooled with ice.
4. Neutralization should be done cautiously, as the liberated gas foams the solution with finely divided $SnCl_2$. The procedure takes much time.
5. On standing in air, 3,4,5,6-tetrafluoro-2-aminophenol is quickly oxidized and darkens in a few days. It is advisable to keep it in sealed ampoules.

Pentafluorophenyl-1,2-epoxyethane

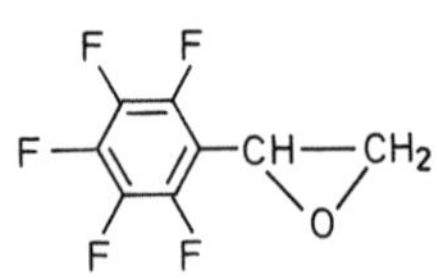

$C_8H_3F_5O$ MW 210

Pentafluorophenyl-1,2-epoxyethane is a liquid, soluble in most organic solvents. Bp 70°–72 °C/15 mm Hg [151].

Method of Preparation

Pentafluorophenyl-1,2-epoxyethane is prepared by the reaction of 2,3,4,5,6-pentafluoro-α-bromoacetophenone with sodium borohydride in an alkaline medium (75% yield) [151].

The following procedure is based on the data reported in [151].

Procedure

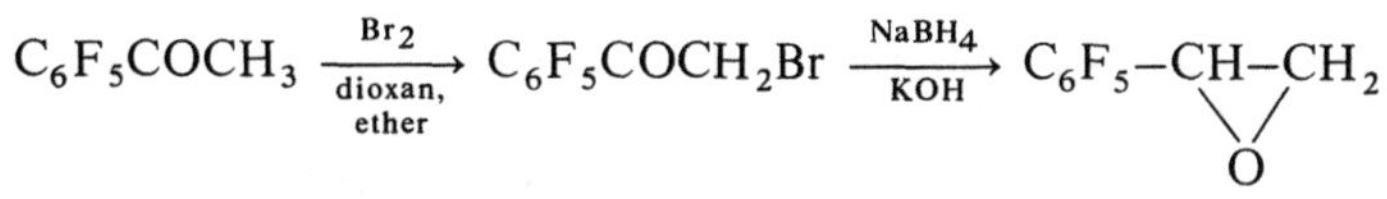

2,3,4,5,6-Pentafluoro-α-bromoacetophenone. In a 350-ml four-necked flask fitted with a stirrer, a reflux condenser with a calcium chloride tube, a thermometer, and a dropping funnel, are placed pentafluoroacetophenone (50 g, 0.24 mol), absolute dioxan (36 ml) and absolute ether (90 ml), and the mixture is heated to 35 °C with stirring. The bath is removed, the mixture is allowed to cool to 25 °C and bromine (38 g, 0.24 mol) is added dropwise during 20–30 min; the temperature is maintained at 25°–27 °C. After all the bromine has been added, the reaction mixture of claret-red colour is allowed to stand at room temperature for further 40 min, then poured into a mixture of 70 ml of ice-water and 230 ml of ether. The ether layer is separated, washed with water (4 × 25 ml), 5% $NaHCO_3$ solution, then again with water, and dried over $MgSO_4$. The ether is distilled off in a water-bath and the residue is distilled on a fractionating column under vacuum. The following fractions are collected: fraction 1 (5–6 g, bp up to 90 °C/10 mm Hg) contains 40% of the starting pentafluoroacetophenone and 60% of pentafluoro-α-bromoacetophenone; fraction 2 (6–8 g, bp 90°–95 °C/10 mm Hg) contains 10% of pentafluoroacetophenone and 90% of pentafluoro-α-bromoacetophenone; fraction 3 (pure pentafluoro-α-bromoacetophenone, 35–38 g (51%–55%), bp 96°–98 °C/10 mm Hg; n_D^{23} 1.4850) (bp 79 °C/5 mm Hg, $n_D^{24.5}$ 1.4820 [154]; mp 21 °C [104]).

Pentafluorophenyl-1,2-epoxyethane. In a 350 ml three-necked flask provided with a stirrer, a thermometer, and a pressure-equalizing dropping funnel with a calcium chloride tube, is placed 35 g (0.12 mol) of pentafluoro-α-bromoacetophenone in 200 ml of dioxan. A solution of $NaBH_4$ (4.8 g, 0.13 mol) in 49 ml of water and 6 ml of an aqueous 2 N KOH solution are then added with stirring during 1 h. The reaction mixture is maintained at 20 °C during 2 h, acidified with 120 ml of 2 N H_2SO_4, poured into 500 ml of water and extracted with diisopropyl ether (3 × 150 ml). An aqueous 2 N KOH solution (150 ml) is then added to the extract and the mixture is distilled into a 1-l two-necked flask equipped with a stirrer and a thermometer. The reaction mixture is heated, with stirring, at 60 °C for 15 min, then at 70 °C for an additional 1 h, cooled and extracted with ether (2 × 70 ml). The ether extract is dried over $MgSO_4$. The solvents are distilled off in a boiling water-bath and the residue is distilled in vacuum. A fraction boiling at 96°–97 °C/17 mm Hg is collected. Then the product is redistilled in vacuum; a fraction boiling at 70°–72 °C/15 mm Hg is collected. The yield of pentafluorophenyl-1,2-epoxyethane is 20–21 g (79%–82%).

2,3,4,5,6-Pentafluoroanisole

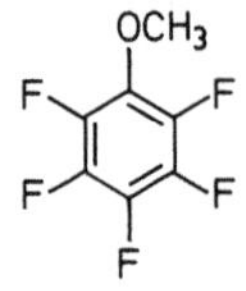

$C_7H_3F_5O$ MW 198

Pentafluoroanisole is a colourless liquid, soluble in common organic solvents. Bp 137°–138 °C/760 mm Hg [230], 138°–139 °C [219], 154°–156 °C [115], 155°–157 °C [10]; d_4^{20} 1.4960 [230]; 1.417 [10]; n_D^{20} 1.4090 [219], 1.4105 [230].

Method of Preparation

Pentafluoroanisole is obtained by the reaction of hexafluorobenzene with sodium methoxide in methanol [10, 219] or in the presence of pyridine (70%–72% yield) [115, 230].

The following procedure is a modification of the one described in [219].

Procedure

$$C_6F_6 \xrightarrow{CH_3ONa;\ CH_3OH} C_6F_5OCH_3$$

In a 3-l three-necked flask equipped with a stirrer and a reflux condenser with a calcium chloride tube, is placed 1700 ml of distilled methanol, and metallic sodium (40 g) is then added in small lumps, with stirring. Hexafluorobenzene (206 g, 1.1 mol, 129 ml) is poured into a solution of sodium methoxide, which has been cooled to room temperature, and the mixture is refluxed for 1 h, then cooled, poured into 2.5 to 3 l of water, and 250 ml of concentrated HCl is added to complete dissolution of the formed precipitate. The mixture is then extracted with ether (four or five times with 300 ml of ether for each 1800 ml of the mixture) (Note 1), the ether extract is washed with 15% Na_2CO_3 solution, with water till it is neutral to litmus, and dried over $MgSO_4$. The ether is removed in a water-bath and the residue distilled on a fractionating column (12–15 theoretical plates). A fraction boiling at 136°–138 °C is collected. The yield of pentafluoroanisole is 130–135 g (60%–62%); n_D^{23} 1.4084 (Note 2).

Notes

1. To ensure good separation of the organic layer from the aqueous one, sodium chloride should be added.
2. The still residue contains a mixture of isomeric tetrafluorodimethoxybenzenes.

Pentafluorothiophenol

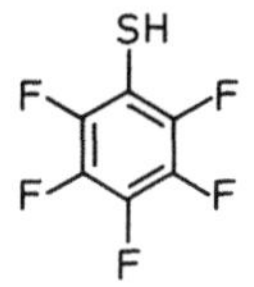

C_6HF_5S MW 200

Pentafluorothiophenol is a colourless liquid with a strong odour of thiophenol; soluble in common organic solvents. Bp 143 °C/760 mm Hg; n_D^{22} 1.4622 [231].

Method of Preparation

Pentafluorothiophenol is prepared by the reaction of hexafluorobenzene with sodium hydrosulphide in ethylene glycol (66% yield) [231].

Procedure

$$C_6F_6 \xrightarrow{KSH} C_6F_5SH$$

In a 350-ml four-necked flask provided with a stirrer, a reflux condenser with a calcium chloride tube, and a pressure-equalizing dropping funnel, is placed 100 ml of 27% KSH solution in anhydrous ethylene glycol (Note 1). Then 40 ml of anhydrous ethylene glycol is added, the solution is heated at 60°–70 °C and a hot (70 °C) solution of 37.2 g (0.2 mol) of hexafluorobenzene in 150 ml of absolute pyridine is added to it with stirring during 10 min. The mixture is heated to 87°–91 °C, allowed to stand for 10 min, poured into 250 ml of water, acidified with concentrated HCl (Note 2) and extracted with ether (5 × 30 ml). The ether extract is dried over $MgSO_4$. The ether is distilled off under a Vigreux column in a water-bath and the residue is distilled at atmospheric pressure. A fraction boiling at 142°–143 °C is collected. The yield of pentafluorothiophenol is 27–28 g (69%–72%); n_D^{22} 1.4622 (Note 3).

Notes

1. KSH solution is obtained by passing hydrogen sulphide into a solution of 56 g of KOH in 155 ml of anhydrous ethylene glycol to saturation (7–8 h).
2. Acidification with concentrated HCl results in the evolution of H_2S and heat, therefore it is necessary to trap the liberated H_2S (in an aqueous 20% NaOH solution) and to cool the flask with an ice-salt bath. A pyridinium salt precipitate may be formed, which is further dissolved in excess HCl.
3. It is advisable to keep the product in sealed ampoules.

Perfluoro(isopropylphenyl)sulphide

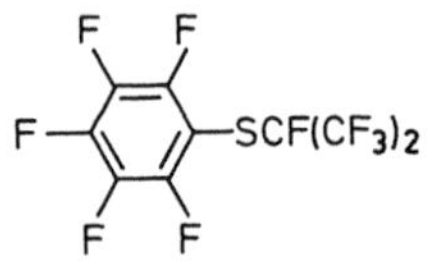

$C_9F_{12}S$ MW 368

Perfluoro(isopropylphenyl)sulphide is a light-yellow liquid, soluble in common organic solvents. Bp 172–173 °C [232].

Method of Preparation

Perfluoro(isopropylphenyl)sulphide is prepared by the reaction of bis(pentafluorophenyl)disulphide with hexafluoropropylene in HF-SbF_5 system (25% yield) [232].

Procedure

$$(C_6F_5S)_2 \xrightarrow{CF_2{=}CFCF_3;\ HF;\ SbF_5} C_6F_5SCF(CF_3)_2$$

In a 200-ml rotating steel autoclave are placed 39.8 g (0.1 mol) of bis(pentafluorophenyl)disulphide and 180 g of SbF_5. The autoclave is cooled to −40° ÷ −50 °C, then anhydrous HF (15 ml) (see p. 1) and hexafluoropropylene (30 g, 0.2 mol) are added and an autoclave is heated at 70 °C for 6 h. The pressure is then reduced to that of the atmosphere, the contents are poured onto 1 kg of ice and extracted with methylene chloride. The extract is dried over $MgSO_4$, methylene chloride is distilled off in a water-bath under a short Vigreux column and the residue is distilled. A fraction boiling at 172°–173 °C is collected. The yield of perfluoro(isopropylphenyl)sulphide is 23–22 g (30%–31%).

Bis(pentafluorophenyl)sulphide

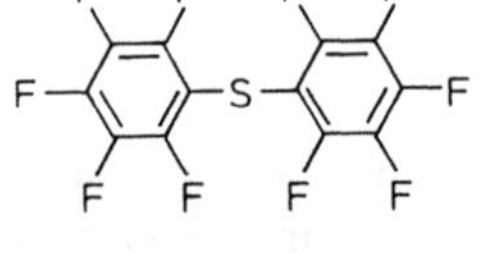

$C_{12}F_{10}S$ MW 366

Bis(pentafluorophenyl)sulphide is a yellowish crystalline substance, soluble in common organic solvents. Mp 84°–86 °C [233], 83°–86 °C [234], 84°–85 °C [235–241], 85° to 86 °C [180], 83 °C [242]; sublimes at 80 °C/0.01 mm Hg [234].

Method of Preparation

Bis(pentafluorophenyl)sulphide is prepared by the reaction of pentafluorophenylmagnesium bromide with SCl_2 in ether (75% yield) [233]; by the reaction of bis(pentafluorophenyl)thallium bromide with sulphur at 190 °C (55% yield) [234]; by treatment of bis(pentafluorophenyl)mercury with sulphur at 250 °C (90% yield) [235]; by the reaction of bromopentafluorobenzene with cuprous pentafluorothiophenoxide in dimethylformamide (84% yield) [180, 236]; by the reaction of iodopentafluorobenzene with sulphur [237, 238]; by the reaction of pentafluorobenzene with sulphur, sulphur chlorides or pentafluorophenylsulphenyl chloride in the presence of SbF_5 (in a quantitative yield) [239, 240]; by the reaction of pentafluorophenylsulphenyl fluoride with pentafluorobenzene at 20 °C in SbF_5 [241]; by the reaction of pentafluorophenylmagnesium bromide with SCl_2 in ether [242]; by the reaction of pentafluorobenzene with bis(pentafluorophenyl)disulphide in SbF_5 (94% yield) [243]; by treatment of bis(pentafluorophenyl)lead with sulphur [244]; by the reaction of pentafluorophenyllithium with SCl_2 (60%–80% yield) [238, 245, 246]; by heating, to the temperature above 0 °C, of tetrakis(pentafluorophenyl)sulphurane formed as an intermediate in the reaction of pentafluorophenyllithium with SF_4 or pentafluorophenylsulphur trifluoride [247]; by the reaction of pentafluorophenyllithium with S_2Cl_2, with bis(pentafluorophenyl)disulphide as a by-product [238]; by heating bis(pentafluorophenyl)selenide with sulphur at 230 °C (in a quantitative yield) [236]; or by the reaction of pentafluorophenylmercuric chloride with sulphur at 230 °C [235, 238].

Procedure

$$C_6HF_5 \xrightarrow[2.\ H_2O]{1.\ \text{sulphur};\ SbF_5} C_6F_5SC_6F_5$$

Finely ground sulphur (3.2 g, 0.1 mol) and freshly distilled SbF_5 (120 g) are placed in a 100-ml three-necked flask provided with a stirrer, a thermometer, and a pressure-equalizing dropping funnel with a calcium chloride tube. The mixture is heated to 90 °C, with stirring for 2 h and cooled to 15 °C. Then pentafluorobenzene (33.6 g, 0.2 mol) is added during 1 h (Note 1) and the mixture is heated at 70 °C for 2 h, cooled and poured onto 0.5 kg of ice (Note 2). The precipitate is filtered off on a Büchner funnel, washed with 350 ml of water, dried, and recrystallized from aqueous ethanol. The yield of bis(pentafluorophenyl)sulphide is 33–34 g (90%–93%); mp 84°–85 °C (Note 3).

Notes

1. The reaction is very exothermic and proceeds extremely violently if the heat is not carried off. It is preferable to add pentafluorobenzene in portions; the temperature of the reaction mixture should not rise above 30 °C.
2. The reaction mixture is a viscous slurry; to transfer it onto ice, a glass blade should be used. When the mixture is added to ice, violent heating occurs, which necessitates vigorous stirring and cooling of the flask with ice-water.
3. Bis(pentafluorophenyl)sulphoxide (~3%) is formed as a by-product.

Pentafluorobenzenesulphonic Acid

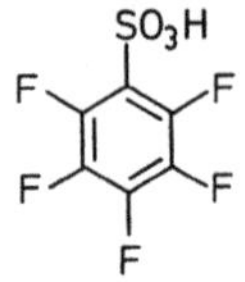

$C_6HF_5O_3S$ MW 248

Pentafluorobenzenesulphonic acid is a white crystalline substance, slightly soluble in common organic solvents, well soluble in water. Mp 32 °C [248].

Method of Preparation

Pentafluorobenzenesulphonic acid is prepared by the reaction of pentafluorobenzene with liquid SO_3 (98% yield) [248]; by the reaction of pentafluorobenzene with 20% oleum [249]; by the hydrolysis of pentafluorobenzenesulphonyl chloride [250].

The following method involves sulphonation of pentafluorobenzene with liquid SO_3 [248].

Procedure

$$C_6HF_5 \xrightarrow{SO_3} C_6F_5SO_3H \cdot H_2O \xrightarrow{-H_2O} C_6F_5SO_3H$$

Pentafluorobenzenesulphonic acid monohydrate. In a 150-ml round-bottom flask provided with an effective reflux condenser, are placed pentafluorobenzene (84 g, 0.5 mol) and freshly distilled SO_3 (48 g, 0.6 mol). The mixture is heated, then cooled, poured onto 400 g of ice, and cautiously neutralized with a saturated aqueous solution of sodium bicarbonate, whereupon $BaSO_4$ is added. Then the mixture is passed through a column (400 × 30 mm) filled with cation exchanger. The eluent is boiled with activated carbon (20 g) during 1 h and the filtrate is evaporated on a rotary vacuum-evaporator. The residue is dried in vacuum at 100 °C/0.1–0.01 mm Hg over a period of 8 h to give 128 g (96%) of pentafluorobenzenesulphonic acid monohydrate, mp 101 °C.

Pentafluorobenzenesulphonic acid. A mixture of pentafluorobenzenesulphonic acid monohydrate (66.5 g, 0.25 mol) and $SOCl_2$ (119 g, 1 mol) is refluxed during 8 h in a 250-ml round-bottom flask fitted with a reflux condenser, whereupon residual $SOCl_2$ is distilled off from the flask. The resulting solid is sublimed at 60 °C/0.1 to 0.01 mm Hg. The yield of pentafluorobenzenesulphonic acid is 61 g (98%); mp 32 °C.

Pentafluorobenzenesulphochloride

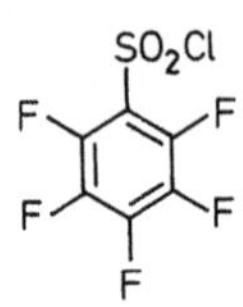

$C_6ClF_5O_2S$ MW 266.5

Pentafluorobenzenesulphochloride is a colourless liquid with a strong odour. Bp 210°–211 °C/760 mm Hg [250].

Method of Preparation

Pentafluorobenzenesulphochloride is prepared by oxidative chlorination of pentafluorothiophenol (60% yield) [250].

The method suggested involves sulphochlorination of pentafluorobenzene with chlorosulphonic acid.

Procedure

Chlorosulphonic acid (157.5 g, 1.35 mol) is placed in a 350-ml four-necked flask provided with a stirrer, a reflux condenser, a thermometer, and a pressure-equalizing dropping funnel; then it is cooled to 5 °C and pentafluorobenzene (37.5 g, 0.225 mol) is carefully dropped in during 45–50 min. By the end of the addition of pentafluorobenzene, the temperature of the reaction mixture has risen to 18 °C. The mixture is then heated at 90 °C for 1 h, at 110 °C for 1 h and at 120 °C for 6 h, cooled, poured onto ice (2 kg) and extracted with methylene chloride (7 × 20 ml). The extract is washed twice with a saturated solution of $NaHCO_3$, then with water until neutral, and dried over $MgSO_4$. The solvent is distilled off in a water-bath and the residue distilled in vacuum. A fraction boiling at 75°–76.5 °C/2 mm Hg is collected. The yield of pentafluorobenzenesulphochloride is 29–30 g (50%–52%); $n_D^{20.5}$ 1.4780.

Pentafluorophenylsulphenyl Chloride

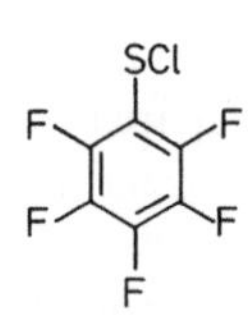

C_6ClF_5S MW 234.5

Pentafluorophenylsulphenyl chloride is a yellow liquid, readily soluble in common organic solvents. Bp 48 °C/1 mm Hg [251], 89°–90 °C/0.1 mm Hg [252]; n_D^{20} 1.4991 [251], n_D^{23} 1.4993 [252].

Method of Preparation

Pentafluorophenylsulphenyl chloride is prepared by chlorination of pentafluorothiophenol in CCl_4 (90% yield) [251], or in autoclave [252]; by chlorination of bis-

(pentafluorophenyl)disulphide in CCl_4 (90% yield) [251], of bis(pentafluorothiophenoxy)lead [252] or mercury [251]. .

The following method is based on the data reported in [251].

Procedure

$$C_6F_5SH \xrightarrow[CCl_4]{Cl_2} C_6F_5SCl$$

Carbon tetrachloride (200 ml) is placed in a 0.5-l three-necked flask fitted with a stirrer, a gas inlet tube, and a pressure-equalizing dropping funnel with a calcium chloride tube; then it is cooled to −10 °C, and a solution of pentafluorothiophenol (36.7 g, 0.184 mol) in 100 ml of CCl_4 is dropped in during 1 h, with flushing of the flask with dry chlorine (30 l/h). After all the pentafluorothiophenol has been added, the flow of chlorine is continued to be let in for an additional 0.5 h. The solvent is then distilled off in a water-bath, in vacuum of a water-jet pump (30–50 mm Hg). The residue, a yellowish oil, is distilled in vacuum and a fraction boiling at 47°–48 °C/1–2 mm Hg is collected. The yield of pentafluorophenylsulphenyl chloride is 38.9 g (90%).

1,4-Bis(pentafluorophenylthio)-2,3,5,6-tetrafluorobenzene

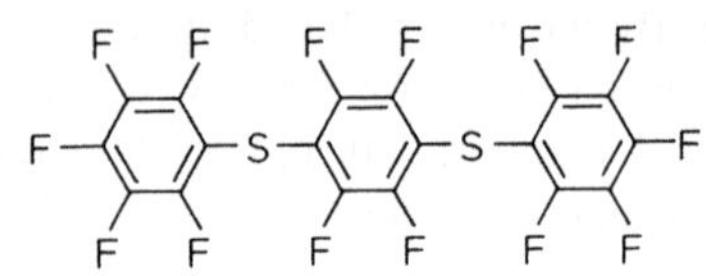

$C_{18}F_{14}S_2$ MW 546

1,4-Bis(pentafluorophenylthio)-2,3,5,6-tetrafluorobenzene is a crystalline substance of pale yellow colour, partially soluble in common organic solvents, well soluble in ether, methylene chloride. Mp 163°–164 °C [239], 164.5°–166 °C [180].

Method of Preparation

1,4-Bis(pentafluorophenylthio)-2,3,5,6-tetrafluorobenzene is prepared by the reaction of bis(pentafluorophenyl)disulphide with 1,2,4,5-tetrafluorobenzene in SbF_5 (92% yield) [230]; by the reaction of cuprous pentafluorothiophenoxide with 2,3,5,6-tetrafluoro-1,4-dibromobenzene (63% yield), or with nonafluoro-4-bromodiphenyl sulphide (80% yield) in dimethylformamide [180].

The following procedure is based on the data reported in [239].

Procedure

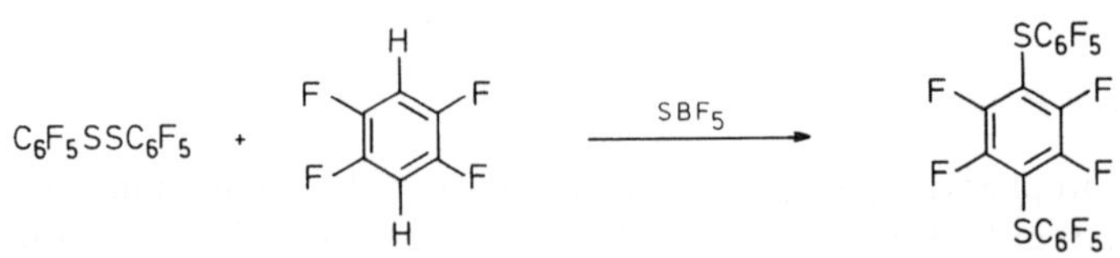

Bis(pentafluorophenyl)disulphide (39.8 g, 0.1 mol), 1,2,4,5-tetrafluorobenzene (16.8 g, 0.1 mol) and freshly distilled SbF_5 (200 g) are placed in a 250-ml two-necked flask fitted with a stirrer and a thermometer. The reaction mixture is stirred at 20 °C for 1 h, then at 70 °C for 2 h, whereupon it is cooled and poured onto 1 kg of ice (Note). The precipitate is filtered off on a Büchner funnel, washed with water, dried and recrystallized from aqueous ethanol. The yield of 1,4-bis(pentafluorophenylthio)-2,3,5,6-tetrafluorobenzene is 46–49 g (84%–90%); mp 163°–164 °C.

Note. Addition of SbF_5 to water results in violent heating, which necessitates vigorous stirring and cooling of the flask with ice.

4,4′-Thiobis(tetrafluorobenzenesulphonyl)Difluoride

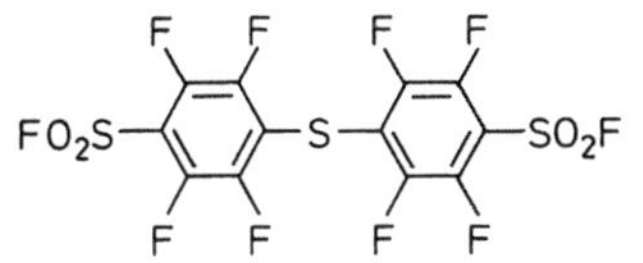

$C_{12}F_{10}O_4S_3$ MW 494

4,4′-Thiobis(tetrafluorobenzenesulphonyl) difluoride is a crystalline substance of light-yellow colour, soluble in common organic solvents. Mp 116.5°–118 °C [253].

Method of Preparation

4,4′-Thiobis(tetrafluorobenzenesulphonyl) difluoride is prepared by the reaction of pentafluorobenzenesulphonyl fluoride with sodium thiosulphate in dimethylformamide (54% yield) [253].

The following procedure is based on the data reported in [253].

Procedure

$$C_6F_5SO_2F \xrightarrow[\text{DMF}]{Na_2S_2O_3 \cdot 5H_2O} S(C_6F_4SO_2F)_2$$

Pentafluorobenzenesulphonyl fluoride (25 g, 0.1 mol), sodium thiosulphate ($Na_2S_2O_3 \times 5\,H_2O$) (37.2 g, 0.15 mol) and dry dimethylformamide (150 ml) are placed in a 250-ml three-necked flask fitted with a stirrer, a thermometer, and a calcium chloride tube. The mixture is stirred at 20 °C for 2 h and poured into 700 ml of water. The precipitate is filtered off on a Büchner funnel, washed with water (2 × 150 ml), dried and recrystsllized from aqueous ethanol. The yield of 4,4′-thiobis(tetrafluorobenzene-sulphonyl) difluoride is 19.8 g (80%); mp 117°–118 °C.

Bis(2,3,5,6-tetrafluoro-4-carbomethoxyphenyl)Sulphone

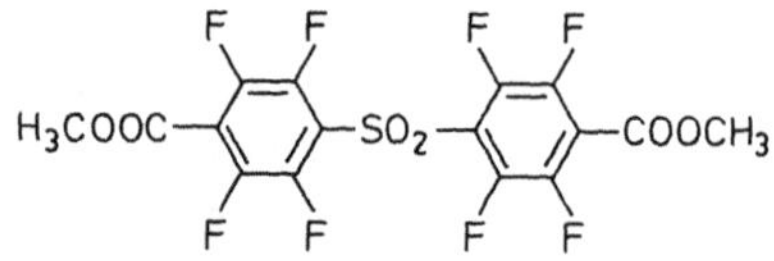

$C_{16}H_6F_8O_6S$ MW 478

Bis(2,3,5,6-tetrafluoro-4-carbomethoxyphenyl)sulphone is a white crystalline substance, partially soluble in common organic solvents, well soluble in ether and methylene chloride. Mp 184°–186 °C [253].

Method of Preparation

Bis (2,3,5,6-tetrafluoro-4-carbomethoxyphenyl)sulphone is prepared by oxidation of bis(2,3,5,6-tetrafluoro-4-carbomethoxyphenyl) sulphide with trifluoroperacetic acid (90% yield) [253].

The following procedure is based on the data reported in [253].

Procedure

$$C_6F_5COOCH_3 \xrightarrow[DMF]{Na_2S_2O_3 \cdot 5\,H_2O} (CH_3OOCC_6F_4)_2S$$

$$(CH_3OOCC_6F_4)_2S \xrightarrow[CH_2Cl_2]{CF_3COOOH} (CH_3OOCC_6F_4)_2SO_2$$

Bis(2,3,5,6-tetrafluoro-4-carbomethoxyphenyl)sulphide. Methyl pentafluorobenzoate (45.2 g, 0.2 mol), sodium thiosulphate (62 g, 0.25 mol) and dimethylformamide (200 ml) are placed in a 350-ml three-necked flask equipped with a stirrer, a thermometer, and a reflux condenser with a calcium chloride tube. The mixture is heated at 100 °C with stirring for 20 h, poured into 1 l of ice-water and extracted with ether (3 × 250 ml). The ether extract is dried over $MgSO_4$. The ether is removed on a rotary vacuum evaporator and the residue recrystallized from aqueous ethanol. The yield of the sulphide is 62–63 g (69%–72%); mp 77°–79 °C.

Bis(2,3,5,6-tetrafluoro-4-carbomethoxyphenyl)sulphone. Trifluoroacetic oxide (200 ml) and methylene chloride (450 ml) are placed in a 1-l three-necked flask provided with a stirrer, an effective reflux condenser, and a dropping funnel; then 100 ml of 95% H_2O_2 is added dropwise vigorous stirring during 0.5 h, at such a rate as to maintain gentle refluxing of the mixture. Then 31.2 g (0.07 mol) of bis (2,3,5,6-tetrafluoro-4-carbomethoxyphenyl)sulphide in 100 ml of methylene chloride is dropped in. The mixture is refluxed for 10 h and poured into 1 l of water. The resulting solution is transferred into a separatory funnel and vigorously shaken. The organic layer is washed with further 500 ml of water and dried over $MgSO_4$. The solvent is removed on a rotary vacuum-evaporator and the residue recrystallized from ethanol. The yield of the sulphone is 32 g (95%); mp 184°–186 °C.

Pentafluorophenylsulphur Trifluoride

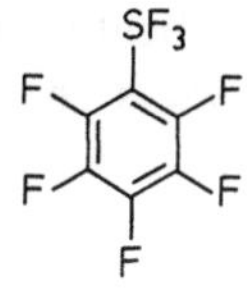

C_6F_8S MW 256

Pentafluorophenylsulphur trifluoride is a colourless liquid, easily hydrolysed by air moisture, soluble in common organic solvents. Bp 41 °–42 °C/5.2 mm Hg [254], 49 °C/6 mm Hg [255].

Method of Preparation

Pentafluorophenylsulphur trifluoride is prepared by the reaction of bis(pentafluorophenyl) disulphide with AgF_2 in 1,1,2-trichloro-1,2,2-trifluoroethane (56% yield) [254], (72% yield) [255].

The following procedure is a modification of the one described in [254].

Procedure

$$C_6F_5S{-}SC_6F_5 \xrightarrow{AgF_2} C_6F_5SF_3$$

A mixture of bis(pentafluorophenyl) disulphide (25.9 g, 0.065 mol), finely divided AgF_2 (40 g, 0.28 g), and 1,1,2-trichloro-1,2,2-trifluoroethane (70 ml) is placed in a 150-ml Teflon (or stainless steel) reactor provided with a stirrer, a gas inlet tube, and a reflux condenser with a calcium chloride tube. Then it is refluxed in a flow of argon over a period of 3 h. The mixture is then cooled, quickly filtered to remove a yellow solid precipitate (AgF), the solvent is distilled off in a water-bath, the residue is transferred to another flask and distilled in vauum. A fraction boiling at 41 °–42 °C/5 mm Hg is collected. The yield of pentafluorophenylsulphur trifluoride is 18.8 g (56%).

Pentafluorobenzenesulphinic Acid Monohydrate

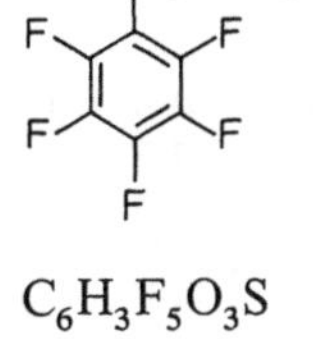

$C_6H_3F_5O_3S$ MW 250

Pentafluorobenzenesulphinic acid monohydrate is a colourless crystalline substance, slightly soluble in common organic solvents, well soluble in ether and dioxan; readily hydrolysed to pentafluorobenzene; decomposes at 120 °C, giving pentafluorobenzene. Mp 62°–63 °C [254].

Method of Preparation

Pentafluorobenzenesulphinic acid monohydrate is prepared by treatment of pentafluorobenzenesulphinyl chloride with equimolar amount of water in ether (46% yield) [254].

The following procedure is based on the data reported in [254].

Procedure

$$C_6F_5Br \xrightarrow[\text{ether, }-80\ ^\circ C]{n\text{-}C_4H_9Li} C_6F_5Li \xrightarrow{SO_2} C_6F_5SO_2Li \xrightarrow{SOCl_2} C_6F_5SOCl;$$

$$C_6F_5SOCl \xrightarrow{H_2O} C_6F_5SO_2H \cdot H_2O$$

Lithium pentafluorobenzenesulphinate. Bromopentafluorobenzene (123.5 g, 0.5 mol) and absolute ether (250 ml) are placed in a 1-l four-necked flask fitted with a stirrer, a thermometer, a gas inlet tube, and a pressure-equalizing dropping funnel with a calcium chloride tube. The mixture is cooled to —70 °C and n-butyllithium in hexane (320 ml, 0.5 mol) is dropped in during 1 h. The temperature is raised to —64 °C and the mixture is allowed to stand for 1.5 h, then cooled to —70 °C and a flow of dry sulphur dioxide is passed over a period of 3 h (25 l/h). A voluminous white precipitate is gradually deposited; it is filtered off and dried in air. The yield of lithium pentafluorobenzenesulphinate is 112 g (94%); decomp.p. 280 °C.

Pentafluorobenzenesulphinyl chloride. Thionyl chloride (400 ml) is placed in a 750-ml two-necked flask equipped with a reflux condenser with a calcium chloride tube; the flask is connected by means of a section of rubber tubing to a 200-ml pear-shaped flask, in which is placed 98 g (0.41 mol) of finely divided lithium pentafluorobenzenesulphinate. The lithium salt is then slowly added to thionyl chloride and an exothermic reaction occurs; the rate of addition is regulated so as to maintain gentle refluxing of the mixture, by cooling of the flask with ice-water. The solution is then filtered to remove LiCl. The filtrate is distilled to remove excess thionyl chloride and the residue distilled in vacuum. A fraction boiling at 62°–63 °C/2–3 mm Hg is collected. The yield of pentafluorobenzenesulphinyl chloride is 86 g (83%); n_D^{25} 1.5042.

Pentafluorobenzenesulphinic acid monohydrate. Pentafluorobenzenesulphinyl chloride (9.53 g, 0.038 mol) and dry ether (230 ml) are placed in a 350-ml three-necked flask fitted with a stirrer, a thermometer, and a pressure-equalizing dropping funnel with a calcium chloride tube. The mixture is cooled to 0 °C and 1 ml of water is slowly dropped in with stirring. The mixture is stirred at 20 °C during 1.5 h, freshly calcinated $MgSO_4$ (~25 g) is added to it, then it is filtered and the solvent is removed on a rotary vacuum-evaporator. The solid product is triturated with petroleum ether (bp 40°–60 °C). The yield of pentafluorobenzenesulphinic acid monohydrate is 4.6 g (46%); mp 62°–63 °C.

Pentafluoroselenophenol

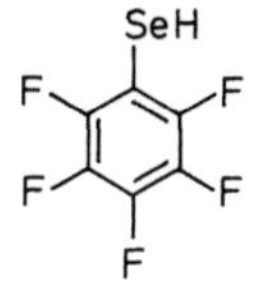

C_6HF_5Se MW 244

Pentafluoroselenophenol is a colourless liquid with an unpleasant odour, soluble in common organic solvents; readily oxidized by atmospheric oxygen to bis(pentafluorophenyl) diselenide. Bp 62°–63 °C/20 mm Hg [242].

Method of Preparation

Pentafluoroselenophenol is obtained by the reaction of hexafluorobenzene with sodium selenide in liquid ammonia [256]; by the reaction of metallic selenium with a Grignard reagent obtained from bromopentafluorobenzene in tetrahydrofuran (43% yield) [242].

The following procedure is based on the data reported in [255].

Procedure

$$Se \xrightarrow[NH_3(liquid)]{NaNH_2} Na_2Se \xrightarrow[2.\ H_2O]{1.\ C_6F_6} C_6F_5SeH$$

In a 1-l three-necked flask provided with a stirrer, an air condenser with a calcium chloride tube, and a dropping funnel, is placed 800 ml of liquid ammonia, then three or four crystals of anhydrous $FeCl_3$ are added. Finely divided sodium (17.5 g, 0.76 mol) is added in portions to the stirred mixture over a period of about 0.5 h, at such a rate as to maintain gentle refluxing of the mixture. Then powdered selenium (30 g, 0.38 mol) is added in portions; the mixture boils due to liberation of ammonia. After all the selenium has been added (~0.5 h), stirring is continued for 1 h, then hexafluorobenzene (68.8 g, 0.37 mol) is added with continuous stirring and ammonia is allowed to escape (Note 1). The residue is allowed to stand for 20 h, whereupon 500 ml of water is carefully added in portions in such a manner as to wash the inner surface of the flask (Note 2). The solution is acidified with 10% HCl solution and extracted with ether. The ether extract is dried over $MgSO_4$. The ether is removed in a water-bath and the residue distilled in vacuum; a fraction boiling at 62°–63 °C/20 mm Hg is collected. The yield of pentafluoroselenophenol is 72–73 g (80%–81%).

Notes

1. The ice is carefully removed from the walls of the flask with a woollen cloth. If ice no longer forms, it means that all the ammonia has evolved.
2. Part of the sodium amide used for the preparation of Na_2Se may remain on the walls of the flask, especially at the top, *so treatment with water should be carried out with great caution*, in view of possible vigorous decomposition of sodium amide.

(Pentafluorophenyl)phenylselenide

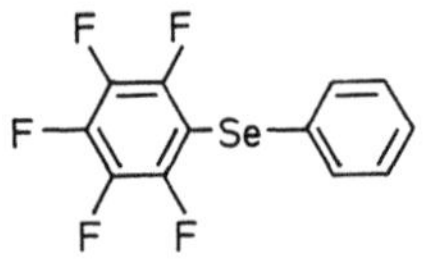

$C_{12}H_5F_5Se$ MW 320

(Pentafluorophenyl)phenylselenide is a colourless crystalline substance, soluble in common organic solvents. Mp 30°–31 °C [257].

Method of Preparation

(Pentafluorophenyl)phenylselenide is prepared by the reaction of benzene with bis-(pentafluorophenyl) diselenide in the presence of SbF_5-in-graphite intercalate (68% yield) [257].

The following procedure is a modification of the one described in [257].

Procedure

$$C_6H_6 + C_6F_5Se\text{—}SeC_6F_5 \xrightarrow{C_nSbF_5} C_6F_5SeC_6H_5$$

SbF_5-in-graphite intercalate. Powdered graphite (40 g; granules of 0.063 to 0.1 mm size) is placed in a horizontal quartz reactor (400 × 30 mm) provided with a heater, gaseous SbF_5 and argon inlet system, and a thermocouple in a well; the reactor is connected by a straight adapter to a two-necked flask fitted with a reflux condenser. Gaseous SbF_5 is then passed at 140°–160 °C and atmospheric pressure, over a period of 2–3 h; this is effected by heating a flask with SbF_5 to 140 °C and passing argon through the hot SbF_5 at the rate of 10 l/h. During that period ~250 g of SbF_5 is passed. The flow of SbF_5 is then cut off and the heater is removed. This procedure gives 140 g of a fluffy product of dark-blue colour.

(Pentafluorophenyl)phenylselenide. SbF_5-in-graphite intercalate (15 g), benzene (36 g, 0.5 mol) and bis(pentafluorophenyl) diselenide (4.86 g, 0.01 mol) are placed in a 150-ml three-necked flask provided with a stirrer, a reflux condenser with a calcium chloride tube and a thermometer. The mixture is stirred at 80 °C for 3 h, then cooled and filtered to remove graphite, using the filter paper. Residual graphite is washed with hot benzene (2 × 30 ml), the extracts are combined with the filtrate, and the mixture is passed through a 1-cm silica gel layer to remove the smallest granules of graphite. The benzene is removed on a rotary vacuum-evaporator and the residue recrystallized from petroleum ether (bp 40°–60 °C). The yield of (pentafluorophenyl)-phenylselenide is 3.9–4.0 g (80%–82%); mp 30°–31 °C.

Pentafluoroaniline

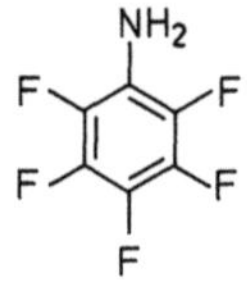

$C_6H_2F_5N$ MW 183

Pentafluoroaniline is a white crystalline substance with an odour of aniline, soluble in common organic solvents. Bp 153°–154 °C [258], 156°–157 °C/760 mm Hg [259]; mp 33 °C [51, 58], 33.5°–35 °C [243], 34 °C [40, 258], 34°–34.5 °C [58], 35°–36 °C [260], 34°–35 °C [261].

Method of Preparation

Pentafluoroaniline is prepared by the reaction of hexafluorobenzene with sodium amide in liquid ammonia [258], with ammonia in aqueous ethanol [259] or with aqueous ammonia [40, 261], in 59%–86% yields; by the pyrolysis of pentafluorophenylhydrazine (42% yield) [57], or of a mixture of hexafluorobenzene with ammonia (~5% yield) [76]; by boiling pentafluorophenylhydrazine [58] or 2-(pentafluorophenylamino)ethanol [260] with 50%–55% aqueous HI (65% yield) [58]. Pentafluoroaniline is obtained in a quantitative yield by the reduction of 4-dimethylamino-2,3,5,6-pentafluoroazobenzene with tin in HCl [51].

The following procedure is a modification of the one described in [261].

Procedure

$$C_6F_6 \xrightarrow{NH_3} C_6F_5NH_2$$

In a 1-l rotating steel autoclave are placed hexafluorobenzene (150 g, 0.8 mol) and 30% aqueous ammonia (700 ml) (d_4^{20} 0.8920). The autoclave is heated at 150°–160 °C for 6 h, then cooled to room temperature. The reaction mixture is removed from the autoclave and extracted with ether (10 × 200 ml). The ether extracts are dried over $MgSO_4$, the ether is distilled off in a water-bath and the residue is distilled on a fractionating column with a glass packing (12–15 theoretical plates). A fraction boiling at 152°–154 °C is collected. The yield of pentafluoroaniline is 105–115 g (70%–77%); mp 33°–34 °C.

2,3,4,5-Tetrafluoroaniline

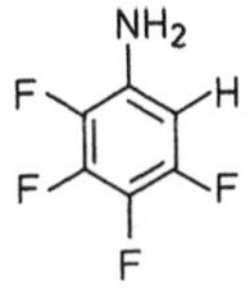

$C_6H_3F_4N$ MW 165

2,3,4,5-Tetrafluoroaniline is a colourless liquid with an odour of aniline, soluble in common organic solvents. Bp 78 °C/13 mm Hg [262], 108°–110 °C/53 mm Hg [263]; mp 27°–28 °C [264]; n_4^{20} 1.4630 [263].

Method of Preparation

2,3,4,5-Tetrafluoroaniline is prepared by the reduction of 1-nitro-2,3,4,5-tetrafluorobenzene with tin in HCl [262], or with hydrogen over palladium-on-charcoal [264]; or by decarboxylation of tetrafluoroanthranilic acid [263]. Yields are 80%–84%.

The following procedure is a modification of the one described in [263].

Procedure

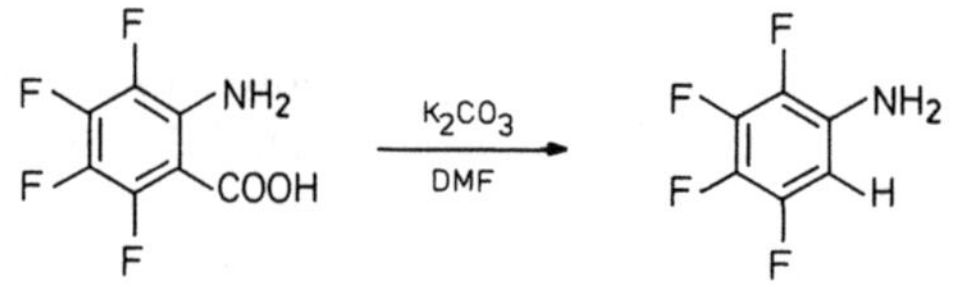

In a 250-ml three-necked flask provided with a stirrer, a thermometer, and a reflux condenser with a calcium chloride tube, are placed 16 g (0.077 mol) of tetrafluoroanthranilic acid, 90 ml of absolute dimethylformamide and 5 g of freshly calcinated K_2CO_3. The reaction mixture is gradually heated to 140°–145 °C with stirring and kept at that temperature for 1.5–2 h. Then it is heated for further 2 h at 150 °C, and cooled to 90°–100 °C. Tetrafluoroaniline is steam-distilled; the amount of the distillate is ~1.5 l. The distillate is slightly acidified with HCl, extracted with ether and dried over $MgSO_4$. The ether is distilled off in a water-bath and the residue distilled in vacuum. A fraction boiling at 72 °C/15 mm Hg is collected. The yield of 2,3,4,5-tetrafluoroaniline is 8–9 g (63%–70%); n_D^{20} 1.4650.

2,3,5,6-Tetrafluoroaniline

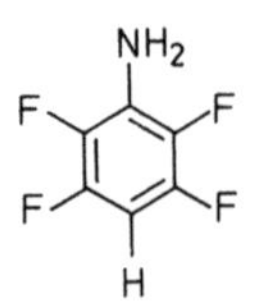

$C_6H_3F_4N$ MW 165

2,3,5,6-Tetrafluoroaniline is a white crystalline substance with an odour of aniline, soluble in common organic solvents. Bp 156 °C [250], 156°–157 °C/740 mm Hg [266], 158°–159 °C [259]; mp 23.5°–26.5 °C [267], 28.5°–29.5 °C [193], 29°–31.5 °C [268], 30–31 °C [269].

Method of Preparation

2,3,5,6-Tetrafluoroaniline is prepared by the reaction of pentafluorobenzene with an aqueous alcoholic solution of ammonia (66% yield) [259]; by decarboxylation of p-aminotetrafluorobenzoic acid (56% yield) [193]; by the reduction of (p-aminotetrafluorophenyl)methylsulphide with skeletal nickel (37% yield) [250], of p-halogenotetrafluorophenylhydrazine with hydroiodic acid (91% yield) [269], of benzaldehyde 2,3,5,6-tetrafluorophenylhydrazone with zinc in hydrochloric acid (79% yield), or of 1,4-dihydrazino-2,3,5,6-tetrafluorobenzene with 3 N solution of NaOH (56% yield) [266].

The following procedure is based on the reaction of pentafluorobenzene with aqueous ammonia.

Procedure

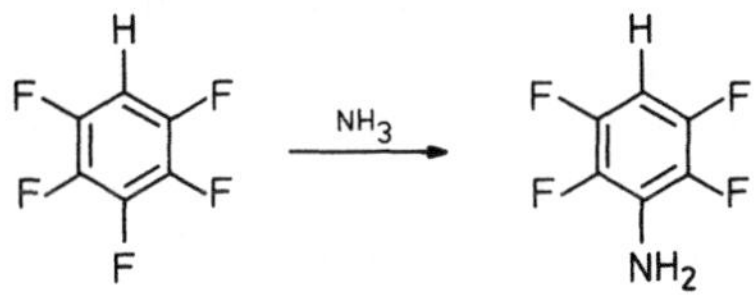

Pentafluorobenzene (100 g, 0.59 mol) and 30% aqueous ammonia (500 ml) are heated at 150 °C during 3 h in a 1-l rotating steel autoclave provided with a manometer, a valve, and a thermocouple in a well. The cooled reaction mixture is extracted with ether (4 × 300 ml). The ether extract is dried over $MgSO_4$, the ether is distilled off in a water-bath and the residue is distilled on a fractionating column with a glass packing (12–15 theoretical plates). A fraction boiling at 160°–161 °C is collected. The yield of 2,3,5,6-tetrafluoroanilinc is 73–78 g (73%–78%); mp 29°–30 °C.

4,4'-Diaminooctafluorobiphenyl

(Octafluorobenzidinc)

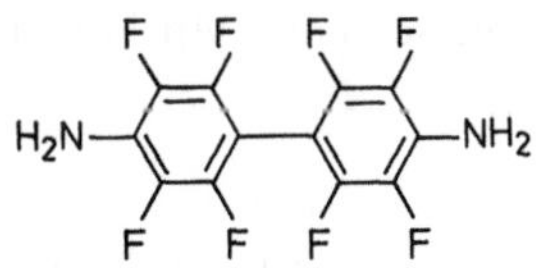

$C_{12}H_4F_8N_2$ MW 328

Octafluorobenzidine is a white crystalline substance, soluble in common organic solvents . Mp 173°–174 °C [55], 175°–176.5 °C [265], 181°–181.5 °C [57].

Method of Preparation

Octafluorobenzidine is prepared by the reaction of decafluorobiphenyl [55] or 4-aminononafluorobiphenyl [57] with 30% aqueous ammonia (73–91%); by the reaction of 4,4'-dihydrazinooctafluorobiphenyl in 55% hydroiodic acid (54% yield) [55], or of bis-hydrazone, formed from 4,4'-dihydrazinooctafluorobiphenyl and acetophenone, with zinc in acetic acid (91% yield) [265].

The following method is a modification of the one described in [55].

Procedure

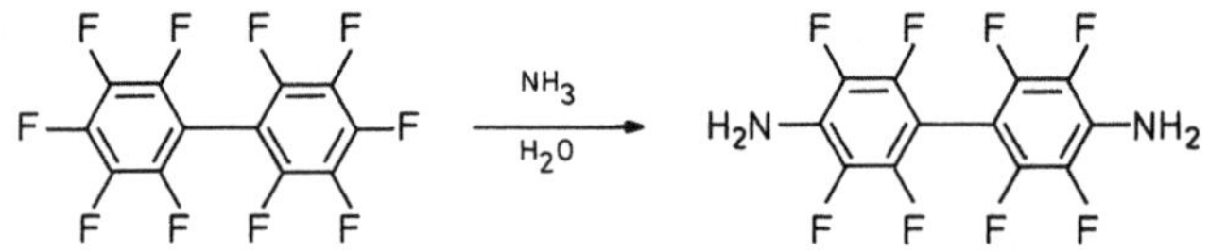

In a 1-l rotating steel autoclave fitted with a manometer, a valve, and a thermocouple in a well, are placed 6 g (0.018 mol) of decafluorobiphenyl and 600 ml of 30% aqueous ammonia. The mixture is heated at 130 °C for 6 h. The autoclave is cooled (excess pressure should be previously released into a hood), the mixture is removed and the solid precipitate is filtered off, washed on a filter and dried. The crystalline residue is washed off the autoclave walls with ether. The aqueous filtrates are extracted with ether (2 × 50 ml). The ethereal solutions are combined (400 ml) and dried over $MgSO_4$. The solid products — the product taken out of the autoclave and that obtained by the removal of ether on a rotary vacuum-evaporator — are combined (the total amount is 5.2–5.4 g). Recrystallization from petroleum ether (bp 40°–60 °C) yields 5.0–5.2 g (83%–87%) of octafluorobenzidine; mp 174°–175.5 °C.

Bis(2,3,5,6-tetrafluoro-4-aminophenyl)Sulphide

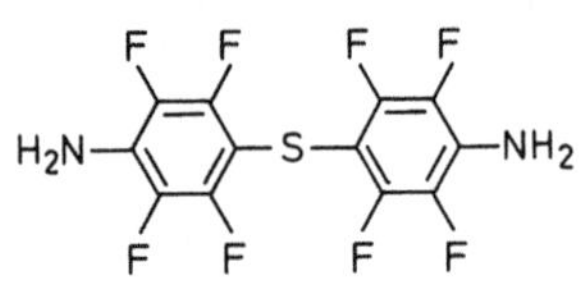

$C_{12}H_4F_8N_2S$ MW 360

Bis(2,3,5,6-tetrafluoro-4-aminophenyl)sulphide is a crystalline substance of slightly yellow colour, partially soluble in common organic solvents. Mp 179°–180 °C [253].

Method of Preparation

Bis(2,3,5,6-tetrafluoro-4-aminophenyl)sulphide is prepared by the reduction of bis-(2,3,5,6-tetrafluoro-4-nitrophenyl)sulphide, obtained by the reaction of pentafluoronitrobenzene with thiourea [245], or with $Na_2S_2O_3$ in dimethylformamide [253]; by treatment of bis(pentafluorophenyl)sulphide with iron turnings in alcohol, in the presence of NH_4Cl (90% yield) [253], or with an aqueous solution of ammonia [253].

The following procedure is based on the data reported in [245, 253].

Procedure

$$C_6F_5NO_2 \xrightarrow{SC(NH_2)_2} (p\text{-}O_2NC_6F_4)_2S \xrightarrow[C_2H_5OH]{Fe} (p\text{-}H_2NC_6F_4)_2S$$

Bis(2,3,5,6-tetrafluoro-4-nitrophenyl)sulphide. In a 300-ml three-necked flask provided with a stirrer, a reflux condenser, and a dropping funnel, are placed 30.4 g (0.4 mol)

Procedure

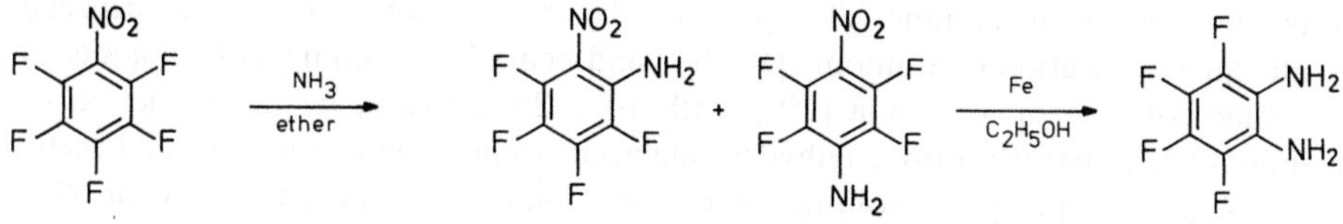

2-Nitro-3,4,5,6-tetrafluoroaniline. In a 2.5-l two-necked flask are placed 70 g (0.33 mol) of nitropentafluorobenzene and 1800 ml of absolute ether. A slow flow of gaseous ammonia is passed during 4 h through the solution. The solution is allowed to stand overnight, then filtered and the ether is removed on a rotary vacuum-evaporator to give 62–65 g of a liquid product, which solidifies on standing.

The product (30 g) is dissolved in benzene, the solution is applied to a chromatographic column (150 cm long, 7 cm in dia.) packed with Al_2O_3 (the 2nd degree of activity) and eluted with benzene. The first zone (of yellow colour) is 2-nitrotetrafluoroaniline, the second zone (of red colour) is a mixture of 4-nitrotetrafluoroaniline and nitrodiaminotrifluorobenzenes. The mixture is continuously separated during 2–3 days. Benzene is distilled off from the fractions to yield 16–18 g (52%–55%) 2-nitrotetrafluoroaniline, mp 42°–44 °C (mp 43°–44 °C [209], 42.5°–43.5 °C [265]), 6–7 g (20%–23%) of 4-nitrotetrafluoroaniline, mp 99°–104 °C (mp 106.5°–108 °C [271]) and 3–5 g of a mixture of isomeric nitrodiaminotrifluorobenzenes.

Tetrafluorophenylenediamine-1,2. In a 100-ml three-necked flask fitted with a stirrer, a reflux condenser, and a dropping funnel, are placed 30 g of cast-iron turnings, 0.2 g of NH_4Cl, 20 ml of water and 30 ml of ethanol. The suspension is heated to reflux and a solution of 5 g (0.024 mol) of 2-nitrotetrafluoroaniline in 20 ml of ethanol is added during 30 min. The mixture is refluxed for 3 h, the hot solution is filtered and thoroughly washed with hot ethanol (3 × 25 ml). The filtrate is poured into 500 ml of water, acidified with 11 ml of concentrated HCl, extracted with ether (3 × 250 ml) and dried over $MgSO_4$. The ether is removed on a rotary vacuum-evaporator at 50 °C/10–15 mm Hg to give 3.7–3.9 g (86%–91%) of the crude product, mp 118°–121 °C. Sublimation of the product in vacuum of a water-jet pump yields 3.4–3.5 g (79%–81%) of 3,4,5,6-tetrafluorophenylenediamine-1,2; mp 129°–130 °C.

2,4,5,6-Tetrafluorophenylenediamine-1,3

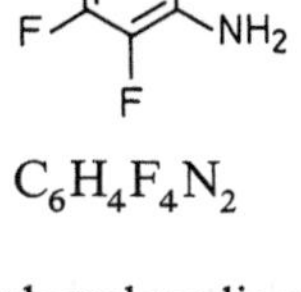

$C_6H_4F_4N_2$ MW 180

Tetrafluorophenylenediamine-1,3 is a white crystalline product, soluble in common organic solvents. Mp 126°–127 °C [272], 127°–130 °C [260], 129°–131 °C [265], 129.5°–131 °C [259], 132°–132.5 °C [261].

Method of Preparation

Tetrafluorophenylenediamine-1,3 is prepared by the reaction of hexafluorobenzene in an aqueous solution of ammonia [40, 261] and pentafluoroaniline with an aqueous alcoholic solution of ammonia [259], with 10%–25% yields; by the reduction of 5-amino-2,3,4,6-tetrafluorophenylhydrazone acetophenone with zinc in glacial acetic acid (54% yield) [265]; by boiling 3-hydrazino-2,4,5,6-tetrafluoroaniline with 50% aqueous solution of HI (39% yield) [263]; or by hydrogenation of 3-imino-1-amino-heptafluorocyclohexene over skeletal nickel (25% yield) [272].

The following procedure is a modification of the preparations described in [40, 261].

Procedure

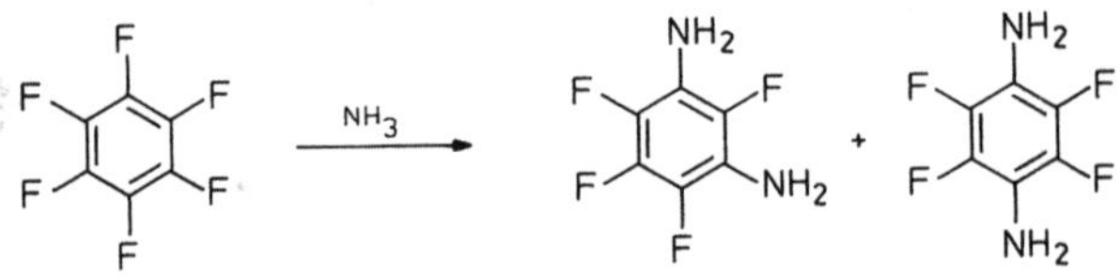

An aqueous 30% solution of ammonia (300 ml) and hexafluorobenzene (50 g, 0.27 mol), placed in a 0.5-l rotating steel autoclave, are heated at 200 °C during 6 h. The reaction mixture is then filtered and the precipitate is recrystallized from petroleum ether (bp 70°–100 °C) to afford 36–40 g (74%–82%) of the product (mp 129°–131 °C) containing about 90% of tetrafluorophenylenediamine-1,3 and 10% of tetrafluoro-phenylenediamine-1, 4. Isomeric tetrafluorophenylenediamines may be separated by the preparative GLC.

2,4,5,6-Tetrafluoro-3-aminophenol

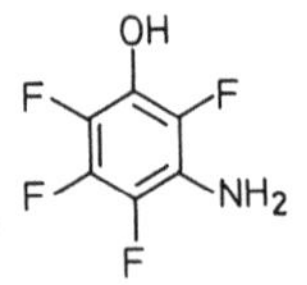

$C_6H_3F_4NO$ MW 181

2,4,5,6-Tetrafluoro-3-aminophenol is a white crystalline substance with an odour of phenol, soluble in common organic solvents. Mp 95 °C [267]; dissociation constant of the neutral form $0.72 \cdot 10^{-7}$ [273, 274].

Method of Preparation

2,4,5,6-Tetrafluoro-3-aminophenol is prepared by the reduction of heptafluoro-3-aminocyclohexen-5-one-1 on a mercury cathode (90% yield) [273, 274], or with zinc in hydrochloric acid (55% yield) [275].

Procedure

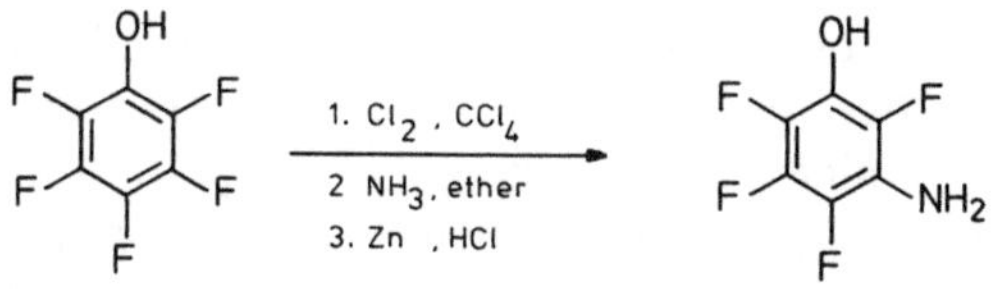

Carbon tetrachloride (400 ml) is placed in a 1-l three-necked flask equipped with a stirrer, a thermometer, and a gas inlet tube, and is cooled to −35 °C. Then 20 g (0.282 mol) of chlorine is condensed into the flask and 50 g (0.225 mol) of potassium pentafluorophenoxide is added with stirring during 0.5 h. The temperature of the mixture is then raised to 20 °C and the mixture is stirred for 1 h. A flow of dry gaseous ammonia is passed during 1 h through the resulting solution (Note 1). A voluminous precipitate formed in the reaction (Note 2) is filtered off on a Büchner funnel and dissolved in 600 ml of ether.

In a 2-l three-necked flask provided with a stirrer and two pressure-equalizing dropping funnels, are placed 130 g (2 mol) of zinc dust and 500 ml of ether. The ethereal solution of the product and 500 ml of concentrated HCl are simultaneously dropped in during 4 h, with vigorous stirring and cooling of the flask in an ice-bath, so as to keep up stirring and gentle boiling of the mixture. The reaction mixture is stirred for further 10 h at 20 °C and filtered. The acidic solution is neutralized with a saturated solution of Na_2CO_3 to pH ~ 6 and extracted with benzene (4 × 250 ml). The extract is washed with 300 ml of water and dried over $MgSO_4$. The solvents are removed on a rotary vacuum-evaporator; this gives 28 g of the crude product, which is then recrystallized from benzene. The yield of 2,4,5,6-tetrafluoro-3-aminophenol is 24–25 g (58%–61%); mp 94°–95 °C.

Notes

1. The solution is a mixture of pentafluoro-6-chlorocyclohexadien-2,4-one-1 and pentafluoro-4-chlorocyclohexadien-2,5-one-1.
2. The bubbler may get clogged; it should be provided with a glass rod inserted through one of two holes by means of a Teflon or rubber gland to dislodge the material which clogs it. The precipitate is a mixture of tetrafluoro-6-chloro-3-aminocyclohexadien-2,4-one-1 and tetrafluoro-4-chloro-3-aminocyclohexadien-2,5-one-1.

2,3,5,6-Tetrafluoro-4-aminophenol

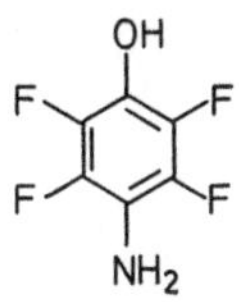

$C_{12}H_8F_6H_4O$ MW 181

2,3,5,6-Tetrafluoro-4-aminophenol is a white crystalline substance, soluble in common organic solvents. Mp 180 °C [229], 180°–181.5 °C [276], 177.5°–178 °C [51], 177° to 178 °C (decompn.) [277]; sublimes at 130 °C/0.24 mm Hg [277].

Method of Preparation

2,3,5,6-Tetrafluoro-4-aminophenol is prepared by the reduction of 2,3,5,6-tetrafluoro-4-nitrophenol with hydrogen over palladium black (82% yield) or over PtO (78% yield) [276]; with cast-iron turnings in the presence of NH_4Cl (70% yield) [276], or with tin in hydrochloric acid [279]; by the reduction of the azo-compound, obtained in the reaction of dimethylaniline with pentafluoroaniline diazonium salt in concentrated H_2SO_4, with tin in hydrochloric acid (24% yield) [51]; by treatment of 4,4'-diethoxyoctafluoroazobenzene with hydroiodic acid (98% yield) [277].

Procedure

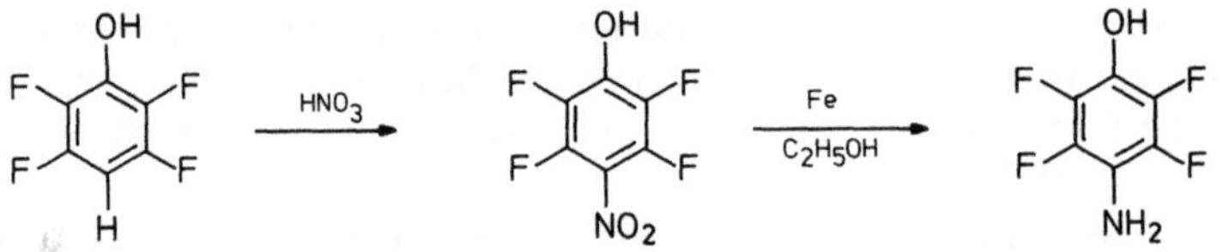

2,3,5,6-Tetrafluoro-4-nitrophenol. Concentrated HNO_3 (d 1.5) (75 ml) is placed in a 150-ml three-necked flask provided with a stirrer, a thermometer, and a pressure-equalizing dropping funnel, and is cooled to $-10° \div -15$ °C. 2,3,5,6-Tetrafluoroaniline (30 g, 0.18 mol) is added in small portions with stirring during 1 h, at such a rate the temperature should not rise above -5 °C. The mixture is stirred for 1 h, whereupon it is transferred, in portions and with vigorous shaking, into a flask filled with 1 kg of ice. The precipitate is filtered off on a Büchner funnel. The aqueous solution is extracted with 250 ml of ether and the ether extract dried over $MgSO_4$. The ether is removed on a rotary vacuum-evaporator, the residue is combined with the precipitate formed in the reaction, dried over P_2O_5 in a vacuum-desiccator and recrystallized from CCl_4. The yield of 2,3,5,6-tetrafluoro-4-nitrophenol is 33–34 g (86%–89%); mp 73°–74 °C (mp 73.5°–74.5 °C [278]).

2,3,5,6-Tetrafluoro-4-aminophenol. In a 0.5-l three-necked flask fitted with a stirrer, a reflux condenser, and a dropping funnel, are placed 180 g (3.2 mol) of cast-iron turnings, 60 ml of water, 0.5 g of NH_4Cl, and 150 ml of ethanol. 2,3,5,6-Tetrafluoro-4-nitrophenol (20 g, 0.095 mol) in 150 ml of ethanol is slowly added during 0.5 h to the stirred and refluxed reaction mixture. The mixture is refluxed for 3 h and filtered while still hot. The cast-iron turnings are washed with hot ethanol and the combined ethanolic solutions are poured into 700 ml of water. The precipitate is filtered off and dried in air. The yield of 2,3,5,6-tetrafluoro-4-aminophenol is 12 to 12.5 g (70%–73%); mp 181°–182 °C.

Bis(2,3,6-trifluoro-4,5-diaminophenyl)Ether

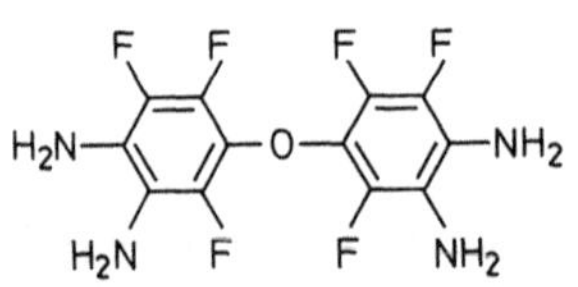

$C_{12}H_8F_6N_4O$ MW 338

Bis(2,3,6-trifluoro-4,5-diaminophenyl) ether is a white crystalline substance, partially soluble in common organic solvents. Mp 184°–185 °C [270, 279].

Method of Preparation

Bis(2,3,6-trifluoro-4,5-diaminophenyl) ether is prepared by the reduction of bis-(trifluoro-4-nitro-3-aminophenyl) ether with iron in ethanol, in the presence of NH_4Cl (95% yield) [270, 279].

Procedure

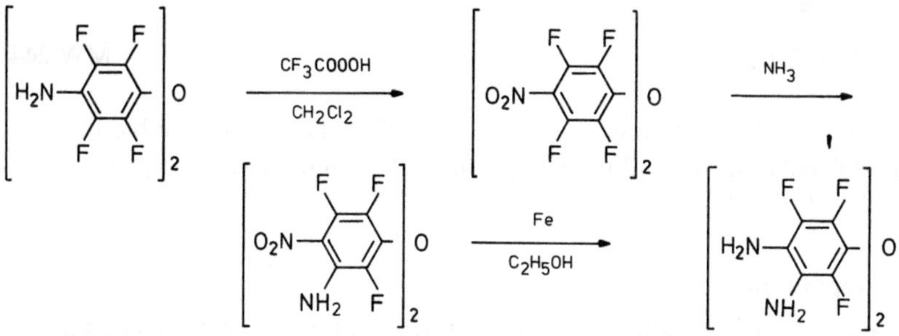

Bis(tetrafluoro-4-nitrophenyl) ether. In a 1-l three-necked flask provided with a stirrer, a reflux condenser, and a dropping funnel, are placed 70 ml of trifluoroacetic oxide and 500 ml of methylene chloride. Then 35 ml of 90% H_2O_2 is added dropwise with vigorous stirring at such a rate as to keep the mixture slightly boiling, whereupon 10 g (0.029 mol) of bis(tetrafluoro-4-aminophenyl) ether in 75 ml of methylene chloride is added (the mixture should gently boil). After 1 h, further 50 ml of trifluoroacetic oxide and 20 ml of 90% H_2O_2 are added. The mixture is boiled for 24 h, cooled and poured into 0.5 l of water. The organic layer is separated, washed with 0.5 l of water and dried over $MgSO_4$. After methylene chloride has been removed in a water-bath, the residue is distilled in vacuum and a fraction boiling at 158°–160 °C/2 mm Hg is collected. The yield of bis(tetrafluoro-4-nitrophenyl) ether is 11 g (94%); bp 158° to 160 °C/2 mm Hg [270].

Bis(trifluoro-4-nitro-3-aminophenyl) ether. In a 0.5-l two necked flask provided with a reflux condenser and a bubbler, are placed 10 g (0.025 mol) of bis(tetrafluoro-4-nitrophenyl) ether and 400 ml of ether. Gaseous NH_3 is passed through the solution during 10 h, at the rate of 13 l/h. The solution is filtered, the filtrate is evaporated on a rotary vacuum-evaporator to remove the ether, and the residue is recrystallized from ethanol. The yield of bis(trifluoro-4-nitro-3-aminophenyl) ether is 9 g (91%); mp 179°–181 °C.

Bis(trifluoro-3,4-diaminophenyl) ether. In a 0.5-l three-necked flask provided with a stirrer, a reflux condenser, and a dropping funnel, are placed 100 g of cast-iron turnings, 1 g of NH_4Cl, 200 ml of ethanol, and 50 ml of water. A hot solution of 9 g (0.022 mol) of bis(trifluoro-4-nitro-3-aminophenyl) ether in 75 ml of ethanol is added to the boiling mixture. The mixture is refluxed for 3 h, filtered while still hot, and the turnings are washed on the filter with 50 ml of hot ethanol. The filtrate is poured into 1.5 l of water, the residue is filtered off on a Büchner funnel, dried and recrystallized from hexane. The yield of bis(2,3,6-trifluoro-4,5-diaminophenyl) ether is 6.8 g (95%); mp 184°–185 °C.

2-Bromo-3,4,5,6-tetrafluoroaniline

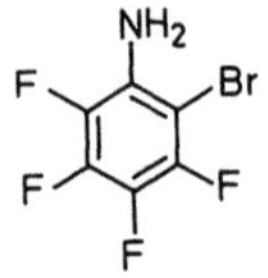

$C_6H_2BrF_4N$ MW 244

2-Bromo-3,4,5,6-tetrafluoroaniline is a white crystalline substance, soluble in common organic solvents. Bp 70 °C/3 mm Hg [259], 65 °C/1 mm Hg [209]; mp 50°–52 °C [209, 262], 52°–53 °C [280].

Method of Preparation

2-Bromo-3,4,5,6-tetrafluoroaniline is prepared by acidolysis of 2-bromo-3,4,5,6-tetrafluoroacetaniline (68% yield) [209, 262], or by bromination of 2,3,4,5-tetrafluoroaniline in acetic acid in the presence of powdered iron and sodium acetate (63% yield) [280].

The following procedure is a modification of the one described in [280].

Procedure

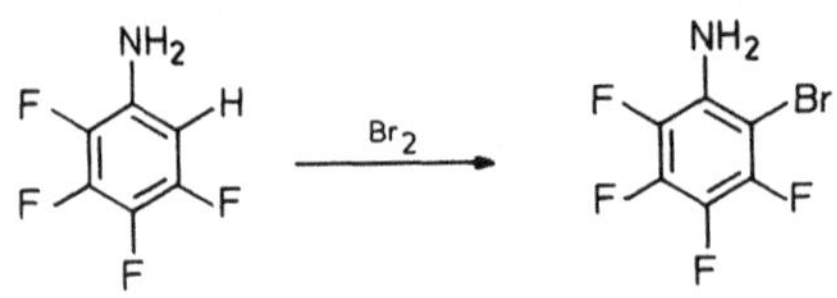

In a 250-ml four-necked flask provided with a stirrer, a reflux condenser, a thermometer, and a dropping funnel, are placed 7 g (0.042 mol) of 2,3,4,5-tetrafluoroaniline, 0.28 g of powdered iron, 5.6 g of anhydrous sodium acetate and 35 ml of glacial acetic acid. The mixture is heated to 45°–50 °C with stirring, then heating is stopped and a solution of 10 g of bromine in 25 ml of glacial acetic acid is added dropwise during 30–40 min. The reaction mixture is stirred for 4 h at 45°–50 °C, then cooled to room temperature. The remaining bromine is reduced by adding a 5% solution of $Na_2S_2O_5$ till the colouring disappears, and bromotetrafluoroaniline formed in the reaction is isolated by steam-distillation of the reaction mixture. The distillate (~2 l) is cooled, the crystallized bromotetrafluoroaniline is filtered off on a glass filter and dried to give 5–6 g of the product. The filtrate is acidified with HCl, extracted with ether, and the extract is dried over $MgSO_4$. The ether is removed to afford additional ~0.5 g of the product. The total yield of 2-bromo-3,4,5,6-tetrafluoroaniline is 5.5–6.5 g (54%–62%); mp 51.5°–52.5 °C.

N,N-Dichloropentafluoroaniline

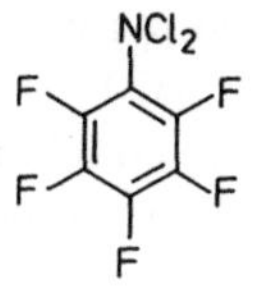

$C_6Cl_2F_5N$ MW 252

N,N-Dichloropentafluoroaniline is a light-yellow liquid, well soluble in common organic solvents, insoluble in water. On standing it is converted to 4-N-chloroimino-1-chloropentafluorocyclohexadiene-2,5. Bp 31°–34 °C/0.8 mm Hg [281]; mp —10 °C [282]; n_D^{20} 1.4012 [281].

Method of Preparation

N,N-Dichloropentafluoroaniline is prepared by the reaction of pentafluoroaniline with tert-butyl hypochlorite in CCl_4 at —30 °C [281, 282].

The following procedure is based on the data reported in [281, 282].

Procedure

$$C_6F_5NH_2 \xrightarrow[CCl_4]{tert\text{-}C_4H_9OCl} C_6F_5NCl_2$$

In a 0.5-l three-necked flask provided with a stirrer, a thermometer, and a pressure-equalizing dropping funnel, are placed 40 g (0.22 mol) of pentafluoroaniline and 200 ml of CCl_4. The mixture is cooled to —20 °C and a solution of 48 g (0.44 mol) of tert-butyl hypochlorite in 200 ml of CCl_4 is dropped in at the above temperature, so that the temperature of the reaction mixture should not rise above —15 °C. The mixture is stirred at that temperature for 1.5 h, whereupon the solution is evaporated in vacuum at 50–100 mm Hg till 70–80 ml is left. The residue is distilled in vacuum and a fraction boiling at 31°–32 °C/1 mm Hg is collected (Note). The yield of N,N-dichloropentafluoroaniline is 41 g (74%).

Note. It is recommended to cool the receiving flask to —15 °C to avoid decomposition of N,N-dichloropentafluoroaniline, which may take place at a temperature above its melting point.

Pentafluorophenylcarbonimidoyl Dichloride

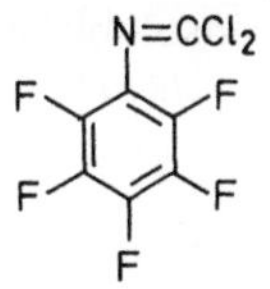

$C_7Cl_2F_5N$ MW 264

Pentafluorophenylcarbonimidoyl dichloride is a colourless liquid, soluble in common organic solvents. Bp 90 °C/40 mm Hg [283].

Method of Preparation

Pentafluorophenylcarbonimidoyl dichloride is prepared by co-pyrolysis of pentafluoroaniline with chloroform or carbon tetrachloride at 620°–670 °C (30%–40%) [283, 284]; by the reaction of pentafluoroaniline with carbon tetrachloride in the presence of $AlCl_3$ (70% yield) [285]; by the reaction of pentafluoroformanilide with sulphuryl chloride [286].

The following procedure is based on the data reported in [286].

Procedure

$$C_6F_5NH_2 \xrightarrow[AlCl_3]{CCl_4} C_6F_5N{=}CCl_2$$

In a 100-ml four-necked flask equipped with a stirrer, a thermometer, a dropping funnel, and a reflux condenser with a calcium chloride tube, are placed 15 ml of absolute carbon tetrachloride and 8.8 g (0.066 mol) of freshly sublimed, powdered aluminium chloride. Then 4 g (0.022 mol) of pentafluoroaniline in 10 ml of absolute CCl_4 is carefully dropped into a vigorously stirred suspension. The reaction mixture is heated to 70°–80 °C with stirring during 12 h, then cooled to room temperature and poured onto 200 g of ice. The organic layer is separated and the aqueous layer extracted with ether (4 × 50 ml). The organic layer and the ether extracts are combined and dried over $MgSO_4$. The solvents are removed on a rotary vacuum-evaporator at 50 °C/10–15 mm Hg and the residue is steam-distilled. The organic layer of the distillate is separated, dried over $MgSO_4$ and distilled in vacuum. A fraction boiling at 93°–94 °C/40 mm Hg is collected. The yield of pentafluorophenylcarbonimidoyl dichloride is 4 g (70%).

Pentafluorophenylhydrazine

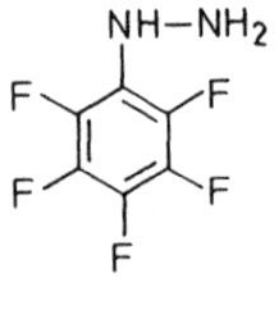

$C_6H_3F_5N_2$ MW 198

Pentafluorophenylhydrazine is a crystalline substance of slightly yellow colour, well soluble in ether, acetone, slightly soluble in water, almost insoluble in petroleum ether; pK_a 9.85 [58]; mp 73°–75 °C [260], 76°–77 °C [259], 77°–78 °C [58].

Method of Preparation

Pentafluorophenylhydrazine is prepared by the reaction of hexafluorobenzene with hydrazine hydrate in aqueous alcoholic solution (50%–71% yield) [58, 259]. The best results are obtained when the reaction is produced in dioxan [58].

The following procedure is based on the data reported in [58, 259].

Procedure

$$C_6F_6 \xrightarrow{H_2N-NH_2 \cdot H_2O} C_6F_5NHNH_2$$

Hexafluorobenzene (40 g, 0.215 mol) and 100% hydrazine hydrate (25 g, 0.5 mol) in 150 ml of absolute dioxan are refluxed during 12 h in a 250-ml two-necked flask provided with a stirrer and a reflux condenser with a calcium chloride tube. The solvent and unchanged hexafluorobenzene are removed on a rotary vacuum-evaporator; the distillate is diluted with water to afford 5 g of hexafluorobenzene. The crystalline precipitate is filtered off and recrystallized from petroleum ether (bp 70° to 100 °C). The yield of pentafluorophenylhydrazine is 20–31 g (69%–73%); mp 75°–77 °C.

Pentafluoronitrosobenzene

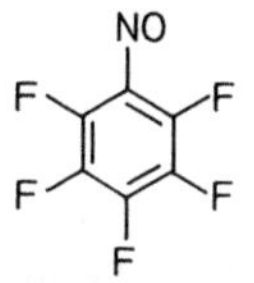

C_6F_5NO MW 197

Pentafluoronitrosobenzene is a crystalline substance of pale-green colour, soluble in common organic solvents, slightly soluble in ethanol and petroleum ether. Mp 44.5°–45 °C [287, 288], 44 °C [289]; bp 65 °C/10 mm Hg [187]; sublimes at 75 °C/10 mm Hg [287]. In a crystalline form it is a dimer, μ 1.93 [290].

Method of Preparation

Pentafluoronitrosobenzene is prepared by oxidation of pentafluoroaniline with performic acid in methylene chloride (48% yield) [287–289]; by the pyrolysis of pentafluorobenzoyl nitrite at 300 °C/2 mm Hg in a platinum tube (the contact time is 5–6 s) (18% yield) [291].

The following procedure is a modification of the one described in [288].

Procedure

$$C_6F_5NH_2 \xrightarrow[HCOOH]{HCOOOH} C_6F_5NO$$

In a 1-l three-necked flask provided with a stirrer, a reflux condenser, and a pressure-equalizing dropping funnel, are placed 200 ml of 98% formic acid and 600 ml of methylene chloride. Then 50 ml of 90% H_2O_2 (d 1.41) is carefully poured in; the mixture should not boil. The reaction mixture is stirred for 0.5 h, then 50 g (0.27 mol) of pentafluoroaniline in 200 ml of methylene chloride is dropped in over a period of 45 min; the mixture should slightly boil. The mixture is refluxed for 5 h, then poured into 500 ml of water, the organic layer is separated, and the dark-green mass is extracted with methylene chloride (2 × 200 ml). The extracts are combined, washed with water (2 × 200 ml) and dried over $MgSO_4$. The mixture is distilled on a fractionating column (high volatility of pentafluoronitrosobenzene may result in losses of the product in distilling off the solvent). This gives 47.6 g of the crude product, which is then distilled in vacuum; a fraction boiling at 50°–60 °C/20 mm Hg is collected. The product is sublimed at 75 °C/10 mm Hg or recrystallized from ethanol. The yield of pentafluoronitrosobenzene is 25.1 g (46.6%); mp 42°–44 °C (the recrystallized product melts at 44.5°–45 °C).

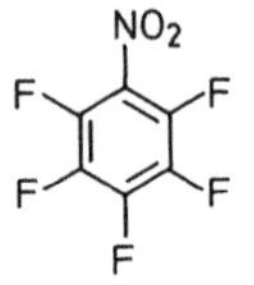

$C_6F_5NO_2$ MW 213

Pentafluoronitrobenzene is a liquid of pale-yellow colour, soluble in common organic solvents; a mild lachrymator. Bp 158°–161 °C [271], 91°–94 °C/52 mm Hg [291].

Method of Preparation

Pentafluoronitrobenzene is prepared by oxidation of pentafluoroaniline with trifluoroperacetic acid (85% yield) [271], or by nitration of pentafluorobenzene with concentrated nitric acid (50%–82%) [271, 291].

The following procedure is a modification of the one described in [271].

Procedure

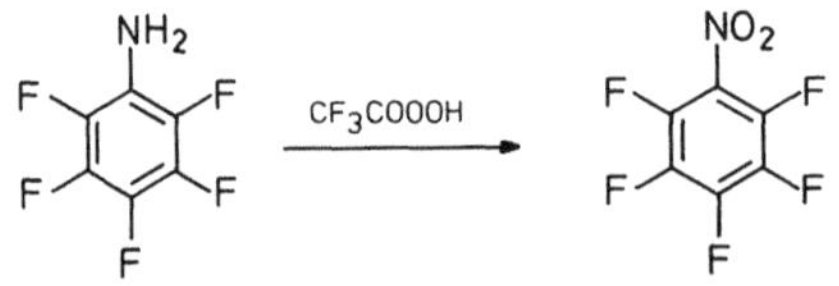

In a 3.5-l three-necked flask provided with a stirrer, an effective reflux condenser with a calcium chloride tube, and a pressure-equalizing dropping funnel, are placed 480 ml of trifluoroacetic oxide (Note 1) and 2 l of methylene chloride. Then 150 ml of 90% H_2O_2 is added dropwise, with vigorous stirring and cooling of the flask with ice-water, at such a rate as to maintain the mixture gently boiling. The mixture is refluxed (gentle refluxing should be maintained) over a period of 1 h (40°–45 °C), whereupon it is cooled to room temperature. Then a solution of 210 g (1.14 mol) of pentafluoroaniline in 300 ml of methylene chloride is gradually added, with vigorous stirring at room temperature; the solution should be added cautiously, so as to maintain gentle boiling of the mixture. The mixture is allowed to boil gently (40°–50 °C) for 1 h, then cooled to room temperature once more, and further 120 ml of trifluoroacetic oxide and 30 ml of 90% H_2O_2 are added to it. Then the mixture is allowed to boil gently, with stirring for 20–30 h (Note 2), whereupon it is cooled, poured into 2 l of cold water, the lower organic layer is separated and washed with cold water (3 × 900 ml). The washings are extracted with methylene chloride (3 × 150 ml), the extracts are combined and dried over $MgSO_4$. After the methylene chloride has been distilled off in a water-bath, the residue is distilled in vacuum and a fraction boiling at 61°–63 °C/17 mm Hg is collected. The yield of pentafluoronitrobenzene is 160–170 g (66%–70%); n_D^{20} 1.4483.

Notes

1. Trifluoroacetic oxide is obtained by the following procedure. In a 2-l flask provided with a reflux condenser with a calcium chloride tube and a dropping funnel, are placed 900 g of P_2O_5, then 300 ml of trifluoroacetic acid is added in two or three portions, with shaking of the flask. The mixture boils up in the flask, so it should be cooled. The mixture is then heated to 40 °C for 4 h and allowed to stand overnight, whereupon it is distilled and a fraction boiling at 40°–49 °C is collected. The yield of trifluoroacetic oxide is 180–200 ml.
2. The end of the oxidation reaction is indicated by the changed colour of the solution (from dark-green to pale-yellow).

Pentafluorophenylazide

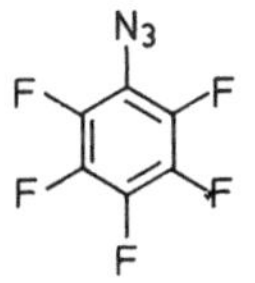

$C_6F_5N_3$ MW 209

Pentafluorophenylazide is a colourless liquid, soluble in common organic solvents; thermally unstable [358]. Bp 52°–54 °C/10 mm Hg [292], 35 °C/5 mm Hg [58].

Method of Preparation

Pentafluorophenylazide is prepared by the reaction of hexafluorobenzene with sodium azide in dimethylformamide (60% yield) [292]; by treatment of pentafluorophenylhydrazine with sodium nitrite in hydrochloric acid (30% yield) [275], or with nitrosyl chloride in acetic acid (52% yield) [275].

The following procedure is based on the data reported in [292].

Procedure

$$C_6F_6 \xrightarrow{NaN_3;\ DMF} C_6F_5N_3$$

In a 0.5-l three-necked flask equipped with a stirrer, a thermometer, and a pressure-equalizing dropping funnel with a calcium chloride tube, are placed 13 g (0.2 mol) of NaN_3 and 200 ml of dry dimethylformamide. The mixture is heated to 60 °C, and 37.2 g (0.2 mol) of hexafluorobenzene is added with stirring, over a period of 1 h. The mixture is then stirred at 65 °C for 2 h, poured into 0.5 l of water and extracted with 250 ml of ether. The ether extract is dried over $MgSO_4$. The ether is distilled off in a water-bath and the residue distilled; a fraction boiling at 52°–54 °C/10 mm Hg is collected (Note). The yield of pentafluorophenylazide is 27–29 g (64%–69%); n_D^{20} 1.4710.

Note. Caution! Pentafluorophenylazide is explosive. Synthesis should be carried out behind the safety screen. Face should be protected with a transparent screen of plastic. Charges should not exceed those indicated above. In distilling off pentafluorophenylazide, the same precautions should be observed as those customarily used when handling explosives. *Overheating of the still should be avoided!* It is preferable to wash off sodium azide from the ether extract, checking it for the azide ion. Explosions have been reported earlier to occur during distillation.

S,S-Difluoro-N-pentafluorophenylsulphimine

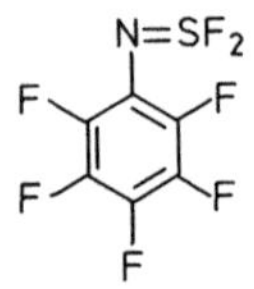

C_6F_7NS MW 251

S,S-Difluoro-N-pentafluorophenylsulphimine is a yellow liquid, soluble in common organic solvents; easily hydrolysed by air moisture. Bp 42°–43 °C/5 mm Hg [293].

Method of Preparation

S,S-Difluoro-N-pentafluorophenylsulphimine is prepared by the reaction of pentafluoroaniline with SF_4 in the presence of NaF (33% yield) [293].

The following procedure is a modification of the one described in [293].

Procedure

$$C_6F_5NH_2 \xrightarrow{SF_4;\ NaF} C_6F_5NSF_2$$

In a 50-ml stainless steel autoclave equipped with a needle valve and a thermocouple in a well, are placed a mixture of 9.2 g (0.05 mol) of pentafluoroaniline and 20 g of NaF ground to homogeneity. The autoclave is cooled to −60 °C with a dry-ice acetone mixture, 10.8 g (0.1 mol) of SF_4 is introduced, and the autoclave is allowed to stand at 0°–3 °C for 3 h, whereupon it is opened and excess pressure is released *(excess SF_4 should be vented off into a good hood)*. The bright-yellow solid mass is transferred from the autoclave to a Teflon bottle with argon (it is not practical to use a glass bottle, as the product is decomposed to bis(pentafluorophenyl)thiodiimide and its yield decreases) and extracted with 1,1,2-trichloro-1,2,2-trifluoroethane. The extract is concentrated in a Teflon vessel and the residue distilled in vacuum; a fraction boiling at 42°–43 °C/5 mm Hg is collected. The yield of S,S-difluoro-N-pentafluorophenylsulphimine is 6.3 g (50%).

N-Sulphinylpentafluoroaniline

C_6F_5NOS MW 229

N-Sulphinylpentafluoroaniline is a yellow liquid, soluble in common organic solvents. Bp 38°–40 °C/4 mm Hg [293].

Method of Preparation

N-Sulphinylpentafluoroaniline is prepared by the reaction of pentafluoroaniline with thionyl chloride in benzene (95% yield) [293].

The following procedure is a modification of the one described in [293].

Procedure

$$C_6F_5NH_2 \xrightarrow[\text{benzene}]{SOCl_2} C_6F_5NSO$$

A mixture of 36.6 g (0.2 mol) of pentafluoroaniline, 35.7 g (0.3 mol) of thionyl chloride, and 40 ml of dry benzene is refluxed during 10 h in a 200-ml round-bottom flask fitted with a reflux condenser with a calcium chloride tube. The solvents are then distilled off in vacuum (30–50 mm Hg) of a water-jet-pump, the residue is distilled on a 25-cm Vigreux column under vacuum and a fraction boiling at 38°–40 °C/4 mm Hg is collected. The yield of N-sulphinylpentafluoroaniline is 43.5 g (95%).

trans-Decafluoroazobenzene

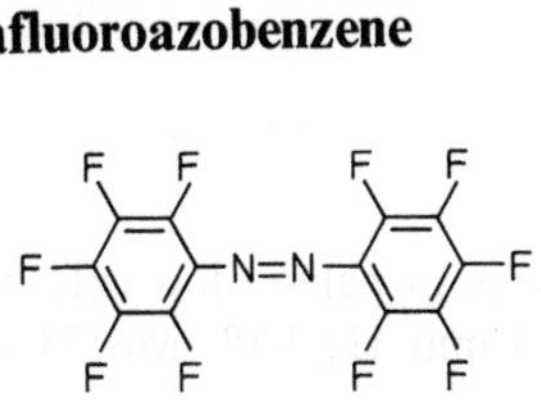

$C_{12}F_{10}N_2$ MW 362

trans-Decafluoroazobenzene is a crystalline substance of red colour, slightly soluble in common organic solvents, partially soluble in acetone, ether and ethanol. Mp 142°–143.5 °C [294], 141°–142 °C [295], 141°–142 °C [281, 289].

Method of Preparation

trans-Decafluoroazobenzene is prepared by the hydrolysis of the solution of decafluorohydrazobenzene in trifluoroacetic acid (47% yield) [294], or of the solution of decafluoroazoxybenzene in SbF_5—HSO_3F (1:1) (90% yield) [294]; by thermolysis (140°–160 °C) of N,N-dichloropentafluoroaniline (60% yield) [282, 295]; by the reaction of N,N-dichloropentafluoroaniline with sulphur [296]; by the reaction of pentafluorophenylazide with pentafluoroaniline (10% yield) [297, 298]; by the reaction of pentafluoronitrosobenzene with pentafluoroaniline (15% yield) [299]; by electrochemical oxidation of pentafluoroaniline on a platinum anode (18% yield) [300], with peracetic acid [40], lead tetraacetate (48% yield) [302], or with chlorinated lime (51% yield) [297].

The following procedure is a modification of the preparations described in [281, 282].

Procedure

$$C_6F_5NCl_2 \xrightarrow{140\ °C} C_6F_5N{=}NC_6F_5$$

In a 150-ml round-bottom flask provided with a reflux condenser with a calcium chloride tube, is placed at −5° to 0 °C 25.2 g (0.1 mol) of N,N-dichloropentafluoroaniline. The flask is then immersed into a Wood's metal bath previously heated to 160°–170 °C. The reaction proceeds with considerable evolution of chlorine and heating of the mixture, and gives a red product, which sublimes onto the inner surface of the flask. After 0.5 h, the hot mixture is pored into water, the precipitate is filtered off, washed with water (2 × 100 ml), dried in air and sublimed in vacuum at 140 °C/5 mm Hg. The yield of *trans*-decafluoroazobenzene is 14.5 g (80%); mp 142° to 143 °C.

Decafluoroazoxybenzene

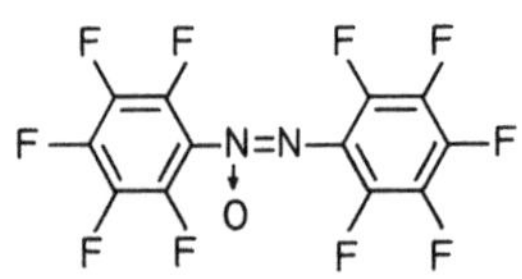

$C_{12}F_{10}N_2O$ MW 378

Decafluoroazoxybenzene is a crystalline substance of pale-yellow colour, soluble in common organic solvents; sublimes at 80 °C/1 mm Hg [40]. Mp 53°–54 °C [40], 54°–55 °C [302].

Method of Preparation

Decafluoroazoxybenzene is prepared by oxidation of decafluoroazobenzene with trifluoroperacetic acid (94% yield) [302]; by oxidation of pentafluoroaniline with peracetic acid (22% yield) [40]; by the reaction of pentafluoronitrosobenzene with triethyl phosphite (26.8% yield), together with some other compounds [303, 304]; by heating pentafluoronitrosobenzene with toluene (32.8% yield) [305]; by the pyrolysis of pentafluoronitrosobenzene [289].

The following procedure is a modification of the one described in [40].

Procedure

$$C_6F_5NH_2 \xrightarrow{CH_3COOOH} C_6F_5\underset{\downarrow \atop O}{N}{=}NC_6F_5$$

In a 250-ml three-necked flask provided with a stirrer, a thermometer, and an air condenser with a calcium chloride tube, are placed 20.1 g (0.11 mol) of pentafluoroaniline, 50 ml of 30% H_2O_2 and 150 ml of acetic acid. The mixture is stirred at 25 °C over a period of 24 h, during which time the colour of the solution has changed from blue to orange. Then it is stirred for an additional 24 h, poured into 600 ml of water and extracted with ether (4 × 150 ml). The extract is washed with water (2 × 100 ml) and dried over $MgSO_4$. The ether is removed on a rotary vacuum-evaporator and the residue recrystallized from petroleum ether (bp 40°–60 °C). The yield of decafluoroazoxybenzene is 12.6 g (60%); mp 53°–54 °C.

4′-Dimethylamino-2,3,4,5,6-pentafluoroazobenzene

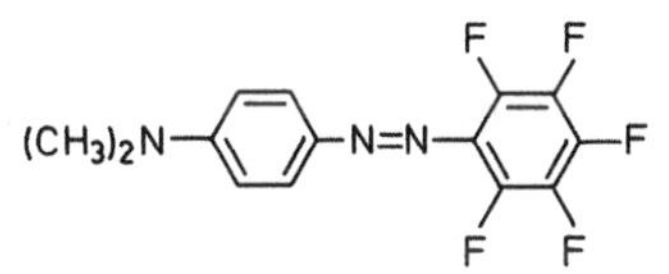

$C_{14}H_{10}F_5N_3$ MW 315

4′-Dimethylamino-2,3,4,5,6-pentafluoroazobenzene is a crystalline substance of orange colour, slightly soluble in common organic solvents, well soluble in acetone and ether. Mp 147.5°–149 °C [51].

Method of Preparation

4′-Dimethylamino-2,3,4,5,6-pentafluoroazobenzene is prepared by the reaction of dimethylaniline with pentafluoroaniline diazonium salt obtained by diazotization of pentafluoroaniline in anhydrous HF (43% yield) [51].

The following procedure is a modification of the one described in [51].

Procedure

$$C_6F_5NH_2 \xrightarrow[HF]{NaNO_2} [C_6F_5N{\equiv}N]^+ \xrightarrow{C_6H_5N(CH_3)_2} C_6F_5N{=}NC_6H_4N(CH_3)_2$$

Anhydrous HF (150 ml) (it is possible to use HF containing up to 20% of water; see p. 1) is placed in a 250-ml Teflon (or polyethylene) reactor equipped with a stirrer, a thermometer, and a pressure-equalizing dropping funnel, and is cooled to −10° to −5 °C. Pentafluoroaniline (18.3 g, 0.1 mol) is then added in portions; the mixture should not boil. Dry sodium nitrite (10.4 g, 0.15 mol) is slowly added in portions during 0.5 h, at the above temperature. The mixture is stirred for 3 h and 12 g of sulphanilic acid is added, followed by 45 ml of dimethylaniline, whereupon the mixture is stirred for 2 h and poured into 600 ml of water. The dark-orange precipitate is filtered off, dried in air and recrystallized from the benzene-petroleum ether (bp 70° to 100 °C) (1:1) mixture. The yield of 4′-dimethylamino-2,3,4,5,6-pentafluoroazobenzene is 15.8 g (50%); mp 148°–149 °C.

Pentafluorophenyltrimethylsilane

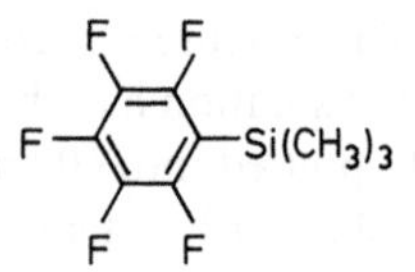

$C_9H_9F_5Si$ MW 240

Pentafluorophenyltrimethylsilane is a colourless liquid, soluble in common organic solvents. Bp 93 °C/57 mm Hg [306], 60 °C/14 mm Hg [307].

Method of Preparation

Pentafluorophenyltrimethylsilane is prepared by the reaction of the Grignard reagent, obtained from bromopentafluorobenzene, with trimethylchlorosilane in ether (90% yield) [307–310], (33% yield) [306]; from pentafluorobenzene, diethyl magnesium and trimethylchlorosilane [309]; by irradiation of the mixture of hexafluorobenzene and trimethylsilane (53% yield) [306, 311].
The following procedure is a modification of the one described in [307].

Procedure

$$C_6F_5MgCl \xrightarrow[\text{ether}]{(CH_3)_3SiCl} C_6F_5Si(CH_3)_3$$

A Grignard reagent (p. 126) is prepared from 20.2 g of chloropentafluorobenzene in a 0.5-l three-necked flask fitted with a stirrer, a reflux condenser with a calcium chloride tube, and a pressure-equalizing dropping funnel. The Grignard reagent is cooled to 0 °C and a solution of 10.9 g (0.1 mol) of trimethylchlorosilane in 100 ml of absolute ether is dropped in; the mixture should not be allowed to boil vigorously. The mixture is then allowed to boil gently with stirring for 5 h, poured onto 0.5 kg of ice, the organic layer is separated and the aqueous one extracted with ether (2 × 50 ml). The ether extracts are combined and dried over $MgSO_4$. The ether is removed in a water-bath and the residue distilled in vacuum; a fraction boiling at 93°–94 °C/57 mm Hg is collected. The yield of pentafluorophenyltrimethylsilane is 21.6 g (90%).

Tris(pentafluorophenyl)phosphine

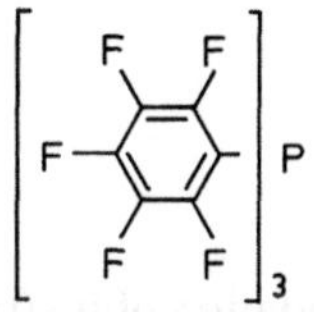

$C_{18}F_{15}P$ MW 532

Tris(pentafluorophenyl)phosphine is a white crystalline substance, soluble in common organic solvents. Mp 116°–117 °C [312–314].

Method of Preparation

Tris(pentafluorophenyl)phosphine is prepared by the reaction of PCl_3 with the Grignard reagent obtained from bromopentafluorobenzene (39.5% yield) [312], (25% yield) [313].

The following procedure is a modification of the one described in [312].

Procedure

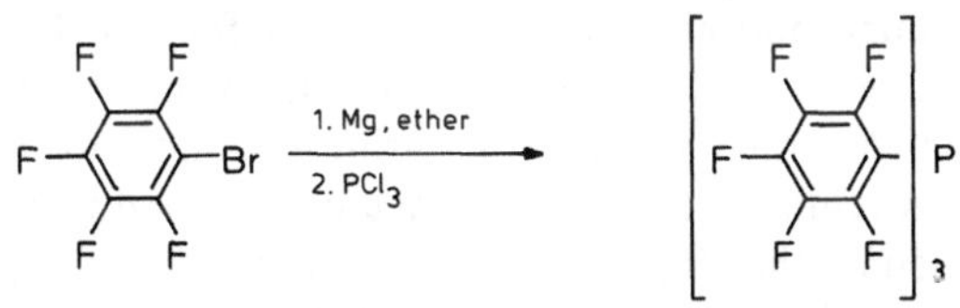

In a 0.5-l three-necked flask provided with a stirrer, a reflux condenser with a calcium chloride tube, and a dropping funnel with a gas inlet tube, are placed 13.8 g (0.1 mol) of freshly distilled PCl_3 in 200 ml of dry ether. A Grignard reagent obtained from 74.1 g (0.3 mol) of bromopentafluorobenzene, and 8.5 g (0.35 mol) of magnesium in 200 ml of ether are added in a flow of argon at —20 °C, with vigorous stirring, in such a way that the temperature should not rise above —18 °C. The mixture is allowed to stay for 20 h under these conditions, poured onto a mixture of 600 g of ice and 200 ml of 10% HCl, and extracted with ether (3 × 200 ml). The ether extract is washed with 500 ml of water and dried over $MgSO_4$. The ether is removed on a rotary vacuum-evaporator, the residue is recrystallized from the pentane-benzene mixture and sublimed at 100°–103 °C/3 mm Hg. The yield of tris(pentafluorophenyl)-phosphine is 32 g (60%); mp 116°–117 °C.

Tris(pentafluorophenyl)phosphine Oxide

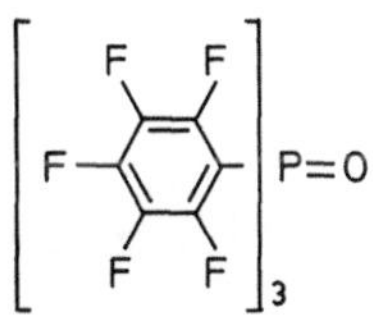

$C_{18}F_{15}OP$ MW 548

Tris(pentafluorophenyl)phosphine oxide is a white crystalline substance, soluble in common organic solvents. Mp 169°–170 °C [312, 315].

Method of Preparation

Tris(pentafluorophenyl)phosphine oxide is prepared by oxidation of tris(pentafluorophenyl)phosphine with sodium chromate in acetic acid (97% yield) [312, 315], or by maintaining it in fluorosulphonic acid with subsequent hydrolysis (98% yield) [316].

The following procedure is a modification of the one described in [312].

Procedure

$$(C_6F_5)_3P \xrightarrow[CH_3COOH;\ H_2SO_4]{Na_2Cr_2O_7} (C_6F_5)_3P{=}O$$

A mixture of tris(pentafluorophenyl)phosphine (26.6 g, 0.05 mol), $Na_2Cr_2O_7$ (100 g), water (350 ml), concentrated H_2SO_4 (25 ml) and acetic acid (350 ml) is refluxed over a period of 6 h in a 1-l round-bottom flask fitted with a reflux condenser. Then it is poured into a 2-l beaker, neutralized with a saturated solution of sodium carbonate, continuously extracted with 500 ml of chloroform during 4 h, and dried over $MgSO_4$. After chloroform has been removed on a rotary vacuum-evaporator, the residue is recrystallized from petroleum ether (bp 70°–100 °C). The yield of tris(pentafluorophenyl)phosphine oxide is 24.7 g (90%); mp 169°–170 °C.

Pentafluorophenyltetrafluorophosphorane

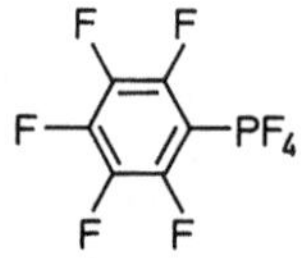

C_6F_9P MW 274

Pentafluorophenyltetrafluorophosphorane is a colourless liquid with a strong odour; *its vapour is choky*. It is extremely sensitive to air moisture; soluble in common organic solvents. Bp 121 °C [317], 122°–125 °C [318].

Method of Preparation

Pentafluorophenyltetrafluorophosphorane is prepared by the reaction of pentafluorophenyldifluorophosphine with chlorine in the presence of SbF_3 (65% yield) [317];

by the reaction of pentafluorophenylmagnesium bromide with PF_5 (10% yield) [318]; by oxidation of pentafluorophenyldifluorophosphine with XeF_2 [319].

The following procedure is based on the data reported in [317].

Procedure

$$C_6HF_5 \xrightarrow[AlCl_3]{PCl_3} C_6F_5PCl_2 \xrightarrow{SbF_3} C_6F_5PF_2 \xrightarrow[SbF_3]{Cl_2} C_6F_5PF_4$$

Pentafluorophenyldichlorophosphine. In a 250-ml three-necked flask provided with a stirrer, a reflux condenser with a calcium chloride tube, and a pressure-equalizing dropping funnel with a gas inlet tube, are placed 50.4 g (0.3 mol) of pentafluorobenzene, 41.2 g (0.3 mol) of freshly distilled PCl_3 and 56 g (0.46 mol) of anhydrous $AlCl_3$. The mixture is heated, with stirring, in a flow of argon at 80 °C for 3 h, then at 150 °C for 1 h, cooled and 64.5 g (0.42 mol) of freshly distilled $POCl_3$ is added to it. A voluminous white precipitate is formed. The mixture is washed by decanting with petroleum ether (4 × 150 ml) and the washings are combined. The solvent is distilled off in a water-bath and the residue is distilled in vacuum; a fraction boiling at 118°–122 °C/18 mm Hg is collected. The yield of pentafluorophenyldichlorophosphine is 47–48 g (59%–60%).

Pentafluorophenyldifluorophosphine. A mixture of 60 g (0.223 mol) of pentafluorophenyldichlorophosphine and 54 g (0.3 mol) of freshly dried SbF_3 (Note 1) is heated in a flow of argon, in a 250-ml two-necked flask provided with a reflux condenser with a calcium chloride tube and a gas inlet system, at 150 °C for 2 h (Note 2). The reaction product is distilled and a fraction boiling at 129°–130 °C is collected. The yield of pentafluorophenyldifluorophosphine is 37–38 g (70%–72%).

Pentafluorophenyltetrafluorophosphorane. A 50-ml stainless steel autoclave is charged with 30 g (0.127 mol) of pentafluorophenyldifluorophosphine and 30 g (0.167 mol) of freshly dried SbF_3. The autoclave is cooled to —120° ÷ —100 °C and 10 g (0.141 mol) of liquid chlorine is then added in portions (Note 3). The autoclave is closed, maintained at 20 °C for 15 h and heated at 110 °C for 6 h, then cooled to —70 °C, the pressure is reduced to that of the atmosphere and the autoclave is opened. The reaction mixture is transferred into a 100-ml flask and distilled; a fraction boiling at 118°–128 °C is collected. This fraction is redistilled, and a fraction boiling at 121°–122 °C is collected. The yield of pentafluorophenyltetrafluorophosphorane is 24–25 g (69%–72%).

Notes

1. Drying of SbF_3 is carried out in a desiccator at 90°–100 °C, during 5 h.
2. The beginning of the reaction is indicated by precipitation of black metallic antimony. The reaction proceeds with evolution of heat, therefore cooling of the flask with an ice-bath is necessary at the start.
3. When chlorine is added to the reaction mixture, a violent reaction starts, therefore the reaction mixture should be vigorously cooled and chlorine should be added in small portions.

Dimethyl Pentafluorobenzenephosphonate

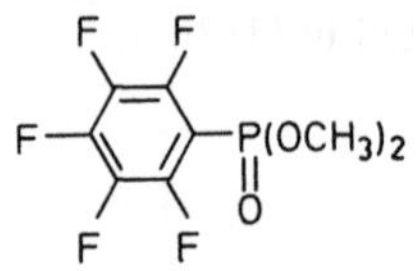

$C_8H_6F_5O_3P$ MW 276

Dimethyl pentafluorobenzenephosphonate is a colourless liquid, soluble in common organic solvents. Bp 46°–47 °C/10 mm Hg [320]; n_D^{23} 1.4443 [320].

Method of Preparation

Dimethyl pentafluorobenzenephosphonate is prepared by treatment of pentafluorobenzenephosphonic acid with diazomethane in ether (74% yield) [320]; by the Michaelis-Becker reaction of hexafluorobenzene with $NaPO(OCH_3)_2$ (15% yield) [321].

The following procedure is a modification of the one described in [321].

Procedure

$$C_6F_6 \xrightarrow{NaPO(OCH_3)_2} C_6F_5PO(OCH_3)_2$$

A mixture of $NaPO(OCH_3)_2$ (13.2 g, 0.1 mol), hexafluorobenzene (14.9 g, 0.08 mol) and absolute tetrahydrofuran (100 ml) is refluxed with stirring for 10 h in a 200-ml round-bottom flask with a stirrer and a reflux condenser with a calcium chloride tube. The mixture is then poured into 500 ml of water, extracted with ether (3 × 100 ml) and dried. The ether is distilled off in a water-bath and the residue distilled in vacuum; a fraction boiling at 46°–47 °C/10 mm Hg is collected. The yield of dimethyl pentafluorobenzenephosphonate is 6 g (27%); n_D^{20} 1.4440.

Tris(pentafluorophenyl)phosphate

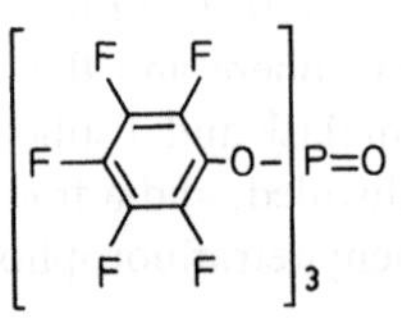

$C_{18}F_{15}O_4P$ MW 596

Tris(pentafluorophenyl)phosphate is a colourless liquid, soluble in common organic solvents. Bp 164°–168 °C/1 mm Hg [322].

Method of Preparation

Tris(pentafluorophenyl)phosphate is prepared by the reaction of pentafluorophenol with $POCl_3$ (25% yield) [322].

The following procedure is a modification of the one described in [322].

Procedure

$$C_6F_5OH \xrightarrow{POCl_3} (C_6F_5O)_3PO$$

In a 250-ml three-necked flask provided with a thermometer and a reflux condenser with a calcium chloride tube, are placed 165 g (0.9 mol) of pentafluorophenol and 46 g (0.3 mol) of freshly distilled $POCl_3$. The mixture is heated, with stirring at a reasonable rate, at 70 °C for 3 h, then at 120 °C for 5 h, and at 170 °C for 6 h. During the reaction HCl is evolved, which is neutralized in a trap with a diluted aqueous solution of an alkali. The reaction mixture is distilled in vacuum and a fraction boiling at 174°–175 °C/2 mm Hg is collected. The yield of tris(pentafluorophenyl)phosphate is 130–140 g (73%–78%).

Bis(pentafluorophenyl)mercury

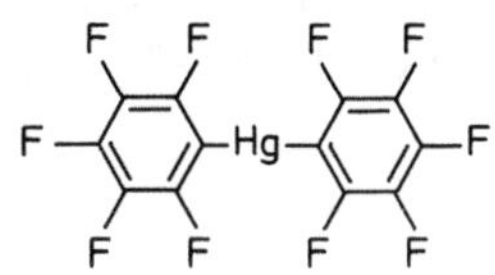

$C_{12}F_{10}Hg$ MW 534

Bis(pentafluorophenyl)mercury is a white crystalline substance, soluble in most organic solvents; sublimes at 140 °C/10 mm Hg [50]. Mp 142.5 °C [323], 138°–139 °C [50, 324], 117°–118 °C [102].

Method of Preparation

Bis(pentafluorophenyl)mercury is prepared by the reaction of iodopentafluorobenzene with mercury at 300 °C (75% yield) [50]; by the reaction of $HgCl_2$ with the Grignard reagent obtained from bromopentafluorobenzene (73% yield) [323]; by the reaction of bromopentafluorobenzene with lithium amalgam (51% yield) [102]; by the reaction of pentafluorobenzene with mercury or its compounds (30%–90% yields) [324]; by the reaction of pentafluorophenyllithium with $HgCl_2$ (94% yield) [108]; by heating pentafluorophenylhydrazine with HgO/Hg (64% yield) [58]; by the reaction of tetrakis(pentafluorophenyl)titanium with $HgCl_2$ in benzene (82% yield) [325]; by the thermolysis of mercury pentafluorobenzoate at 212 °C (56% yield) [326].

The following procedure is a modification of the one described in [324].

Procedure

$$C_6HF_5 \xrightarrow[SbF_5]{Hg_2F_2} C_6F_5HgC_6F_5$$

A mixture of 3.36 g (0.02 mol) of pentafluorobenzene, 4.38 g (0.01 mol) of Hg_2F_2, and 50 g of SbF_5 is heated at 70 °C over a period of 2 h, in a 100-ml three-necked fask provided with a stirrer, a thermometer, and a pressure-equalizing dropping funnel. The mixture is then poured onto 0.5 kg of ice, the precipitate is filtered off on a Büchner funnel, washed with water till neutral, and recrystallized from CCl_4. The yield of bis(pentafluorophenyl)mercury is 9.8 g (90%); mp 138°–139 °C.

Pentafluorophenyliodine Tetrafluoride

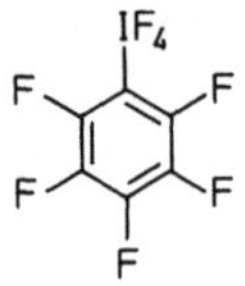

C_6F_9I MW 370

Pentafluorophenyliodine tetrafluoride is a white crystalline substance, soluble in common organic solvents, hydrolysed by air moisture. Mp 74°–75 °C [328, 327].

Method of Preparation

Pentafluorophenyliodine tetrafluoride is prepared by the reaction of pentafluorophenyliodine difluoride with XeF_2 (100% yield) [327], or by the reaction of iodopentafluorobenzene with ClF_3 [328].

The following procedure is a modification of the one described in [327].

Procedure

$$C_6F_5I \xrightarrow[HNO_3]{(CF_3CO)_2O;} C_6F_5I(OCOCF_3)_2 \xrightarrow{SF_4} C_6F_5IF_2 \xrightarrow{XeF_2} C_6F_5IF_4$$

Di(trifluoroacetoxy)iodopentafluorobenzene. In a 100-ml four-necked flask fitted with a stirrer, a reflux condenser, a thermometer, and a pressure-equalizing dropping funnel, are placed 14.7 g (0.05 mol) of iodopentafluorobenzene and 45 ml of trifluoroacetic oxide. The reaction mixture is cooled to —35° ÷ —30 °C, and 5 ml (0.12 mol) of nitric acid (d 1.5) is slowly dropped in with vigorous stirring; the temperature of the mixture should not rise above —25 °C. After all the nitric acid has been added, the mixture is heated to 20 °C and stirred during 3 h, whereupon it is evaporated to dryness at 30°–40 °C/50–100 mm Hg on a rotary vacuum-evaporator. To remove residual nitric acid, the product is maintained at 40 °C/1 mm Hg for 2.5 h. The yield of di(trifluoroacetoxy)iodopentafluorobenzene is 25.5 g (98%); mp 119°–120 °C (mp 118°–119 °C [329]).

Pentafluorophenyliodine difluoride. A 100-ml autoclave fitted with a thermocouple in a well and a needle valve, is charged with 26 g (0.05 mol) of di(trifluoroacetoxy)-iodopnetafluorobenzene. The autoclave is cooled to —78 °C with dry-ice acetone mixture and 12 g (0.11 mol) of SF_4 is condensed into it (see p. 2). The autoclave is closed and heated at 70 °C for 3 h, then cooled to —10 °C and excess pressure is carefully released. It should be borne in mind that part of the mixture may be ejected if the temperature is sharply raised to 20 °C. The autoclave is opened and heated to 70 °C, in order to facilitate the discharge. The contents of the autoclave are transferred into a flask and distilled in vacuum; a fraction boiling at 88°–90 °C/0.1 mm Hg is collected. The yield of pentafluorophenyliodine difluoride is 15 g (90 %); mp 47° to 48 °C (mp 47°–48 °C [329]).

Pentafluorophenyliodine tetrafluoride. In a 50-ml two-necked flask provided with a thermometer and an air condenser connected with a bubble counter filled with H_2SO_4, are placed 3.2 g (0.01 mol) of pentafluorophenyliodine difluoride and 2.2 g (0.013 mol) of XeF_2. The temperature of the mixture is then slowly raised during 0.5 h to 60° to 65 °C; bubbling should not be vigorous. The mixture is maintained at that temperature for 3 h; the flask should be intermittently shaken to make XeF_2, sublimed to the top of the flask, drop to the bottom. The mixture is then allowed to stand at 20 °C for 12 h, whereupon the contents of the flask are pumped out at 20 °C during 2 h and the residue is distilled in vacuum at 1 mm Hg. The yield of pentafluorophenyliodine tetrafluoride is 3.8 g (100 %); mp 74°–75 °C.

5,6,7,8-Tetrafluorobenzobicyclo[2,2,2]octatriene

(Tetrafluorobenzobarrelene)

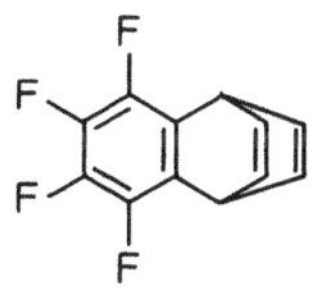

$C_{12}H_6F_4$ MW 226

Tetrafluorobenzobarrelene is a white crystalline product, stable to air moisture and light. Bp 108°–110 °C/2 mm Hg [145, 330]; mp 70°–71 °C [145, 330, 332], 70°–72 °C [333], 71°–72 °C [331], 72°–72.5 °C [334].

Method of Preparation

Tetrafluorobenzobarrelene is prepared by the reaction of tetrafluorodehydrobenzene with benzene (32 %–55 % yield). Yields of the product depend to a certain extent on the method of generation of tetrafluorodehydrobenzene [145, 331–334].

The following procedure is a modification of the one described in [331].

Procedure

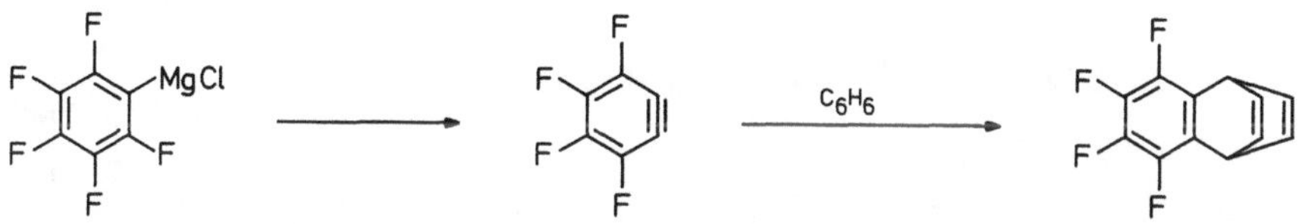

Absolute benzene (400 ml) is added to pentafluorophenylmagnesium chloride prepared from 90 g of chloropentafluorobenzene (p. 126) and the ether-benzene mixture is distilled off at 60°–80 °C (the temperature of the bath). Then 200 ml of absolute benzene is added, the mixture is cooled to —20 °C and, after 15 min, a solution of 46 ml of freshly distilled ethylene oxide in 100 ml of absolute benzene is added in small portions. The mixture is allowed to stand for 15 min, then refluxed during 5 h, cooled to 0 °C, poured into 500 ml of 10% HCl solution and extracted with ether. The ethereal solution is washed with 10% Na_2CO_3 solution, then with water till neutral, and dried over $MgSO_4$. The ether is evaporated and the residue is distilled to remove α- and β-pentafluorophenylethanols; a fraction boiling at 43°–45 °C/5 mm Hg is collected. The still-residue is cooled, dissolved in 250 ml of ether, passed through an Al_2O_3 column and eluted with 200 ml of benzene. The solvent is distilled off, the residue is recrystallized from ethanol and sublimed under vacuum (5 mm Hg) (the temperature of a bath should not rise above 100 °C). The yield of tetrafluorobenzobarrelene is 9–11 g (9%–11%); mp 71°–71 °C.

4,5,6,7-Tetrafluoroindole

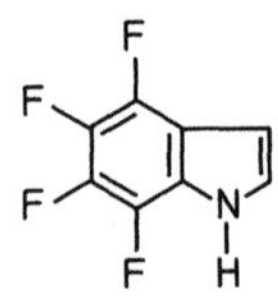

$C_8H_3F_4N$ MW 189

4,5,6,7-Tetrafluoroindole is a white crystalline substance with a specific indole odour; soluble in common organic solvents. Mp 92°–93 °C [335], 92.5°–93 °C [336], 93° to 93.5 °C [335].

Method of Preparation

4,5,6,7-Tetrafluoroindole is prepared by dehydration of 4,5,6,7-tetrafluoroindoline (35% yield); by cyclization of 1-pentafluorophenyl-2-aminoethanol by heating it in dimethylformamide (60% yield) [335]; or by the hydrolysis and subsequent decarboxylation of 2,3-dicarbethoxy-4,5,6,7-tetrafluoroindole, formed in cyclization of diethyl N-acetyl-N-(2,3,4,5,6-pentafluorophenyl)aminofumarate (~3% yield) [336].

The following procedure is a modification of the one described in [335].

Procedure

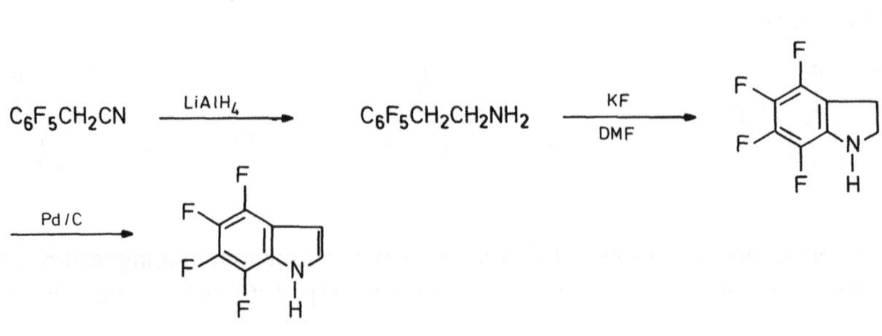

2-(Pentafluorophenyl)ethylamine. In a 1-l four-necked flask equipped with a hermetically sealed stirrer, a reflux condenser with a calcium chloride tube, a pressure-equalizing dropping funnel, and a gas inlet tube, is placed 9.5 g of $LiAlH_4$ in 380 ml of absolute ether. Then 50 g (0.24 mol) of pentafluorophenylacetonitrile in 75 ml of absolute ether is added dropwise to the stirred suspension in a flow of dry argon during 40–60 min. The rate of addition is regulated so as to maintain gentle boiling of the mixture. If the mixture boils vigorously, the flask is cooled with cold water. The reaction mixture is then stirred at room temperature for 1 h, and at 40 °C for 4 h. Then 57 ml of water and 15 ml of 20% NaOH solution are very slowly dropped into the mixture cooled with ice-water *(the reaction proceeds vigorously, with considerable evolution of heat)*. The mixture is allowed to settle and is filtered. The precipitate and the aqueous layer are washed with ether, the ether is separated by decantation. The ethereal solutions are combined and dried over $MgSO_4$. The ether is removed ona rotary vacuum-evaporator. The residue is immediately (to avoid decomposition) distilled in vacuum of a water-jet pump; a fraction boiling at 80°–90 °C/20 mm Hg is collected. The yield of 2-(pentafluorophenyl)ethylamine is 15–16 g (30%–31%) (Note 1) (bp 80°–90 °C/20 mm Hg [103]).

4,5,6,7-Tetrafluoroindoline. In a 1-l three-necked flask fitted with a stirrer, a reflux condenser, and a thermometer, are placed 14.2 g (0.067 mol) of 2-(pentafluorophenyl)-ethylamine in 375 ml of anhydrous dimethylformamide and 4.3 g of freshly calcinated anhydrous KF. The mixture is refluxed during 10 h, whereupon it is poured into 400 ml of water and steam-distilled. First a white product is distilled off, then a slightly yellow one, which crystallizes in a condenser. The solid product is filtered off and the aqueous layer is extracted with three or four portions of ether. The ether extracts are combined and dried over $MgSO_4$. The ether is distilled off and the oily residue crystallizes upon standing at 0° ÷ −5 °C. Sublimation at 70 °C/30 mm Hg gives 5–6 g (39%–47%) of 4,5,6,7-tetrafluoroindoline; mp 60°–61 °C.

4,5,6,7-Tetrafluoroindole. In a 100-ml round-bottom flask provided with a long air condenser, are placed 3.5 g of 10% palladium-on-charcoal (Note 2) and 8 g (0.42 mol) of 4,5,6,7-tetrafluoroindoline. The mixture is heated at 200 °C (bath) for 5 h, then allowed to cool, treated with ether and filtered to remove the catalyst. The residue is washed with ether till the odour of tetrafluoroindole disappears. The ethereal solutions are combined and dried over $MgSO_4$. The ether is distilled off and the residue is sublimed at 90 °C/20 mm Hg or recrystallized from hexane. The yield of 4,5,6,7-tetrafluoroindole is 3–3.5 g (38%–44%); mp 92°–93 °C.

Notes

1. Removal of the ether and distillation of the residue in vacuum should be carried out only before further use of 2-(pentafluorophenyl)ethylamine, as it is unstable. Care should be taken to avoid solidification of the product in the condenser during distillation; distillation should preferably be carried out in a flow of dry argon.
2. Preparation of 10% palladium-on-charcoal is as follows: 3 g of $PdCl_2$ is dissolved in a mixture of 2 ml of concentrated HCl and 14.4 ml of water, in a 0.5-l two-necked flask. The mixture is heated at 50 °C for 1 h, then a solution of 48.6 g of crystalline sodium acetate ($CH_3COONa \cdot 3\,H_2O$) in 180 ml of water is added, together with 16.2 g of freshly activated charcoal. To obtain freshly activeted

charcoal, 150 ml of 15% HCl is poured onto 50 g of activated charcoal and the mixture is boiled for 0.5 h. The hot solution is filtered on a Büchner funnel, the charcoal is washed with hot water till it is no longer acid to litmus and dried at 110 °C for 1 h. Testing of the sample of this charcoal for Fe^{3+} by treatment with a mixture of diluted HCl and HNO_3, followed by the addition of $K_3[Fe(CN)_6]$, should be negative.
The mixture is placed in a shaker for hydration and hydrated during 3 h with hydrogen from a cylinder. The catalyst is filtered off, washed with water, dried in air at room temperature, then over KOH in a desiccator. The product should be kept in a hermetically sealed container.

4,5,6,7-Tetrafluoro-2-methyl-3-carbethoxycoumarone

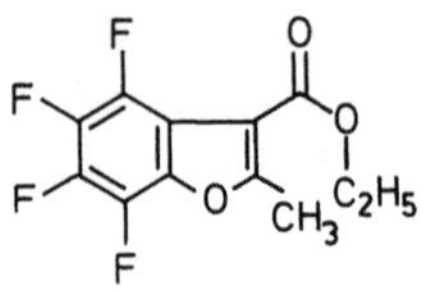

$C_{12}H_8F_4O_3$ MW 276

4,5,6,7-Tetrafluoro-2-methyl-3-carbethoxycoumarone is a colourless or pale-yellow crystalline substance, soluble in common organic solvents. Mp 69.7°–70 °C [226].

Method of Preparation

4,5,6,7-Tetrafluoro-2-methyl-3-carbethoxycoumarone is prepared by the reaction of hexafluorobenzene with sodium acetoacetate ether (30% yield) [226].

Procedure

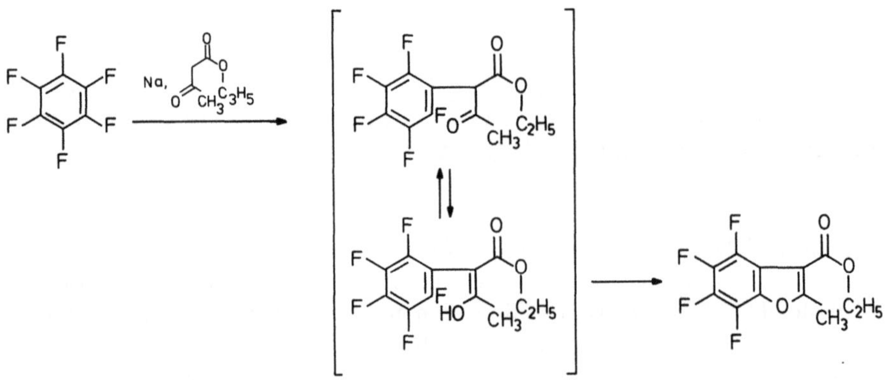

In a 0.5-l four-necked flask provided with a stirrer, a reflux condenser with a calcium chloride tube, a thermometer, and a pressure-equalizing dropping funnel, are placed 7 g (0.29 mol) of sodium hydride and 200 ml absolute dimethylformamide. Acetoacetate ether (34.9 g, 0.269 mol) is gradually added, with continuous stirring; the temperature in the flask is maintained at 25°–28 °C. The reaction proceeds with evolution of hydrogen and the mixture assumes a yellow colour. Stirring is continued for 2 h till hydrogen has ceased to evolve. Then 50 g (0.27 mol) of hexafluorobenzene is poured in and the mixture is refluxed at 130 °C for 5 h. The dark-brown mixture is cooled, poured into 600 ml of water and extracted with chloroform (4 × 200 ml). The solvents are distilled off to give 27–28 g of brown solid product. After recrystallization from methanol the yield of 4,5,6,7-tetrafluoro-2-methyl-3-carbethoxycoumarone is 20–22 g (30%–33%); mp 67°–69 °C.

References

1. Organic Reactions, vol. 5. R. Adams (ed). Wiley, New York, 1949.
2. Organic Synthesis, Coll. Vol. 2. A. Blatt (ed). Wiley, New York, 1946.
3. USSR Pat. 279,608; C.A. *74*, 64044 (1971). Brit. Pat. 1,130,263; C.A. *70*, 47063 (1969)
4. Grakauskas, V.: J. Org. Chem. *34*, 2835 (1964)
5. Shaw, M. J., Hyman, H. H., Filler, R.: J. Am. Chem. Soc. *92*, 6498 (1970)
6. Schiemann, G.: J. Prakt. Chem. *140*, 97 (1934)
7. Nad, M. M., Talalaeyeva, T. V., Kazennikova, G. V., Kocheshkov, K. K.: Izv. Akad. Nauk SSSR Otd. Khim. Nauk *1959*, 65
8. Chambers, R., Heyes, J., Musgrave, W.: Tetrahedron *19*, 891 (1963)
9. Desirant, Y.: Bull. Soc. Chim. Belg. *67*, 676 (1958)
10. Godsell, J., Stacey, M., Tatlow, J.: Nature (Lond.) *178*, 199 (1956)
11. Godsell, J., Stacey, M., Tatlow, J.: Tetrahedron *2*, 193 (1958)
12. Yakobson, G. G., Platonov, V. E., Vorozhtsov, N. N.: Zh. Obshch. Khim. *35*, 1158 (1965)
13. Parshall, G.: J. Org. Chem. *27*, 4649 (1962)
14. Evans, F., Tiley, P.: J. Chem. Soc. (Lond.) C *1966*, 134
15. Counsell, I., Green, I., Hales, I., Martin, I.: Trans. Farad. Soc. *61*, 212 (1965)
16. Birchall, J., Haszeldine, R.: J. Chem. Soc. (Lond.) *1959*, 13
17. McBee, E., Lindgrem, V., Ligett, V.: Ind. Eng. Chem. *39*, 378 (1947)
18. Vorozhtsov, N. N., Platonov, V. E., Yakobson, G. G.: Izv. Akad. Nauk SSSR Otd. Khim. Nauk *1963*, 1524
19. Fuller, G.: J. Chem. Soc. (Lond.) *1965*, 6264
20. Gething, B., Patrick, C., Tatlow, J., Banks, R., Barbour, A., Tipping, A.: Nature (Lond.) 183,586 (1959)
21. Brit. Pat. 917,412; C.A. *59*, 3831 (1963)
22. US Pat. 3,004,077; C.A. *59*, 2711 (1963)
23. Brit. Pat. 917,413; C.A. *59*, 3852 (1963)
24. Coe, P., Patrick, C., Tatlow, J.: Tetrahedron *9*, 240 (1960)
25. Brooke, G., Chambers, R., Heyes, J., Musgrave, W.: Proc. Chem. Soc. Lond. *1963*, 94
26. Brooke, G., Chambers, R., Heyes, J., Musgrave, W.: J. Chem. Soc. (Lond.) *1964*, 729
27. Mobbs, R., Musgrave, W.: Chem. Ind. (Lond.) *1961*, 1268
28. Belg. Pat. 617,337; C.A. *58*, 13841 (1963)
29. Chambers, R., Heaton, C., Musgrave, W.: J. Chem. Soc. (Lond.) C *1968*, 1933
30. Desirant, Y.: Bull. Sci. Acad. R. Belg. *41*, 759 (1955)
31. Hellmann, L., Peters, E., Pummer, W., Wall, L.: J. Am. Chem. Soc. *79*, 5654 (1957)
32. US Pat. 2,927,138; C.A. *54*, 14187 (1960)
33. Wall, L., Fearn, J., Pummer, W., Lowry, R.: J. Res. NBS *65 A*, 239 (1961)
34. US Pat. 3,033,905; C.A. *57*, 11099 (1962)
35. Birchall, J., Haszeldine, R., Parkinson, A.: J. Chem. Soc. (Lond.) *1961*, 2204
36. Banks, R., Birchall, J., Haszeldine, R., Sim, J., Sutcliffe, H.: Proc. Chem. Soc. Lond. *1962*, 281
37. Christe, K., Pavlath, A.: Chem. Ber. *97*, 2092 (1964)

38. Yakobson, G. G., Steingarts, V. D., Vorozhtsov, Jr, N. N.: Izv. Akad. Nauk, Ser. Khim. *1964*, 1551
39. US Pat. 2,488,216; C.A. *44*, 1628 (1950)
40. Wall, L., Pummer, W., Fearn, J., Antonucci, J.: J. Res. NBS *67 A*, 481 (1963)
41. Sokolov, S. V., Tataurov, G. P., Yuminov, V. S., Udilov, G. P., Nirenburg, N. Ya.: Zh. Obshch. Khim. *36*, 362 (1966)
42. Gething, B., Patrick, C., Stacey, M., Tatlow, J.: Nature (Lond.) *183*, 588 (1959)
43. Gething, B., Patrick, C., Tatlow, J.: J. Chem. Soc. (Lond.) *1962*, 186
44. Fr. Pat. 1,393,516; (1965); Chem. Abstr., *63*, 6916 (1965)
45. Banks, R., Ginsberg, A., Haszeldine, R.: Proc. Chem. Soc. Lond. *1960*, 211
46. Banks, R., Haszeldine, R., Datham, J., Young, I.: Chem. Ind. (Lond.) *1964*, 835; J. Chem. Soc. (Lond.) *1965*, 594
47. Burdon, J., Gilman, D., Patrick, C., Stacey, M., Tatlow, J.: Nature (Lond.) *186*, 231 (1960)
48. Chambers, R., Hutchinson, J., Musgrave, W.: J. Chem. Soc. (Lond.) *1964*, 3573; Proc. Chem. Soc. Lond. *1964*, 83
49. Brit. Pat. 914,933; C.A. *59*, 1600 (1963)
50. Birchall, J., Hazard, R., Haszeldine, R., Wakalski, P.: J. Chem. Soc. (Lond.) C *1967*, 47
51. Brooke, G., Forbes, E., Richardson, R., Stacey, M., Tatlow, J.: J. Chem. Soc. (Lond.) *1965*, 2088
52. Pummer, W., Wall, L.: J. Res. NBS *63 A*, 167 (1959)
53. Callander, D., Coe, P., Tatlow, J.: Tetrahedron *22*, 419 (1966)
54. Belf, L., Buxton, M., Fuller, G.: J. Chem. Soc. (Lond.) *1965*, 3372
55. Yakobson, G. G., Shteingarts, V. D., Miroshnikov, A. I., Vorozhtsov, Jr, N. N.: Dokl. Akad. Nauk SSSR *159*, 1109 (1964)
56. US Pat. 3,046,313; C.A. *57*, 15003 (1962)
57. Brooke, G., Musgrave, W.: J. Chem. Soc. (Lond.) *1965*, 1864
58. Birchall, J., Haszeldine, R., Parkinson, A.: J. Chem. Soc. (Lond.) *1962*, 4966
59. Fr. Pat. 1,360,917; C.A. *61*, 13235 (1964)
60. Nield, E., Stephens, R., Tatlow, J.: J. Chem. Soc. (Lond.) *1959*, 166
61. Shchegoleva, G. S., Barkhash, V. A., Vorozhtsov, Yr, N. N.: Izv. Akad. Nauk SSSR Ser. Khim. *1967*, 708
62. Comaggi, G., Campbell, S., Perry, D. Stephens, R., Tatlow, J.: Tetrahedron *22*, 1755 (1965)
63. Platonov, V. E., Senchenko, T. V., Malyuta, N. G., Yakobson, G. G.: Izv. Akad. Nauk SSSR Ser. Khim. *1973*, 2827
64. Platonov, V. E., Senchenko, T. V., Yakobson, G. G.: Zh. Org. Khim. *12*, 816 (1976)
65. Brovko, V. V., Sokolenko, V. A., Yakobson, G. G.: Zh. Org. Khim. *10*, 2385 (1974)
66. Platonov, V. E., Furin, G. G., Malyuta, N. G., Yakobson, G. G.: Zh. Org. Khim. *8*, 430 (1972)
67. Malyuta, N. G., Platonov, V. E., Furin, G. G., Yakobson, G. G.: Tetrahedron *31*, 1201 (1975)
68. Harrison, D., Stacey, M., Stephens, R., Tatlow, J.: Tetrahedron *19*, 1893 (1963)
69. US Pat. 3,511,883; Off. Gaz. *874*, 606 (1970)
70. Banks, R., Harrison, A., Haszeldine, R.: J. Chem. Soc. (Lond.) C *1966*, 2102
71. Platonov, V. E., Ermolenko, N. V., Yakobson, G. G., Vorozhtsov, Yr, N. N.: Izv. Akad. Nauk SSSR Ser. Khim. *1968*, 2752
72. Yakhlakova, O. M., Mazalov, S. F., Tataurov, G. P., Ivanova, E. P., Platonov, V. E.: USSR Pat. 341,786; C.A. *78*, 15360 (1973)
73. Kolenko, I. P., Lundin, B. N., Abezgauz, F. I., Marek, E. I., Strizhevsky, I. L.: Zh. Obshch. Khim. *37*, 1686 (1967)
74. Vorozhtsov, Yr, N. N., Yezersky, S. N., Kollegov, V. F. et al.: Izv. Akad. Nauk SSSR Ser. Khim. *1966*, 1291
75. Vorozhtsov, Yr, N. N., Yermolenko, N. V., Mazalov, S. A., Osina, O. I., Platonov, V. E., Tyurin, V. S., Yakobson, G. G.: Izv. Akad. Nauk SSSR Ser. Khim. *1969*, 196
76. Antonucci, J., Wall, L.: J. Res. NBS *70 A*, 473 (1966)
77. Letchford, B., Patrick, C., Tatlow, J.: Tetrahedron *20*, 1381 (1964)
78. Florin, R., Pummer, W., Wall, L.: J. Res. NBS *1959*, 107
79. US Pat. 2,586,364; C.A. 8675 (1952)
80. Lock, G.: Chem. Ber. *69*, 2553 (1936)
81. Pozdnyakovich, Yu. V., Shteingarts, V. D.: J. Fluor. Chem., *4*, 317 (1974)

82. Fokin, E. P., Gerasimova, T. N., Vasilyevskaya, T. N., Baturina, I. I., Matoshina, K. I.: Izv. Sib. Otd. Akad. Nauk SSSR Ser. Khim. Nauk. *1971*, vyp. 4, N 9, 135
83. Platonov, V. E., Gatilova, V. P., Dvornikova, K. V., Yakobson, G. G.: Izv. Akad. Nauk SSSR Ser. Khim. *1974*, 1668
84. Mobbs, R.: J. Fluor. Chem. *1*, 361 (1972)
85. USSR Pat. 290,899; C.A. *74*, 125121 (1971)
86. US Pat. 3,000,976; C.A. *57*, 9736 (1962)
87. Brit. Pat. 917,413; C.A. *59*, 3852 (1963)
88. Patrick, C. R., Pedler, A. E., Seabra, A., Stephens, R., Tatlow, J.: Chem. Ind. (Lond.) *1963*, 1557
89. USSR Pat. 370,195; C.A. *79*, 18337 (1973)
90. USSR Pat. 255,230; C.A. *75*, 88290 (1971)
91. USSR Pat. 405,891; C.A. *80*, 146038 (1974)
92. Chambers, R., Corbally, R., Holmes, T., Musgrave, W.: J. Chem. Soc. (Lond.) Perkin Trans. 1, *1974*, 108
93. Delyagina, N. I., Pervova, E. Ya., Dyatkin, B. L., Knunyants, I. L.: Zh. Org. Khim. *8*, 851 (1972)
94. Yakobson, G. G., Shteingarts, V. D., Mironova, N. E., Vorozhtsov, Yr, N. N.: Zh. Obshch. Khim. *36*, 145 (1966)
95. Burdon, J., Hollyhead, W., Tatlow, J.: J. Chem. Soc. (Lond.) *1965*, 6336
96. Stephens, R., Tatlow, J.: Chem. Ind. (Lond.) *1957*, 821
97. Johncock, P., Mobbs, R., Musgrave, W.: Ind. Eng. Chem. Proc. Des. Dev. *1*, 267 (1962)
98. Coe, P., Croll, B., Patrick, C.: Tetrahedron *29*, 2097 (1964)
99. Florin, R., Pummer, W., Wall, L.: J. Res. NBS *62*, 119 (1959)
100. Tilney-Bassett, I.: Chem. Ind. (Lond.) *1965*, 693
101. Vorozhtsov, Yr, N. N., Barkhash, B. A., Ivanova, N. G., Anichkina, S. A., Andreevskay, O. I.: Dokl. Akad. Nauk SSSR *159*, 125 (1964)
102. Coe, P., Stephens, R., Tatlow, J.: J. Chem. Soc. (Lond.) *1962*, 3227
103. Barbour, A., Buxton, M., Coe, P., Stephens, R., Tatlow, J.: J. Chem. Soc. (Lond.) *1961*, 808
104. Coe, P., Plevey, R., Tatlow, J.: J. Chem. Soc. (Lond.) C *1969*, 597
105. Vorozhtsov, Yr, N. N., Barkhash, V. A., Gerasimova, T. N., Lokshina, E. G.: Izv. Akad. Nauk SSSR Ser. Khim. *1965*, 2086
106. Yakobson, G. G., Moronova, N. E., Petrov, A. K., Vorozhtsov, Yr, N. N.: Zh. Obshch. Khim. *36*, 147 (1966)
107. Vorozhtsov, Yr, N. N., Barkhash, V. A., Gerasimova, T. N., Lokshina, E. G., Ivanova, N. G.: Zh. Obshch. Khim. *37*, 1293 (1967)
108. Cohen, S., Reddy, M., Massey, A.: J. Chem. Soc. Chem. Commun. *1967*, 451; J. Organometal. Chem. *11*, 563 (1968)
109. Pummer, W., Wall, L.: J. Res. NBS *63 A*, 167 (1959)
110. Bardin, V. V., Furin, G. G., Yakobson, G. G.: Zh. Org. Khim. *17*, 999 (1981)
111. Birchall, J., Clarke, T., Haszeldine, R.: J. Chem. Soc. (Lond.) *1962*, 4977
112. Ambrose, D.; J. Chem. Soc. (Lond.) A *1968*, 1381
113. Birchall, J., Haszeldine, R.: J. Chem. Soc. (Lond.) *1961*, 3719
114. Filler, R., Ayyangar, N., Gustowski, W., Kang, H.: J. Org. Chem. *34*, 534 (1969)
115. Pummer, W., Wall, L.: Science, (Wash DC) *127*, 643 (1958)
116. Harper, R., Soloski, E., Tamborski, Ch.: J. Org. Chem. 29, 2385 (1964)
117. Sayers, D., Stephens, R., Tatlow, J.: J. Chem. Soc. (Lond.) *1964*, 3035
118. USSR Pat. 386,896; C.A. 79, 126055 (1973)
119. Pasquale, R., Tamborski, C.: J. Org. Chem. 34, 1736 (1969)
120. Letchford, B., Patrick, C., Stacey, M., Tatlow, J.: Chem. Ind. (Lond.) *1962*, 1472
121. Tamborski, C., Soloski, E., Pasquale, R.: J. Organometal. Chem. *15*, 494 (1968)
122. Soloski, E., Ward, W., Tamborski, C.: J. Fluor. Chem. 2, 361 (1972)
123. Stacey, M., Tatlow, J. in: Advances in Fluorine Chemistry, Vol. 1. Stacey, M., Tatlow, J., Sharpe, A. (eds). Butterworths, London 1960, p. 203
124. US Pat. 3,513,206; C.A. *73*, 26956 (1970)
125. Fr. Pat. 1,544,617; C.A. *71*, 112689 (1969)
126. US Pat. 3,265,746; C.A. *65*, 13602 (1966)

127. Antonucci, J., Brown, D., Wall, L. Abstr. 156th Natl. Am. Chem. Soc. Meeting, Div. Fluorine Chem. New York, paper 20 (1968)
128. Pedler, A., Tatlow, J., Uff, A.: Tetrahedron *25*, 1597 (1969)
129. Brovko, V. V., Sokolenko, V. A., Yakobson, G. G.: Zh. Org. Khim. *10*, 300 (1974)
130. Dvornikova, K. V., Platonov, V. E., Yakobson, G. G.: Zh. Org. Khim. *11*, 2383 (1975)
131. Waugh, E., Walton, D.; J. Organometal. Chem. *39*, 275 (1972)
132. Bogoradovsky, E. T., Zavgorodnii, V. S., Mingaleva, K. S., Petrov, A. A.: Zh. Obshch. Khim. *44*, 142 (1974)
133. Barrow, I., Pedler, A.: Tetrahedron *39*, 1829 (1976)
134. Molosnova, V. P., Vysochin, V. I., Barkhash, V. A., Vorozhtsov, Yr, N. N.: Izv. Akad. Nauk SSSR Ser. Khim. *1969*, 146
135. Filler, R., Rao, Y.: J. Org. Chem. *39*, 3421 (1974)
136. Birchall, J., Bowden, F., Haszeldine, R., Lever, A.: J. Chem. Soc. (Lond.) A *1967*, 747
137. Filler, R., Heffern, E.: J. Org. Chem. 32, 3249 (1967)
138. Brooke, J.; J. Chem. Soc. Perkin Trans. 1, *1974*, 233; Tetrahedron Lett. *1971*, 2377
139. Callander, D., Coe, P., Tatlow, J., Terrell, R.: J. Chem. Soc. (Lond.) C *1971*, 1542
140. Webb, A., Gilman, H.: J. Organometal. Chem. *20*, 281 (1969)
141. Dvornikova, K. V. Ph. D. Dissertation, Novosibirsk, 1982
142. Brit. Pat. 985,610
143. Molosnova, V. P., Barkhash, V. A., Vorozhtsov, Yr, N. N.: Zh. Obshch. Khim. *39*, 1774 (1969)
144. McBee, E., Rapkin, E.: J. Am. Chem. Soc. *73*, 1366 (1951)
145. Brewer, I., Eckhard, I., Heaney, H., Marples, B.: J. Chem. Soc. (Lond.) C *1968*, 664
146. Filler, R., Woods, S., Frendenthal, A.: J. Org. Chem. *38*, 811 (1973)
147. Vlasov, V. M., Yakobson, G. G.: Zh. Obshch. Khim. *39*, 2071 (1969)
148. Fuller, G., Warwick, D.: Chem. Ind. (Lond.) *1965*, 651
149. Petrova, T. D., Savchenko, T. I., Ardyukova, T. F., Yakobson, G. G.: Izv. Sib. Otd. Akad. Nauk SSSR Ser. Khim. Nauk *1970*, vyp. 3, 119
150. Ger. Pat. 1,168,419; C.A. 61, 3027 (1964)
151. Chapman, N., Clarke, K., Pinder, R., Sawhney, S.: J. Chem. Soc. (Lond.) C *1967*, 293
152. Vysochin, V. I., Barkhash, V. A., Vorozhtsov, Yr, N. N.: Zh. Obshch. Khim. *39*, 1607 (1969)
153. Brit. Pat. 923,115; C.A. *59*, 9902 (1963)
154. Barkhash, V. A., Andreevskaya, O. I., Vorozhtsov, Yr, N. N.: Dokl. Akad. Nauk SSSR *166*, 1343 (1966)
155. Prudchenko, A. T., Shchegoleva, G. S., Barkhash, V. A., Vorozhtsov, Yr, N. N.: Zh. Obshch. Khim. *37*, 2487 (1967)
156. Chambers, R., Spring, D.: J. Chem. Soc. (Lond.) C *1968*, 2394
157. Vorozhtsov, Yr, N. N., Barkhash, V. A., Anichkina, S. A.: Zh. Org. Khim. *2*, 1903 (1966)
158. Pozdnyakovich, Y. V., Shteigarts, V. D.: J. Fluor. Chem. *4*, 317 (1974)
159. Anichkina, S. A., Barkhash, V. A., Vorozhtsov, Yr, N. N.: Zh. Org. Khim. *38*, 2493 (1968)
160. Prudchenko, A. T., Barkhash, V. A., Vorozhtsov, Yr, N. N.: Izv. Akad. Nauk SSSR Ser. Khim. *1965*, 1798; Vorozhtsov, Yr, N. N., Barkhash, V. A., Prudchenko, A. T., Shchegoleva, G. S.: Zh. Obsch. Khim. *35*, 1501 (1965)
161. Filler, R., Rao, Y., Biezais, A., Miller, F., Beaucaire, V.: J. Org. Chem. *35*, 930 (1970)
162. Furin, G. G., Yakobson, G. G.: Izv. Sib. Otd. Akad. Nauk SSSR Ser. Khim. Nauk *1974*, vyp. 1, 78
163. Coe, P., Whitingham, A.: J. Chem. Soc. Perkin Trans. 1, *1974*, 917
164. Brit. Pat. 1,102,692; C.A. *68*, 115904 (1968)
165. Brit. Pat. 1,102,693; C.A. *69*, 43602 (1968)
166. Shteingarts, V. D., Budnik, A. G.: Izv. Sib. Otd. Akad. Nauk SSSR Ser. Khim. Nauk *1967*, vyp. 3, 124
167. Sah, P., Peoples, S.: Arzneim.-Forsch. *11*, 27 (1961); C.A. *57*, 16466 (1962)
168. Burdon, J., Damodaran, V., Tatlow, J.: J. Chem. Soc. (Lond.) *1964*, 763
169. Burdon, J., Castander, J., Tatlow, J.: J. Chem. Soc. (Lond.) *1964*, 5017
170. Shteingarts, V. D., Budnik, A. G., Yakobson, G. G., Vorozhtsov, Yr, N. N.: Zh. Obshch. Khim. *37*, 1537 (1967)
171. Wallenfels, K., Draber, W.: Chem. Ber. *90*, 2819 (1957)

172. Yakobson, G. G., Shteingarts, V. D., Vorozhtsov, Yr, N. N.: Zh. Vses. Khim. Obshch. Im D.I. Mendeleeva *1964*, 701
173. Niel, E., Tatlow, J.: Tetrahedron *8*, 38 (1960)
174. Yakobson, G. G., Odinokov, B. N., Vorozhtsov, Yr, N. N.: Tetrahedron Lett. *1965*, 4473
175. Odinokov, V. N., Yakobson, G. G., Vorozhtsov, Yr, N. N.: Zh. Obshch. Khim. *37*, 176 (1967)
176. Christe, K., Pavlath, A.: Chem. Ber. *96*, 2537 (1963)
177. Harper, R., Tamborski, C.: Chem. Ind. (Lond.) *1962*, 1824
178. Letchford, B., Patrick, C., Stacey, M., Tatlow, J.: Chem. Ind. (Lond.) *1962*, 1472
179. Belg. Pat. 623,680; C.A. *60*, 1657 (1964)
180. Belf, L., Buxton, M., Fuller, G.: J. Chem. Soc. (Lond.) *1965*, 3372
181. Felstead, E., Feilding, H., Wakefield, B.: J. Chem. Soc. (Lond.) *1966*, 708
182. Birchall, J., Haszeldine, R., Jone, M.: J. Chem. Soc. (Lond.) C *1971*, 1343
183. US Pat. 3,150,163; C.A. *61*, 16010 (1964)
184. US Pat. 3,046, 313; C.A. *57*, 15003 (1962)
185. Belg. Pat. 630,479; C.A. *61*, 3033 (1964)
186. US Pat. 3,429,935; C.A. *70*, 97376 (1969)
187. Yakobson, G. G., Odinokov, V. N., Vorozhtsov, Yr, N. N.: Zh. Obshch. Khim. *36*, 139 (1966)
188. Yakobson, G. G., Shteingarts, V. D., Vorozhtsov, Yr, N. N.: Zh. Vses. Khim. Obshch. Im D.I. Mendeleeva *1964*, 702
189. Gething, B., Patrick, C., Tatlow, J.: J. Chem. Soc. (Lond.) *1961*, 1574
190. Belg. Pat. 615,347; C.A. *58*, 473 (1963)
191. Brit. Pat. 986,892; C.A. *62*, 16142 (1965)
192. Sartori, P., Golloch, A.: Chem. Ber. *101*, 2004 (1968)
193. Yakobson, G. G., Odinokov, V. N., Petrova, T. D., Vorozhtsov, Yr, N. N.: Zh. Obshch. Khim. *34*, 2953 (1964)
194. Tamborski, C., Soloski, E.: J. Org. Chem. *31*, 743 (1966)
195. Rupp, E.: Ber. *29*, 1629 (1896); Rabjohn, N.: J. Chem. Soc. (Lond.) *1948*, 3518
196. Ueda, K., Hashiguchi, T.: Bull. Chem. Soc. Jpn *40*, 688 (1967)
197. Jap. Pat. 6,928,493; C.A. *72*, 66669 (1970)
198. Hellmann, N., Bilbo, A.: J. Am. Chem. Soc. *75*, 4590 (1953)
199. Prudchenko, A. T., Barkhash, V. A., Vorozhtsov, Yr, N. N.: Izv. Akad. Nauk SSSR Ser. Khim. *1965*, 1798
200. Shchegoleva, G. S., Petrov, A. K., Barkhash, V. A., Vorozhtsov, Yr, N. N.: Khim. Geterotsikl. Soedin. *1970*, 278
201. Prudchenko, A. T., Vovdenko, L. P., Barkhash, V. A., Vorozhtsov, Yr, N. N.: Khim. Geterotsikl. Soedin, *1968*, 967
202. Vlasov, V. M., Yakobson, G. G.: Izv. Akad. Nauk SSSR Ser. Khim. *1969*, 893
203. Petrova, T. D., Kann, L. I., Barkhash, V. A., Yakobson, G. G.: Khim. Geterotsikl. Soedin. *1969*, 778
204. Vorozhtsov, Yr, N. N., Barkhash, V. A., Prudchenko, A. T., Khomenko, T. I.: Zh. Obshch. Khim. 35,1501 (1965); Dokl. Akad. Nauk SSSR *164*, 1046 (1965)
205. USSR Pat. 270,718; C.A. *73*, 98622 (1970)
206. Alsop, D., Burdon, J., Tatlow, J.: J. Chem. Soc. (Lond.) *1962*, 1801
207. Tamborski, C., Soloski, E.: J. Org. Chem. *31*, 746 (1966); US Pat. 3,459,794; C.A. *71*, 80959 (1969)
208. Ryan, T., Berner, K.: Spectrochim. Acta Part a Mol Spectrosc *25*, 1155 (1969)
209. Belf, L., Buxton, M., Tilney-Bassett, J.: Tetrahedron *23*, 4719 (1967)
210. Gerasimova, T. N., Semikolenova, N. V., Orlova, N. A., Fomenko, T. V., Fokin, E. P.: Izv. Sib. Otd. Akad. Nauk SSSR Ser. Khim. Nauk *1975*, vyp. 6, 54
211. Petrov, V. P., Koptjug, V. A.: Reakt. Sposobnost Org. Soedin. *3*, 135 (1966)
212. Brit. Pat. 901, 880 (1962); Chem. Abstr. *58*, 1411 (1962)
213. Germ. Pat. 1,146,890; Chem. Abstr. *59*, 11331 (1963)
214. Connet, J.; Chem. Ind. (Lond.) *1965*, 1695
215. Vlasov, V. M., Yakobson, G. G.: Zh. Obshch. Khim. *39*, 2071 (1969)
216. Vlasov, V. M., Yakobson, G. G.: Zh. Obshch. Khim. 39, 785 (1969)
217. Brit. Pat. 977,961; C.A. *63*, 8266 (1965)

218. Markovsky, L. N., Furin, G. G., Shermolovich, Yu. G., Tychkina, O. N., Yakobson, G. G.: Zh. Obshch. Khim. *49*, 710 (1979)
219. Forbes, E., Richardson, R., Stacey, M., Tatlow, J.: J. Chem. Soc. (Lond.) *1959*, 2019
220. Pummer, W., Wall, L.: J. Res. NBS *68 A*, 277 (1964)
221. Antonucci, J., Wall, L.: J. Res. NBS *71 A*, 33 (1967)
222. US Pat. 3,394,190; C.A. *69*, 58939 (1968)
223. Brit. Pat. 1,110,232; C.A. *68*, 114245 (1968)
224. Budnik, A. G.: Ph. D. Dissertation, Novosibirsk, 1969
225. Budnik, A. G., Rusakov, S. L., Khomenko, E. I., Yakobson, G. G.: Zh. Org. Khim. *14*, 1281 (1978)
226. Yakobson, G. G., Petrova, T. D., Kann, L. I., Savchenko, T. I., Petrov, A. K., Vorozhtsov, Yr, N. N.: Dokl. Akad. Nauk SSSR *158*, 926 (1964)
227. Denivelle, L., Chesneau, R.: C.R. Acad. Sci. *254*, 1646 (1962)
228. Birchall, J., Haszeldine, R., Nikokavouras, J., Wilks, E.: J. Chem. Soc. (Lond.) C *1971*, 562
229. Furin, G. G.: Ph. D. Dissertation, Novosibirsk, 1969
230. Tataurov, G. P., Sokolov, S. V.: Zh. Obshch. Khim. *36*, 537 (1966)
231. Robson, P., Stacey, M., Stephens, M., Tatlow, J.: J. Chem. Soc. (Lond.) *1960*, 4754
232. Belenky, G. G., German, L. S., Knunyants, I. L., Furin, G. G., Yakobson, G. G.: Zh. Org. Khim. *12*, 1183 (1976)
233. Furin, G. G., Terentieva, T. V., Yakobson, G. G.: Izv. Sib. Otd. Akad. Nauk SSSR Ser. Khim. Nauk *1972*, vyp. 6, 78
234. Deacon, G., Parratt, J.: J. Organometal. Chem. *22*, 287 (1970)
235. Burdon, J., Coe, P., Fulton, M.: J. Chem. Soc. (Lond.) *1965*, 2094
236. Belf, L., Buxton, M., Woiton, D.: Chem. Ind. (Lond.) *1966*, 238
237. Cohen, S., Reddy, M., Massey, A.: Chem. Commun. *1967*, 451
238. Cohen, S., Reddy, M., Massey, A.: J. Organometal. Chem. *11*, 563 (1968)
239. Yakobson, G. G., Furin, G. G., Terentieva, T. V.: Zh. Org. Khim. *10*, 799 (1974)
240. Yakobson, G. G., Furin, G. G., Terentieva, T. V.: Izv. Akad. Nauk SSSR Ser. Khim. *1972*, 2128
241. Furin, G. G., Yakobson, G. G.: Izv. Sib. Otd. Akad. Nauk SSSR Ser. Khim. Nauk *1976*, vyp. 6, 120
242. Furin, G. G., Terentieva, T. V., Yakobson, G. G.: Izv. Sib. Otd. Akad. Nauk SSSR Ser. Khim. Nauk *1972*, vyp. 6, 78
243. Furin, G. G., Shchegoleva, L. N., Yakobson, G. G.: Zh. Org. Khim. *11*, 1290 (1975)
244. Peach, M., Spinney, H. Can. J. Chem. *49*, 644 (1971)
245. Chambers, R., Cunningham, J., Pyke, D.: Tetrahedron *24*, 2783 (1968)
246. Chambers, R., Cunningham, J.: Chem. Commun. *1966*, 469
247. Sheppard, W.: J. Am. Chem. Soc. *93*, 5597 (1971)
248. Sartori, P., Bauer, G.: J. Fluor. Chem. *12*, 203 (1978)
249. Nield, E., Stephens, R., Tatlow, J.: J. Chem. Soc. (Lond.) *1959*, 166
250. Robson, P., Smith, T. A., Stephens, R., Tatlow, J.: J. Chem. Soc. (Lond.) *1963*, 3692
251. Sartori, P., Golloch, A.: Chem. Ber. *103*, 3936 (1970)
252. Niel, R., Peach, M., Spinney, H.: Inorg. Nucl. Chem. Lett. *6*, 509 (1970)
253. Furin, G. G., Krupoder, S. A., Yakobson, G. G.: Izv. Sib. Otd. Akad. Nauk SSSR Ser. Khim. Nauk *1976*, vyp. 5, 146
254. Sheppard, R., Foster, S.: J. Fluor. Chem. *2*, 53 (1972/1973)
255. Furin, G. G., Andreevskaya, O. I., Yakobson, G. G.: Izv. Sib. Otd. Akad. Nauk Ser. Khim. Nauk *1980*, vyp. 2, 100
256. Clase, H., Poole, R.: 6th Int. Symp. Fluor. Chem., Durham, England, A 14 (1971)
257. Furin, G. G., Avramenko, A. A., Nikonorov, Yu. I., Yakobson, G. G.: Zh. Org. Khim. *17*, 1505 (1981)
258. Forbes, E., Richardson, R., Tatlow, J.: Chem. Ind. (Lond.) *1958*, 630
259. Brooke, G., Burdon, J., Stacey, M., Tatlow, J.: J. Chem. Soc. (Lond) *1960*, 1768
260. Holland, D., Moore, G., Tamborski, C.: J. Org. Chem. *29*, 1562 (1964)
261. Yakobson, G. G., Shteingarts, V. D., Furin, G. G., Vorozhtsov, Yr, N. N.: Zh. Obshch. Khim. *34*, 3514 (1964)
262. Belf, L., Buxton, M., Wotton, D.: Chem. Ind. (Lond.) *1966*, 238

263. Petrova, T. D., Mamaev, V. P., Yakobson, G. G.: Izv. Akad. Nauk SSSR Ser. Khim. *1969*, 679
264. Coe, P., Jukes, A., Tatlow, J.: J. Chem. Soc. (Lond.) *C 1966*, 2323
265. Holland, D., Moore, C., Tamborski, C.: Chem. Ind. (Lond.) *1965*, 1376
266. Holland, D., Moore, C., Tamborski, C.: J. Org. Chem. *29*, 3042 (1964)
267. Jukes, A., Gilman, H.: J. Organometal. Chem. *17*, 145 (1969)
268. Brooke, G., Burdon, J., Tatlow, J.: J. Chem. Soc. (Lond.) *1962*, 3253
269. Burdon, J., Coe, P., Marsh, C.: Tetrahedron *22*, 1183 (1966)
270. Kobrina, L. S., Furin, G. G., Yakobson, G. G.: Zh. Obshch. Khim. *38*, 514 (1968)
271. Brooke, J., Burdon, J., Tatlow, J.: J. Chem. Soc. (Lond.) *1961*, 802
272. Robson, P., Roylance, J., Stephens, R., Tatlow, J., Worthington, R.: J. Chem. Soc. Suppl. 1, *1964*, 5748
273. Doyle, A., Pedler, A.: J. Chem. Soc. (Lond.) *C 1972*, 282
274. Brit. Pat. 1,232,285; C.A. *75*, 44321 (1971)
275. Akhmetova, N. E.: Ph. D. Dissertation, Novosibirsk, 1978
276. Budnik, A. G., Kalinichenko, N. V., Shteingarts, V. D.: Zh. Org. Khim. *10*, 1923 (1974)
277. Birchall, J., Haszeldine, R., Kemp, J.: J. Chem. Soc. (Lond.) *C 1970*, 449
278. USSR Pat. 225,210; C.A. *70*, 19797 (1969)
279. USSR Pat. 242,907; C.A. *71*, 80937 (1969)
280. Yakobson, G. G., Furin, G. G., Kobrina, L. S., Vorozhtsov, Yr, N. N.: Zh. Obshch. Khim. *37*, 1289 (1967)
281. Shermolovich, Yu. G., Polumbrik, O. M., Markovsky, L. N., Saenko, E. P., Furin, G. G., Yakobson, G. G.: Zh. Org. Khim. *13*, 2589 (1977)
282. Banks, R., Noakes, T.: J. Chem. Soc. Perkin Trans. 1 *1976*, 143
283. Savchenko, T. I., Petrova, T. D., Platonov, V. E., Yakobson, G. G.: J. Fluor. Chem. *9*, 505 (1977)
284. Savchenko, T. I., Petrova, T. D., Platonov, V. E., Yakobson, G. G.: Zh. Org. Khim. *15*, 1018 (1979)
285. Savchenko, T. I., Petrova, T. D., Kolesnikova, I. V., Platonov, V. E.; Izv. Akad. Nauk SSSR Ser. Khim. *1980*, 1213
286. Koppang, R.: J. Fluor. Chem. *16*, 479 (1980)
287. Brooke, G., Burdon, J., Tatlow, J.: Chem. Ind. (Lond.) *1961*, 832
288. Castellano, J., Green, J., Kaufman, J.: J. Org. Chem. *31*, 821 (1966)
289. Burdon, J., Morton, C., Thomas, D.: J. Chem. Soc. (Lond.) *1965*, 2621
290. Pront, C., Coda, A., Forder, R., Kammer, B.: Cryst. Struct. Commun. *3*, 39 (1974)
291. Birchall, J., Clarke, T., Haszeldine, R.: J. Chem. Soc. (Lond.) *1962*, 4977
292. Kashkin, A. V., Bakhmutov, Yu. L., Marchenko, N. N.: Zh. Vses. Khim. Obshch. Im D.I. Mendeleeva *15*, 591 (1970)
293. Zibarev, A. V., Golding, I. R., Sladkov, A. M., Furin, G. G., Yakobson, G. G.: Izv. Akad. Nauk SSSR Ser. Khim. *1979*, 2536
294. Andreevskaya, O. I., Furin, G. G., Yakobson, G. G.: Zh. Org. Khim. *13*, 1684 (1977)
295. Glemser, O., Halasz, S., Biermann, U.: Inorg. Nucl. Chem. Lett. *4*, 591 (1978)
296. Zibarev, A. V., Dolenko, G. N., Krupoder, S. A., Mazalov, L. N., Poleshchuk, O. Kh., Furin, G. G., Yakobson, G. G.: Zh. Org. Khim. *16*, 390 (1980)
297. Banks, R., Prakash, A.: Tetrahedron Lett. *1973*, 99
298. Banks, R., Prakash, A.: J. Chem. Soc. Perkin Trans. 1 *1974*, 1365
299. Brokke, M., Goodall, B., Sheppard, G., Stone, F.: J. Chem. Soc. Dalton Trans. *1975*, 591
300. Hudson, A., Pedler, A., Tatlow, J.: Tetrahedron *26*, 3791 (1970)
301. Korobeinicheva, I. K., Andreevskaya, O. I., Podgornaya, M. I., Furin, G. G.: Izv. Sib. Otd. Akad. Nauk SSSR Ser. Khim. Nauk *1982*, vyp. 5, 103
302. Birchall, J., Haszeldine, R., Kemp, J.: J. Chem. Soc. (Lond.) *C 1970*, 1519
303. Abramovich, R., Challand, S.: J. Chem. Soc. Chem. Commun. *1972*, 1160
304. Abramovich, R., Challand, S., Yamada, Y.: J. Org. Chem. *40*, 1541 (1975)
305. Abramovich, R., Challand, S., Scriven, E.: J. Am. Chem. Soc. *94*, 1374 (1972)
306. Birchall, J., Daniewski, W., Haszeldine, R., Holfen, D.: J. Chem. Soc. (Lond.) *1965*, 6702
307. Fild, M., Glemser, O., Christoph, G.: Angew. Chem. Int. Ed. Engl. *3*, 801 (1964)
308. Janzen, A., Rodesiler, P., Willis, C.: J. Chem. Soc. Chem. Commun. *1966*, 672

309. Respers, W., Tamborski, C.: J. Org. Chem. *11*, 619 (1968)
310. Whittingham, A., Jarvie, A.: J. Organometal. Chem. *13*, 125 (1968)
311. Brit. Pat. 1,101,972; C.A. *68*, 69119 (1968)
312. Wall, L., Donadio, R., Pummer, W.: J. Am. Chem. Soc. *82*, 4846 (1960)
313. Fild, M., Glemser, O., Christoph, G.: Angew. Chem. Int. Ed. Engl. *23*, 953 (1964)
314. Kemmitt, R., Nichols, D., Peacock, R.: Chem. Commun. *1967*, 599
315. Emeleus, H., Miller, J.: J. Inorg. Nucl. Chem. *28*, 662 (1966)
316. Andreevskaya, O. I., Krupoder, S. A., Furin, G. G., Yakobson, G. G.: Izv. Sib. Otd. Akad. Nauk SSSR Ser. Khim. Nauk *1980*, vyp. 6, 97
317. Fild, M., Schmutzler, R.: J. Chem. Soc. (Lond.) *A 1969*, 840
318. Yakobson, G. G., Furin, G. G., Terentieva, T. V.: Zh. Org. Khim. *9*, 1707 (1973)
319. Gibson, J., Janzen, A.: 6th Int. Symp. Fluor. Chem, Durham, England, C 26 (1971)
320. Furin, G. G., Terentieva, T. V., Rezvukhin, A. I., Yakobson, G. G.: Zh. Obshch. Khim. *45*, 1473 (1975)
321. Markovsky, L. N., Furin, G. G., Shermolovich, Yu. G., Yakobson, G. G.: Izv. Akad. Nauk SSSR Ser. Khim. *1981*, 867
322. US Pat. 3,341,631; C.A. *68*, 29321 (1968)
323. Chambers, R., Coates, G., Livingstone, J., Musgrave, W.: J. Chem. Soc. (Lond.) *1962*, 4367
324. Furin, G. G., Andreevskaya, O. I., Yakobson, G. G.: Zh. Org. Khim. *12*, 2627 (1976)
325. Cohen, S., Massey, A.: J. Organometal. Chem. *10*, 471 (1967)
326. Connett, J., Davies, A., Deacon, G., Green, J.: Chem. Ind. (Lond.) *1965*, 512; J. Chem. Soc. (Lond.) *1966*, 106
327. Maletina, I. I., Orda, V. V., Aleinikov, N. N., Korsunsky, B. L., Yagupolsky, L. M.: Zh. Org. Khim. *12*, 1371 (1976)
328. Berny, J., Oates, G., Winfield, J.: J. Chem. Soc. Dalton Trans. *1974*, 509
329. Lyalin, V. V., Orda, V. V., Alekseeva, L. A., Yagupolsky, L. M.: Zh. Org. Khim. *6*, 329 (1970)
330. Brewer, I., Heaney, H.: Tetrahedron Lett., *1965*, 4709
331. Vorozgtsov, Yr, N. N., Ivanova, N. G., Barkhash, V. A.: Zh. Org. Khim. *3*, 220 (1967); Izv. Akad. Nauk SSSR Ser. Khim. *1967*, 1514
332. Cohen, S., Fenton, D., Tomlunson, A., Massey, A.: J. Organometal. Chem. *6*, 301 (1966)
333. Callander, D., Coe, P., Tatlow, J.: Tetrahedron *25*, 25 (1969)
334. Petrova, T. D., Savchenko, T. I., Yakobson, G. G.: Zh. Obshch. Khim. *37*, 1171 (1967)
335. Petrov, V. P., Barkhash, V. A., Shchegoleva, G. S., Petrova, T. D., Savchenko, T. I., Yakobson, G. G.: Dokl. Akad. Nauk SSSR *178*, 864 (1968)

Aromatic Compounds with Fluorinated Side Chains

Yu. A. Fialkov, L. M. Yagupolski

1,2,3-Tris(trifluoromethyl)Benzene

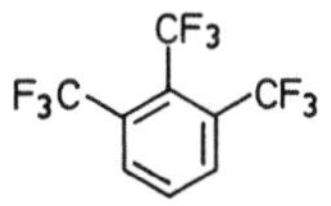

$C_9H_3F_9$ MW 282

1,2,3-Tris(trifluoromethyl)benzene is a colourless crystalline substance, well soluble in organic solvents, insoluble in water. Mp 28.5°–29 °C; bp 162°–163 °C/760 mm Hg; n_D^{30} 1.3786 [1].

Method of Preparation

1,2,3-Tris(trifluoromethyl)benzene is prepared by fluorination of 1,2,3-benzenetricarboxylic acid with SF_4 in anhydrous HF solution (71%) [1].

Procedure [1]

$$C_6H_3(COOH)_3 \xrightarrow{SF_4;\ HF} C_6H_3(CF_3)_3$$

Caution! Sulphur tetrafluoride is toxic. Its toxicity is comparable to that of phosgene, therefore is should be handled only under an efficient hood (see p. 2).

In a 140-ml rotating stainless steel autoclave provided with a needle valve, is placed 16 g (0.076 mol) of 1,2,3-benzenetricarboxylic acid. The autoclave is closed, evacuated to 1 mm Hg residual pressure and cooled with liquid nitrogen (Note 1), then 100 g of 85% (0.8 mol of 100%) SF_4 and 45 ml (2.25 mol) of anhydrous HF are condensed into it through the valve. The valve is closed, the autoclave is heated at 300 °C for 120 h, whereupon it is cooled and the gaseous products (SOF_2, HF, excess SF_4) are led off into an alkaline trap. The autoclave is heated in a water-bath at 25°–30 °C for 30 min to ensure complete removal of HF. Then the autoclave is cooled, opened, 25 ml of 25% aqueous NH_3 is carefully placed in it, the autoclave is closed and heated at 150° 170 °C for 1.5 h (Note 2). After that it is again cooled, opened and its contents are steam-distilled. 1,2,3-Tri(trifluoromethyl)-benzene is extracted from the distillate with ether (3 × 100 ml), the extract is dried over Na_2SO_4, the ether is distilled off and the residue distilled. Yield is 20 g (93%); mp 28.5°–29 °C, bp 162°–163 °C/760 mm Hg, n_D^{30} 1.3786 [1].

Notes

1. The autoclave may be cooled with dry-ice chloroform mixture. In that case condensation of the gas into the autoclave is more sluggish.
2. During treatment of the reaction products with aqueous NH_3, benzenecarboxylic acid fluorides (partial fluorination products) form nonvolatile amides. 1,2,3-Tris(trifluoromethyl)benzene can be easily isolated from the latter by steam-distillation or by extraction with pentane, in which the amides are insoluble.

Pentakis(trifluoromethyl)Benzene

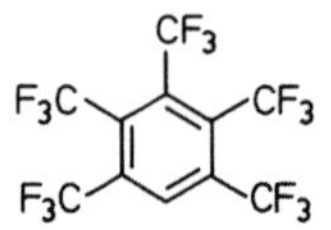

$C_{11}HF_{15}$ MW 418

Pentakis(trifluoromethyl)benzene is a colourless crystalline substance, well soluble in organic solvents, insoluble in water. Mp 55°–56 °C [2].

Method of Preparation

Pentakis(trifluoromethyl)benzene is prepared by fluorination of 1-trifluoromethyl-2,3,5,6-benzenetetracarboxylic acid with SF_4 in anhydrous HF solution (53% yield) [2].

Procedure

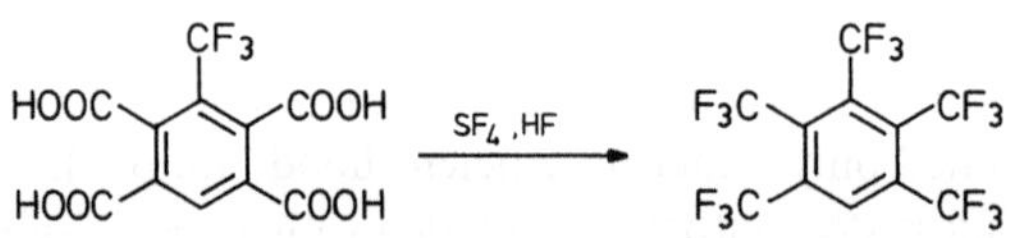

In a 100-ml rotating stainless steel autoclave provided with a needle valve, is placed 4.5 g (0.014 mol) of 1-trifluoromethyl-2,3,5,6-benzenetetracarboxylic acid (see p. 246). The autoclave is closed, evacuated to 1 mm Hg residual pressure and cooled with liquid nitrogen (see Note 1 on p. 234), then 55 g of 85% (0.43 mol) of 100% SF_4 and 15 g (0.75 mol) of anhydrous HF are condensed into it through the valve (see pp. 1, 2). The autoclave is heated at 300 °C for 50 h and at 330 °C for 50 h, whereupon it is cooled, the gaseous products are let out into an alkaline trap and HF is removed by heating the autoclave at 25°–30 °C for 30 min. Then the autoclave is opened and 25 ml of 25% aqueous NH_3 is carefully added; the autoclave is closed and heated at 150°–170 °C for 2 h. The products are steam-distilled. The distillate is extracted with ether (3 × 20 ml) and the ether extracts are dried over Na_2SO_4. The ether is removed and the product is recrystallized from pentane. The yield of pentakis-(trifluoromethyl)benzene is 3.1 g (53%); mp 55°–56 °C [2].

p-(Perfluoroisopropyl)Benzotrifluoride

(ω-Decafluorocumol)

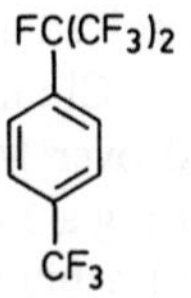

$C_{10}H_4F_{10}$ MW 314

p(-Perfluoroisopropyl)benzotrifluoride is a colourless liquid, well soluble in organic solvents, insoluble in water. Bp 135°–135.5 °C/760 mm Hg; n_D^{26} 1.3632; d_{26}^{26} 1.5255 [3].

Method of Preparation

p-(Perfluoroisopropyl)benzotrifluoride is prepared by fluorination with SF_4 of p-trifluoromethylphenylbis(trifluoromethyl)-carbinol obtained by the reaction of p-trifluoromethylphenylmagnesium bromide with hexafluoroacetone (total yield is 70%) [3].

Procedure

$$p\text{-}CF_3C_6H_4Br \xrightarrow{Mg} [p\text{-}CF_3C_6H_4MgBr] \xrightarrow{(CF_3)_2CO} p\text{-}CF_3C_6H_4COH(CF_3)_2$$

$$p\text{-}CF_3C_6H_4COH(CF_3)_2 \xrightarrow{SF_4} p\text{-}CF_3C_6H_4CF(CF_3)_2$$

p-Trifluoromethylphenylbis(trifluoromethyl)carbinol. Magnesium (7.5 g, 0.3 mol) in 50 ml of anhydrous ether is placed in a 250-ml three-necked flask, previously flushed with dry, oxygen-free nitrogen and provided with a stirrer, a dropping funnel, and a reflux condenser with a bubble counter filled with H_2SO_4. Then 67.5 g (0.3 mol) of p-bromobenzotrifluoride [4] in 150 ml of anhydrous ether is added dropwise with stirring and the mixture is heated in a water-bath. A mildly exothermic reaction starts in 30–40 min after the addition of about 10% of the volume of ethereal solution, with heating to 40 °C in a water-bath. During the reaction the mixture becomes cloudy and coloured. The rate of addition of p-bromobenzotrifluoride is regulated so that the solution should gently reflux all the time. After the addition of p-bromobenzotrifluoride the contents of the flask are refluxed with stirring for another 1 h. The reaction mixture is then allowed to cool to room temperature, the dropping funnel is replaced with a bubbler and the flask is cooled with dry-ice chloroform mixture. Hexafluoroacetone (50 g, 0.3 mol) is bubbled through the Grignard reagent during 6–7 h at −30 °C, with stirring; the temperature should be gradually increased to −12° ÷ −15 °C. Then the temperature is brought to 20 °C during 4 h and the mixture is allowed to stand for 12 h. The contents of the reactor is poured onto a mixture of 500 g of ice and 100 ml of 10% HCl, the ether layer is separated and the aqueous one extracted with ether (3 × 150 ml). The combined ether extracts are washed with 15% HCl (2 × 100 ml), with water, with 5% $NaHCO_3$ solution (2 × 50 ml), with 10% Na_2SO_3 solution (2 × 50 ml), with water, and dried over $MgSO_4$. The ether is removed and the residue is distilled twice in vacuum using a 12-cm column packed with glass helices. The yield of p-trifluoromethylphenylbis(trifluoromethyl)-carbinol is 40 g (43%); bp 169°–170.5 °C/760 mm Hg (~98% purity by GLC).

In order to purify the product, 6.3 g of the distillate is dissolved in 30 ml of 2 N NaOH and stirred at 50 °C for 3 h, then acidified with 15% HCl and extracted with ether (3 × 25 ml), the ether is washed with water and dried over $MgSO_4$. The ether is distilled off and the product is distilled twice in vacuum using a 5-cm Vigreux column. The yield of p-trifluoromethylphenylbis(trifluoromethyl)carbinol of 99.3% purity (as shown by GLC) is 4.5 g (71.5%); bp 168°–169 °C/755 mm Hg, n_D^{25} 1.3850; d_{25}^{25} 1.5630 [5].

p-(Perfluoroisopropyl)benzotrifluoride. In a 50-ml stainless steel autoclave equipped with a needle valve, is placed 6.3 g (0.02 mol) of p-trifluoromethylphenylbis(trifluoro-methyl)carbinol. The autoclave is cooled with liquid nitrogen and evacuated to 1–2 mm Hg residual pressure, then 6.4 g of 85% (0.06 mol of 100%) SF_4 is condensed into the autoclave through the valve. The temperature in the autoclave is brought during 3–4 h to that of the room and the gaseous products (SF_4, SOF_2, HF) are led off through the valve into an alkaline trap. The autoclave is then opened, its contents are extracted with 25 ml of methylene chloride and kept for 12 h over 2–3 g of NaF in a polyethylene bottle (to remove HF), whereupon the solvent is distilled off and the residue distilled. The yield of p-(perfluoroisopropyl)benzotrifluoride is 4.4 g (70%); bp 135°–135.5 °C/760 mm Hg; n_D^{26} 1.3632; d_{26}^{26} 1.5255 [3].

Similar fluorination of the corresponding hydroxy compounds with SF_4 gives (perfluoroisopropyl)benzene (yield 86%; bp 121°–122 °C; n_D^{23} 1.3832) and 1-methyl-4-(perfluoroisopropyl)benzene (yield 70%; bp 145°–147 °C; n_D^{20} 1.3940) [3, 6].

1,1,2,2-Tetrafluoroacenaphthene

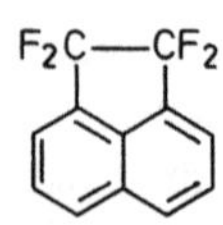

$C_{12}H_6F_4$ MW 226

1,1,2,2-Tetrafluoroacenaphthene is a white crystalline substance, well soluble in common organic solvents, insoluble in water; sublimes in vacuum at 150 °C/10–15 mm Hg; mp 91°–92 °C [7].

Method of Preparation

1,1,2,2-Tetrafluoroacenaphthene is prepared by fluorination of acenaphthenequinone with SF_4, with 40% [7] and 85% [8] yields.

Procedure [8]

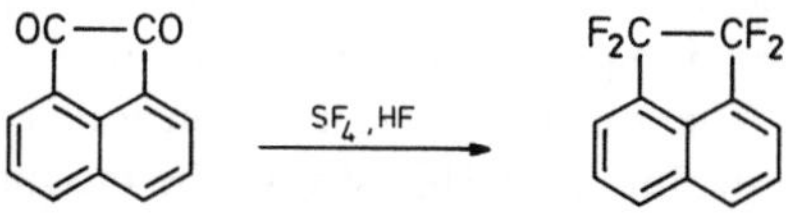

In a 100-ml stainless steel autoclave fitted with a needle valve, are placed 6 g (0.033 mol) of acenaphthenequinone, 0.3 ml of water (Note 1), then 18 g of 85% (0.14 mol of 100%) SF_4 is introduced (Note 2). The autoclave is heated at 150 °C for 2 h, at 180 °C for 2 h and at 200 °C for 2 h, then cooled and gaseous products are led off through the valve into alkaline traps. The contents of the autoclave are extracted with 100 ml of ether, then kept over 5 g of NaF for 10 h, and the ether is distilled off. 1,1,2,2-Tetrafluoroacenaphthene is extracted with petroleum ether and purified by sublimation in vacuum at 150 °C/10–15 mm Hg and crystallization from petroleum ether (bp 40°–60 °C). Yield is 6.3 g (85%); mp 92.5°–93.5 °C [8].

Notes

1. Water is introduced to obtain a catalyst, HF, formed in hydrolysis of SF_4.
2. SF_4 should be free from chlorine and sulphur chlorides, otherwise the yield of 1,1,2,2-tetrafluoroacenaphthene is considerably decreased (see p. 2).

2,4,6-Trimethylbenzotrifluoride

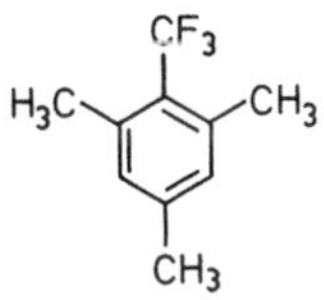

$C_{10}H_{11}F_3$ MW 188

2,4,6-Trimethylbenzotrifluoride is a clear colourless liquid, well soluble in organic solvents, insoluble in water. Bp 179°–180 °C, n_D^{19} 1.4600; d_4^{19} 1.140 [9].

Method of Preparation

2,4,6-Trimethylbenzotrifluoride is prepared by fluorination with SbF_3 of 2,4,6-trimethylbenzotrichloride (63% yield) [9], obtained by the reaction of mesitylene with CCl_4 [10].

Procedure

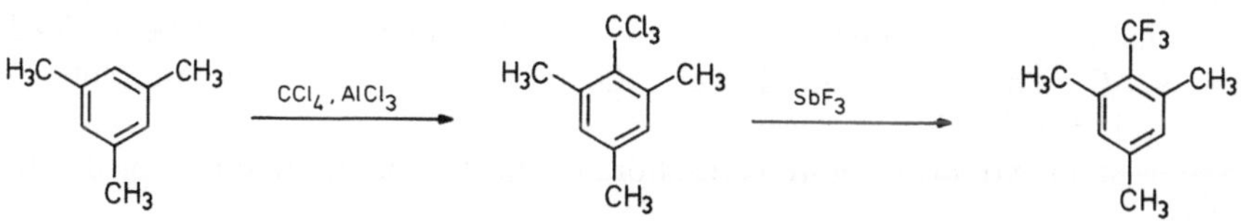

2,4,6-Trimethylbenzotrichloride. In a 0.5-l four-necked flask fitted with a stirrer, a thermometer, a dropping funnel, and a reflux condenser with a bubble counter

(Note 1), are placed 67 g (0.5 mol) of sublimed $AlCl_3$ and 100 ml of anhydrous CCl_4. To a vigorously stirred suspension at 20 °C is gradually added 30 g (0.25 mol) of mesitylene in 50 ml of CCl_4. The reaction proceeds with evolution of heat (Note 2) and HCl; the mixture turns dark-red. After all mesitylene has been added, the mixture is heated at 33°–35 °C for 3 h till the evolution of HCl stops. The mixture cooled to 10 °C is then poured, with vigorous stirring, onto 0.5 kg of crushed ice and 50 ml of concentrated HCl. The organic layer is separated, and the aqueous one is extracted with CCl_4 (3×75 ml). The extract is combined with the organic layer, washed with water and dried over $CaCl_2$. The solvent is distilled off in vacuum of a water-jet pump and the residue distilled. The yield of 2,4,6-trimethylbenzotrichloride is 48 g (80%); bp 126°–127 °C/5 mm Hg; n_D^{25} 1.5640 [10].

2,4,6-Trimethylbenzotrifluoride. In a 200-ml Claisen flask provided with a thermometer, a dropping funnel, and a receiver with a calcium chloride tube, is placed 35.7 g (0.2 mol) of calcinated SbF_3 (Note 3), then 42 g (0.18 mol) of 2,4,6-trimethylbenzotrichloride is carefully poured in. An exothermic reaction immediately starts (Note 4). When it has stopped, the dropping funnel is replaced with a capillary, the mixture is heated during several minutes with a burner, and cooled (Note 5). The product is distilled off in vacuum of a water-jet pump; a fraction boiling at 86°–90 °C/70 mm Hg is collected, treated with 20% HCl and extracted with ether (3×80 ml). The ether extracts are washed 5–6 times with 20% HCl, with water, with 5% Na_2CO_3 solution, then again with water and dried over Na_2SO_4. The ether is distilled off and the residue distilled in vacuum. The yield of 2,4,6-trimethylbenzotrifluoride is 20.9 g (63%); bp 179°–180 °C [9].

Notes

1. The bubble counter is a glass U-tube. Its arms (50 mm long, 15 mm i.d.) are linked by a bended tube (5–6 mm i.d.) filled with 0.5–0.7 ml of concentrated H_2SO_4. Rectangular bends (side 20–25 mm, 5–6 mm i.d.) are soldered onto the arms; they serve to connect the counter to the apparatus and to the absorbing system. The bubble counter is intended for controlling the evolution of gases during the process and for keeping out air moisture. Its design prevents H_2SO_4 from entering the flask.
2. The solution of mesitylene in CCl_4 is added at such a rate that the temperature of the mixture should not rise above 35 °C, since overheating results in resinification of the product and a decreased yield of the trichloromethyl derivative.
3. In a platinum or copper dish is placed, to the height of 7–8 mm, SbF_3 powder, which is then heated at 200°–250 °C for 30 min, with constant stirring and crushing of the lumps with a stainless steel spatula. Melting of SbF_3 should be avoided (mp 292 °C). The dish is cooled over P_2O_5 in a desiccator; before use, SbF_3 is thoroughly powdered.
4. To prevent an extremely vigorous reaction, the flask is intermittently cooled with water.
5. Cooling is necessary since connection of vacuum to the hot flask may result in vigorous boiling up of the mixture and its transfer to the receiver.

trans,trans,trans,trans-1,8-Diphenylperfluoroocta-1,3,5,7-tetraene

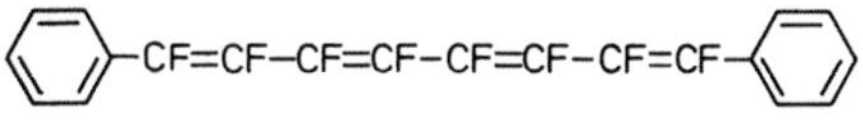

$C_{20}H_{10}F_8$ MW 402

trans,trans,trans,trans-1,8-Diphenylperfluoroocta-1,3,5,7-tetraene is a crystalline substance of pale-yellow colour, soluble in common organic solvents, insoluble in water. Mp 75°–77 °C [11] (see UV spectrum in [11], [^{19}F]NMR spectrum in [12], X-ray structural analysis data in [13]).

Method of Preparation

trans,trans,trans,trans-1,8-Diphenylperfluoroocta-1,3,5,7-tetraene is prepared by the reaction of α,β-difluorostyryllithium with perfluorobuta-1,3-diene in ether-tetrahydrofuran-pentane mixture at —90° ÷ —80 °C (22% yield) [11].

Procedure [11]

$$C_6H_5{-}CF{=}CFCl \xrightarrow{n\text{-}C_4H_9Li} C_6H_5{-}CF{=}CFLi \xrightarrow{CF_2{=}CF{-}CF{=}CF_2}$$

$$\rightarrow C_6H_5{-}(CF{=}CF)_4{-}C_6H_5$$

In a 150-ml four-necked flask, which has previously been flushed with dry, oxygen-free argon and provided with a stirrer, a dropping funnel coated with a jacket cooled with dry-ice methylene chloride mixture, a thermometer, and a reflux condenser with a calcium chloride tube, are placed 6.48 g (0.04 mol) of α,β-difluoro-β-chlorostyrene [14] and a mixture of anhydrous 30 ml of tetrahydrofuran and 30 ml of pentane or hexane. A solution of n-butyllithium cooled to —78 °C, prepared from 0.56 g (0.08 mol) of lithium turnings and 3.68 g (0.04 mol) of n-butyl chloride in 39 ml of anhydrous ether, is added during 20 min from the dropping funnel, with stirring at —90° ÷ —80 °C. Stirring is continued at —90° ÷ —80 °C for 1 h, then 3.24 g (0.02 mol) of perfluorobuta-1,3-diene [15] is added to the reaction mixture, which is stirred at —80 °C for an additional 2 h, then poured into 150 ml of 10% HCl and ice. The product is extracted with ether (4 × 30 ml), the combined ether extracts are washed with water, with 5% solution of Na_2CO_3, with water, and dried over $MgSO_4$. The solvent is removed in vacuum, methanol is added to the residue and the solution is cooled to —78 °C. The precipitate is crystallized from hexane. The yield of *trans,trans-trans,trans*-1,8-diphenylperfluoroocta-1,3,5,7-tetraene is 6.22 g (22%); mp 75°–77 °C [11].

1,1,3,3-Tetrafluoro-1,3-dihydroisobenzofuran

(1,1,3,3-Tetrafluorophthalane)

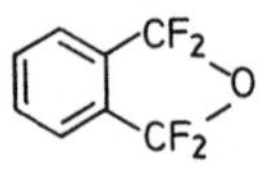

$C_8H_4F_4O$ MW 192

1,1,3,3-Tetrafluoro-1,3-dihydroisobenzofuran is a volatile colourless liquid, well soluble in most organic solvents, insoluble in water; stable during long-term storage. Bp 153 °C; n_D^{20} 1.4155; d_{20}^{20} 1.387 [16].

Method of Preparation

1,1,3,3-Tetrafluoro-1,3-dihydroisobenzofuran is prepared by fluorination of 1,1,3,3-tetrachloro-1,3-dihydroisobenzofuran with SbF_3 (57% yield) [16] or with HF (70%) [17]. The first process gives also o-trifluoromethylbenzoyl fluoride, the product of isomerization of the substrate.

The substrate, 1,1,3,3-tetrachloro-1,3-dihydroisobenzofuran, is obtained by chlorination of the phthalide with PCl [18].

Procedure [16, 17]

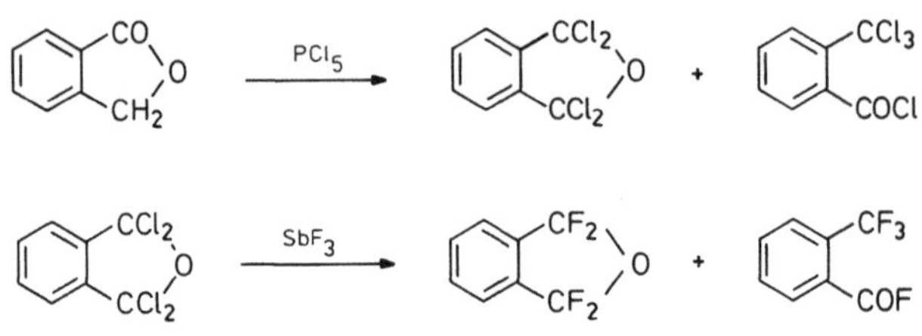

1,1,3,3-Tetrachloro-1,3-dihydroisobenzofuran. In a 100-ml Claisen flask fitted with a thermometer, a capillary, and a receiver connected with a bubble counter filled with concentrated H_2SO_4 (see p. 238), are placed 16 g (0.12 mol) of phthalide [19], 75 g (0.36 mol) of PCl_5 and 1 ml of $POCl_3$. The mixture is heated to 110 °C and kept at that temperature till HCl has ceased to evolve, the temperature is raised to 120 °C and heating is continued for 2 h. $POCl_3$ and PCl_3 are distilled off in vacuum of a water-jet pump, with heating in a water-bath. The residue is vacuum-distilled; a fraction boiling at 110°–112 °C/2 mm Hg (Note 1) is collected. The yield of 1,1,3,3-tetrachloro-1,3-dihydroisobenzofuran is 27 g (87%); mp 84°–85 °C. After crystallization (from petroleum ether, bp 40°–60 °C) mp 86°–87 °C [18].

1,1,3,3-Tetrafluoro-1,3-dihydroisobenzofuran.

a) Fluorination with SbF_3. In a 100-ml Claisen flask equipped with a thermometer, a capillary, and an ice-cooled receiver, are placed 15.5 g (0.06 mol) of 1,1,3,3-tetrachloro-1,3-dihydroisobenzofuran (Note 2) and 14.5 g (0.08 mol) of SbF_3. The mixture is cautiously heated with a burner to gentle boiling, then distilled (Note 3) in vacuum of a water-jet pump at 40–50 mm Hg. Distillation is conducted till the temperature of vapours has reached 160 °C. To remove the antimony salts, the condensate is treated with a mixture of petroleum ether (bp 40°–60 °C) and 20% HCl. The organic layer is separated, washed 4 or 5 times with 20% HCl, then with water and dried over Na_2SO_4. The solution is filtered, placed in a flask provided with a reflux condenser and a bubbler, and a flow of dry ammonia is passed in, with cooling with ice, till the precipitate is not deposited any longer. The precipitate (a mixture of o-trifluoromethylbenzamide, mp 161°–162 °C, and NH_4F) is filtered off and washed on the filter with petroleum ether. The washings are combined with the filtrate, transferred to a flask carrying a dephlegmator and petroleum ether is distilled off in a water-bath. The residue is distilled. The yield of 1,1,3,3-tetrafluoro-1,3-dihydroisobenzofuran is 6.5 g (57%); b.p. 152–153 °C; n_D^{20} 1.4155 [16].

b) Fluorination with anhydrous HF. In a 100-ml polyethylene reactor provided with a copper reflux condenser with a water jacket and a polyethylene bubble counter filled with H_2SO_4, is placed 20 g (1.0 mol) of anhydrous HF (Note 4). The reactor is cooled with ice-water, then 1,1,3,3-tetrachloro-1,3-dihydroisobenzofuran (7.75 g, 0.03 mol) is added in small portions. The reaction starts immediately and proceeds with considerable evolution of HCl. In 20 min after the addition is complete, the ice-bath is removed and the reactor is allowed to stand at 20 °C for 1 h. Excess HF is then distilled off, while the reactor is being heated in a water-bath to 30 °C. The residue is dissolved in ether, washed with water, with 1% K_2CO_3 solution, and dried over Na_2SO_4. The ethereal solution is placed in a flask provided with a reflux condenser and a bubbler, and a flow of dry ammonia is passed in during 20 min, with cooling of the flask. The NH_4F precipitate and a small amount of phthalodiamide (Note 5) are filtered off. The precipitate is washed with ether. The ethereal solutions are combined with the filtrate and ether is distilled off. The yield of 1,1,3,3-tetrafluoro-1,3-dihydroisobenzofuran is 4 g (70%); bp 151–153 °C [17].

Notes

1. o-Trichloromethylbenzoyl chloride, formed in the reaction, boils at a higher temperature. 1,1,3,3-Tetrachloro-1,3-dihydroisobenzofuran can be easily isolated from that impurity by treating it with the ethereal solution of ammonia, with cooling. Under these conditions, ammonia does not react with tetrachlorodihydroisobenzofuran.
2. Since heating of 1,1,3,3-tetrachloro-1,3-dihydroisobenzofuran with SbF_3 results in isomerization of the former, producing o-trichloromethylbenzoyl chloride, a mixture of these two products, obtained in the reaction of phthalide with PCl_5, may also be taken for fluorination.
3. Prolonged heating of the reaction mixture during fluorination of tetrachlorodihydroisobenzofuran with SbF_3 should be avoided, as this decreases the yield of the cyclic product.
4. *Caution*! For the safety measures when handling anhydrous HF see p. 1.
5. A small amount of phthalic anhydride, removed as an amide, is formed in the hydrolysis of the starting tetrachlorophthalane with water (0.3%) contained in HF.

1,2-Difluoromethylenedioxybenzene

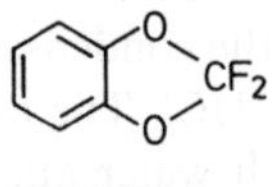

$C_7H_4F_2O_2$ MW 158

1,2-Difluoromethylenedioxybenzene is a colourless liquid, well soluble in organic solvents, insoluble in water. Bp 129°–130 °C; n_D^{20} 1.4429; d_4^{20} 1.3079 [20].

Method of Preparation

1,2-Difluoromethylenedioxybenzene is prepared by fluorination of 1,2-dichloromethylenedioxybenzene with SbF_3 in dioxan (52% yield) [20].

Procedure

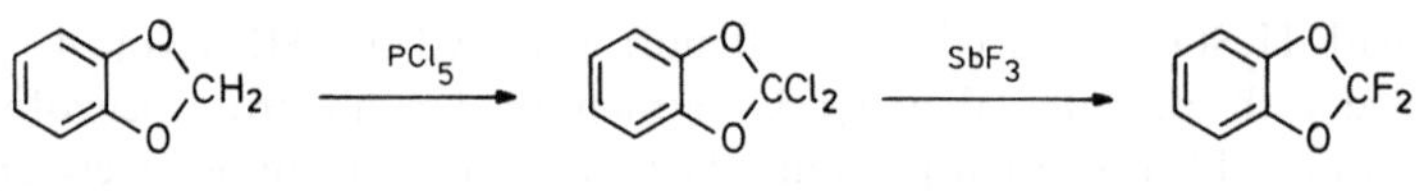

1,2-Dichloromethylenedioxybenzene. 1,2-Methylenedioxybenzen (24.4 g, 0.2 mol) [21, 22] and PCl_5 (84.4 g, 0.4 mol) are mixed *(carefully, as the reaction starts immediately)* in a 250-ml round-bottom flask with a reflux condenser, with cooling. The mixture is heated at 70 °C for 2 h, then the temperature is raised to 90 °C and the mixture is allowed to stand till HCl has ceased to evolve. PCl_3 is distilled off at atmospheric pressure and the reaction product is vacuum-distilled; a fraction boiling at 99°–100 °C/20 mm Hg is collected. The yield of 1,2-dichloromethylenedioxybenzene is 27 g (71%).

1,2-Difluoromethylenedioxybenzene. In a 250-ml round-bottom flask fitted with a downward condenser, are placed 26.5 g (0.15 mol) of sublimed SbF_3 and 90 ml of anhydrous dioxan. To remove traces of moisture, 30 ml of dioxan is distilled off. The mixture is cooled, then a solution of 28.6 g (0.15 mol) of 1,2-dichloromethylenedioxybenzene in 30 ml of dioxan is poured in and the mixture is refluxed during 7 h. Then it is cooled, poured into 300 ml of 20% HCl and extracted with ether (2 × 50 ml). The extract is washed with 20% HCl (4–5 × 20 ml), with water, and dried over Na_2SO_4. The ether is distilled off and the residue is distilled with a dephlegmator. The yield of 1,2-difluoromethylenedioxybenzene is 12.3 g (52%); bp 129°–130 °C; n_D^{20} 1.4429 [20].

2,2,9,9-Tetrafluoro-5-nitronaphtho[1,8-c,d]Pyrane

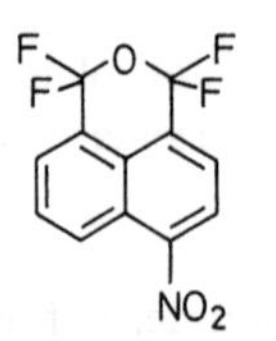

$C_{12}H_5F_4NO_3$ MW 287

2,2,9,9-Tetrafluoro-5-nitronaphtho[1,8-c,d]pyrane is a pale-yellow crystalline substance, soluble in ethanol, benzene, insoluble in water. Mp 84°–86 °C [23].

Method of Preparation

2,2,9,9-Tetrafluoro-5-nitronaphtho[1,8-c,d]pyrane is prepared by fluorination of 4-nitronaphthalic acid with SF_4 (85% yield) [23].

Procedure

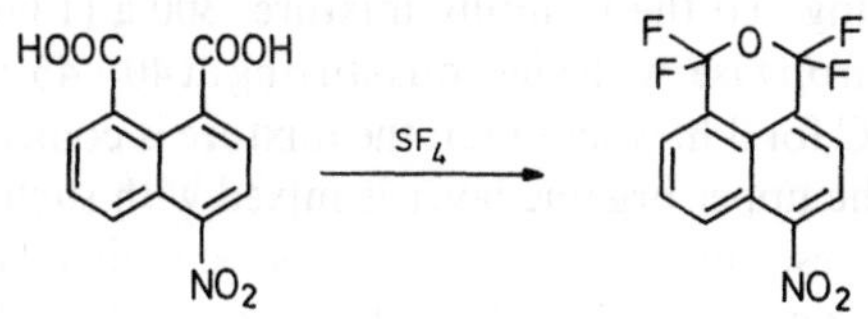

A 10-ml stainless steel autoclave provided with a needle valve, is charged with 14.3 g (0.05 mol) of 4-nitronaphthalic acid (Note), closed and cooled with liquid nitrogen, whereupon 38 g of 85% (0.3 mol of 100%) SF_4 is condensed into it through the valve (see p. 2). The autoclave is heated at 250 °C for 12 h, then cooled, and the gaseous products are led off through the valve into an alkaline trap. The contents of the autoclave are boiled during 30 min with 100 ml of 10% KOH solution, and cooled. The products is filtered off, washed with water, then squeezed and crystallized from aqueous methanol. The yield of 2,2,9,9-tetrafluoro-5-nitronaphtho[1,8-c,d]-pyrane is 13.3 g (85%); mp 84°–86 °C [23].

Note. Freshly prepared 4-nitronaphthalic acid should be used. It is obtained from 4-nitronaphthalic anhydride by settling it in a cool alkaline solution. The acid is thoroughly washed and squeezed on a filter, and dried in vacuum without heating.

4-Chloro-3,5-dinitrobenzotrifluoride

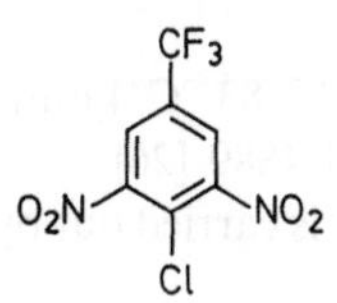

$C_7H_2ClF_3N_2O_4$ MW 270.5

4-Chloro-3,5-dinitrobenzotrifluoride is a yellowish crystalline substance, soluble in organic solvents, insoluble in water. Mp 58 °C [24], 57–59 °C [25].

Method of Preparation

4-Chloro-3,5-dinitrobenzotrifluoride is prepared by nitration of p-chlorobenzotrifluoride with a mixture of concentrated HNO_3 and H_2SO_4, and then with HNO_3 in the presence of oleum [24, 25].

Procedure [24]

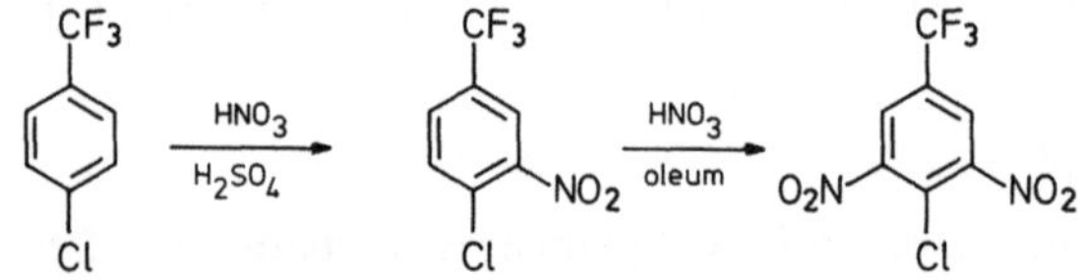

4-Chloro-3-nitrobenzotrifluoride. In a 700-ml four-necked flask provided with a stirrer, a thermometer, a reflux condenser, and a dropping funnel, is placed 150 ml of 94% (2.7 mol of 100%) H_2SO_4. Then 74 ml of 98% (1.73 mol of 100%) HNO_3 is added, with stirring and cooling. To the resulting mixture, 300 g (1.66 mol) of p-chlorobenzotrifluoride is added dropwise, with vigorous stirring at 40°–45 °C (Note 1). Stirring is continued at 40°–45 °C for 3 h, whereupon the mixture is cooled and transferred to a separatory funnel. The upper organic layer is mixed with such an amount of ether that the density of the resulting solution should be <1, then washed with water, with 0.5% Na_2CO_3 solution and with water till it is neutral, and dried over Na_2SO_4. The ether is removed and the product vacuum-distilled. The yield of 4-chloro-3-nitrobenzotrifluoride is 345–355 g (92%–96%); bp 113°–115 °C/20 mm Hg (Note 2).

4-Chloro-3,5-dinitrobenzotrifluoride. In a 100-ml four-necked flask fitted with a stirrer, a thermometer, a reflux condenser, and a dropping funnel, is placed 8 ml of 98% (0.185 mol of 100%) HNO_3, then 33.5 ml of 22% oleum is gradually added, with stirring and cooling. After that 22.5 g (0.1 mol) of 4-chloro-3-nitrobenzotrifluoride is dropped into a vigorously stirred mixture. The mixture is gradually heated to 110° to 115 °C and stirred at that temperature for 3 h, whereupon it is cooled and poured onto ice. The crystalline precipitate is filtered off, washed thoroughly with water, then squeezed and dried. The yield of 4-chloro-3,5-dinitrobenzotrifluoride is 23.5 g (87%); mp 51°–54 °C (Note 3).

Notes

1. Nitration proceeds with considerable evolution of heat, therefore the reactor should be cooled and the rate of addition of p-chlorobenzotrifluoride regulated so as to achieve the required temperature.
2. Boiling point of 4-chloro-3-nitrobenzotrifluoride is 81°–83 °C/4 mm Hg, 94° to 95 °C/10 mm Hg [24]; 110°–111 °C/12 mm Hg; n_D^{24} 1.4889 [26].
3. The product is useful in most syntheses. Its purification is carried out by recrystallization from ethanol; mp 58 °C [24].

3-Trifluoromethylphthalic Anhydride

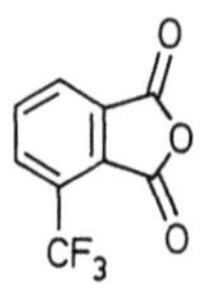

$C_9H_3F_3O_3$ MW 216

3-Trifluoromethylphthalic anhydride is a white crystalline substance with a strong, sweet odour, soluble in benzene, ether, ethanol, slightly soluble in hexane and water. Mp 66°–67 °C [27].

Method of Preparation

3-Trifluoromethylphthalic anhydride is prepared by partial fluorination of hemellitic acid with SF_4 [27, 28], with subsequent hydrolysis of 3-trifluoromethylphthalyl difluoride and dehydration of 3-trifluoromethylphthalic acid with acetic anhydride (65% yield) [27] (Note 1).

Procedure [27]

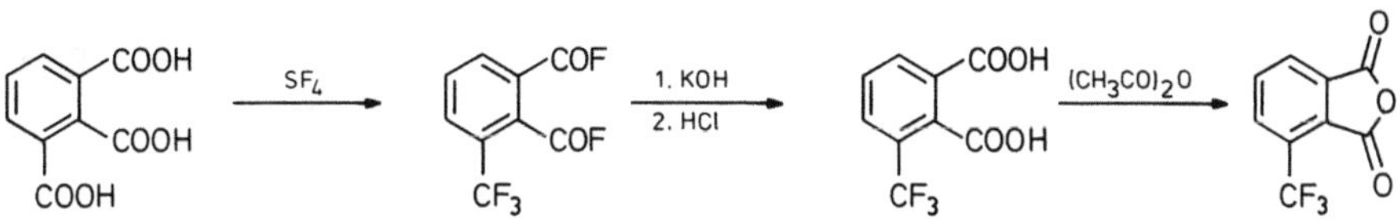

3-Trifluoromethylphthalyl difluoride. In a 100-ml stainless steel autoclave fitted with a needle valve is placed 10.5 g (0.05 mol) of hemellitic acid. The autoclave is closed, cooled with liquid nitrogen, evacuated to 1 mm Hg residual pressure, and 27 g of 85% (0.21 mol of 100%) SF_4 is condensed into it through the valve (see p. 2).
The autoclave is heated at 100 °C for 1 h, at 120 °C for 2 h, and at 150 °C for 8 h, then cooled, and the gaseous products (SF_4, SOF_2, HF) are led off through the valve into an alkaline trap. Then the autoclave is opened, its contents are extracted with 100 ml of methylene chloride, maintained over 5 g of NaF in a polyethylene bottle for 12 h, and the solvent is distilled off. The residue is vacuum-distilled; a fraction boiling at 107°–108 °C/15 mm Hg is collected. The yield of 3-trifluoromethylphthalyl difluoride is 8.2 g (69%); n_D^{20} 1.4408; d_4^{20} 1.5480 [27].

3-Trifluoromethylphthalic acid. A mixture of 7.15 g (0.03 mol) of 3-trifluoromethylphthalyl difluoride and 50 ml of 10% KOH solution is refluxed to complete dissolution of the difluoride. The solution is cooled and acidified with concentrated HCl. The precipitate is filtered off, washed with 10% HCl and dried at 100 °C. The filtrate and the washings are extracted with ether (3 × 100 ml), the extract is dried over Na_2SO_4. The ether is removed and the residue is combined with the major portion of the product and crystallized from 10% HCl. The yield of 3-trifluoromethylphthalic acid (Note 2) is 6 g (85%); mp 169°–170 °C [27].

3-Trifluoromethylphthalic anhydride. A mixture of 3 g (0.013 mol) of 3-trifluoromethylphthalic acid and 4.5 g (0.045 mol) of acetic anhydride is refluxed for 1 h. Acetic acid and excess acetic anhydride are distilled off in vacuum and the residue crystallized from a benzene-hexane mixture. The yield of 3-trifluoromethylphthalic anhydride is 2.6 g (93.5%); mp 66°–67 °C [27].

Notes

1. Likewise, 3,6-bis(trifluoromethyl)phthalic anhydride is obtained from prenitic acid; yield 60%–65% [27].

2. 3-Trifluoromethylphthalic acid is a colourless crystalline substance, soluble in water and ether, slightly soluble in benzene; when melted it is converted to 3-trifluoromethylphthalic anhydride.

1-Trifluoromethyl-2,3,5,6-benzenetetracarboxylic Acid

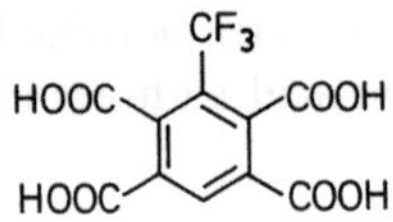

$C_{11}H_5F_3O_8$ MW 322

1-Trifluoromethyl-2,3,5,6-benzenetetracarboxylic acid is a colourless crystalline substance, soluble in ether, slightly soluble in water. Mp 280°–281 °C [2].

Method of Preparation

1-Trifluoromethyl-2,3,5,6-benzenetetracarboxylic acid is prepared by substitution of iodine in iododurene with trifluoromethyl group, followed by oxidation of methyl groups with HNO_3 (58% yield) [2].

Procedure

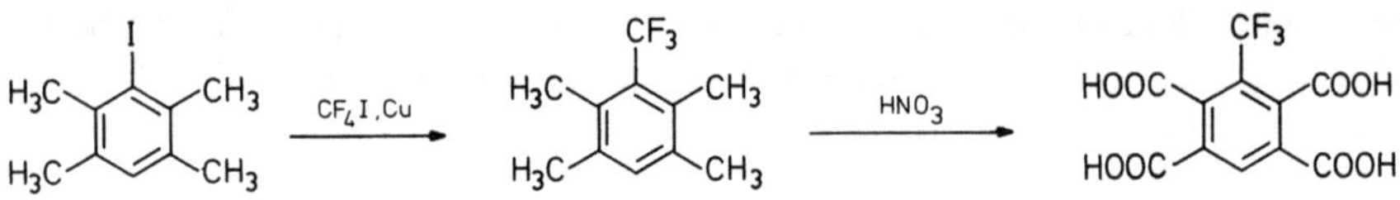

2,3,5,6-Tetramethylbenzotrifluoride. In a 300-ml rotating stainless steel autoclave provided with a needle valve, are placed 30 g (0.115 mol) of iododurene [29, 30], 20 g of copper powder, 10 g of bronze powder, and 150 ml of freshly distilled dimethylformamide. The autoclave is closed, cooled with liquid nitrogen and evacuated to 1 mm Hg residual pressure, then 40 g (0.2 mol) of trifluoroiodomethane is condensed into it through the valve. The autoclave is heated at 150 °C for 50 h. Then it is cooled, the gaseous products are let out through the valve, the autoclave is opened, the copper powder is filtered off and washed on the filter with dimethylformamide. The filtrate is poured onto 100 g of ice mixed with 200 ml of water. The precipitate is filtered off, dried and distilled in vacuum. A fraction boiling at 62°–63 °C/3 mm Hg is collected. The yield of 2,3,5,6-tetramethylbenzotrifluoride is 15 g (65%); mp 37°–38 °C (from pentane) (Note) [31].

1-Trifluoromethyl-2,3,5,6-benzenetetracarboxylic acid. A 300-ml stainless steel autoclave is charged with 5 g (0.025 mol) of 2,3,5,6-tetramethylbenzotrifluoride, 100 ml of 25% (0.5 mol of 100%) HNO_3 and heated at 170 °C for 15 h. Then the autoclave is cooled and the pressure is brought to that of the atmosphere. The autoclave is opened and the product extracted with ether (3 × 60 ml). The ether is distilled off, the residue is dissolved in 100 ml of 6% aqueous KOH solution and heated in a boiling water-bath, then solid $KMnO_4$ is added in portions with stirring to complete disappearance of the violet colour. Excess $KMnO_4$ is decomposed with 5 ml of methanol. The solvent is filtered off from MnO_2 and the precipitate is washed on the filter with boiling water. The filtrate is acidified with 35% HCl to pH ~4–5, evaporated till 30 ml has been left and cooled. The precipitate is filtered off, washed with ice-water, dried and crystallized from water. The yield of 1-trifluoromethyl-2,3,5,6-benzenetetracarboxylic acid is 7.1 g (89%); mp 280°–281 °C [2].

Note. Previously, tetramethylbenzotrifluoride was generally prepared from bromodurene in 35% yield [9].

4-Aminobenzotrifluoride

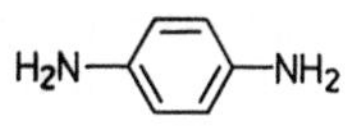

$C_7H_6F_3N$ MW 161

4-Aminobenzotrifluoride is a colourless crystalline substance with a specific odour, soluble in common organic solvents, insoluble in water. Mp 8 °C [32], 38 °C [33]. Bp 83 °C/12 mm Hg, 91 °C/19 mm Hg, 107 °C/39 mm Hg, 117.5 °C/60 mm Hg [32], 73°–74 °C/5–6 mm Hg [34], 85 °C/14 mm Hg [35] 86°–88 °C/15 mm Hg, 85°–88 °C/16 mm Hg [36], 90°–91 °C/20 mm Hg [37], 102°–102.5 °C/30 mm Hg [38], 117 °C/40 mm Hg [39]; n_D^{22} 1.4831 [38], n_D^{25} 1.4815 [32]; d_{27}^{27} 1.283 [32].

Method of Preparation

4-Aminobenzotrifluoride is prepared by the reduction of 4-nitrobenzotrifluoride with tin, in the mixture of HCl and acetic acid (75% yield) [33], with $SnCl_2$ in HCl (90% yield) [32], with hydrogen on Raney nickel at 2.5 atm (80% yield) [38], in petroleum ether at 100 °C/100 atm (70% yield) [36], in ethanol at a low pressure (80% yield) [39]; by fluorination of 4-phthaloimidobenzotrichloride with SF_4, with subsequent saponification (total yield 10%) [36]; by fluorination of 4-aminobenzoic acid with SF_4 in the presence of HF (37% yield) [35]; or by amination of 4-chlorobenzotrifluoride with $NaNH_2$ in liquid ammonia (yield of 1:1 mixture of 4- and 3-aminobenzotrifluorides is 25%) [40].

Procedure [32]

$$\text{4-}O_2NC_6H_4CHCl_2 \xrightarrow{NaOBr} \text{4-}O_2NC_6H_4CBrCl_2 \xrightarrow[SbCl_5]{SbF_3;}$$

$$\rightarrow \text{4-}O_2NC_6H_4CF_3 \xrightarrow{SnCl_2;\ HCl} \text{4-}H_2NC_6H_4CF_3$$

4-Nitro-α-bromo-α,α-dichlorotoluene. In a 10-l three-necked reactor fitted with a stirrer and a thermometer, are placed 6 l of water, 700 g (17.5 mol) of NaOH, the reactor is cooled to 0 °C and 250 ml (4.9 mol) of bromine is introduced in to it. To the resulting NaOBr solution is added, with vigorous stirring, 250 g (1.22 mol) of 4-nitrobenzal chloride, and the mixture is stirred at 50 °C for 50–60 h. The mixture is then cooled, the solid product is filtered off, washed with water till it is neutral, dried and crystallized from methanol. The yield of 4-nitro-α-bromo-α,α-dichlorotoluene is 235 g (67%); mp 54°–56 °C [41].

4-Nitrobenzotrifluoride. In a 250-ml Claisen flask provided with an air condenser and a receiver which is kept cool, are placed 150 g (0.52 mol) of 4-nitro-α-bromo-α,α-dichlorotoluene and 120 g (0.66 mol) of SbF_3, then 5 ml of $SbCl_5$ is added. The reaction mixture evolves heat. The flask is carefully heated with a burner for several minutes, during which time the bromine vapours are evolved. After the vigorous reaction has stopped, the mixture is boiled for 10 min and fluorination product is distilled off (at the end in 40–50 mm Hg vacuum). The product is mixed with 200 ml of 20% HCl and extracted with ether (2 × 250 ml). The extract is washed with 20% HCl to remove the antimony salts (Note 1), with water, with 5% Na_2CO_3 solution, again with water, and dried. The ether is distilled off. The yield of 4-nitrobenzotrifluoride is 90 g (90%); mp 40°–41 °C (Notes 2, 3) [41].

4-Aminobenzotrifluoride. In a 1-l three-necked flask provided with a stirrer, a thermometer, and a funnel, are placed 160 ml of ethanol and 240 ml of HCl (d 1.19), then 57.5 g (0.3 mol) of 4-nitrobenzotrifluoride is introduced, whereupon 295 g (1.3 mol) of $SnCl_2 \cdot 2\,H_2O$ is added at 60 °C (Note 4), with stirring during 1 h. The mixture is stirred at 60 °C for another 1 h, then cooled and poured with stirring onto a mixture of 1 kg of ice and 800 ml of 40% NaOH solution (Note 5). The amine is extracted with ether (3 × 200 ml), the extract is washed with 40% NaOH solution to complete dissolution of the stannates, then with water till it is neutral, and dried over $MgSO_4$. The ether is distilled off and the residue vacuum-distilled. Yield is 41.5–43 g (86%–89%); bp 83°–84 °C/12 mm Hg [32].

Notes

1. The ethereal solution is washed to remove the antimony salts till the HCl solution (3–5 ml) stops to develop cloudiness when poured into water.
2. The product is sufficiently pure to be used in the next step. It may be also purified by vacuum-distillation (bp 81°–82 °C/10 mm Hg [32]), or by crystallization from pentane.
3. 4-Nitrobenzotrifluoride may be obtained also by the reaction of 4-nitrobenzoic acid with SF_4 in anhydrous HF at 100 °C, which proceeds for 2 h (82% yield) [42].
4. The temperature of the reaction may be regulated by the rate of addition of $SnCl_2$ and by cooling of the flask. Overheating should be avoided as it leads to the decreased yield of the amine.
5. Decomposition of the reaction mixture should be conducted with constant stirring and cooling, since in the presence of excess alkali any local overheating will lead to the hydrolysis of the trifluoromethyl group. The temperature of the reaction mixture should not exceed +5 °C; the flask is cooled with an ice-salt mixture and more lumps of ice are added to it when necessary.

2-Nitro-4,5-bis(trifluoromethyl)Aniline

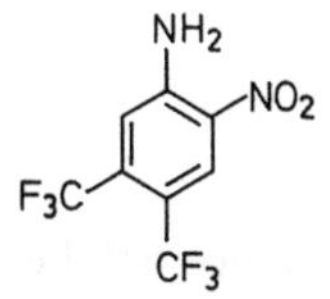

$C_8H_4F_6N_2O_2$ MW 274

2-Nitro-4,5-bis(trifluoromethyl)aniline is a crystalline substance, soluble in ethanol, insoluble in water. Mp 89°–90 °C [43].

Method of Preparation

2-Nitro-4,5-bis(trifluoromethyl)aniline is prepared by fluorination of 5-nitro-4-chlorophthalic acid with SF_4, with subsequent substitution of chlorine by the amino group (70% yield) [43].

Procedure

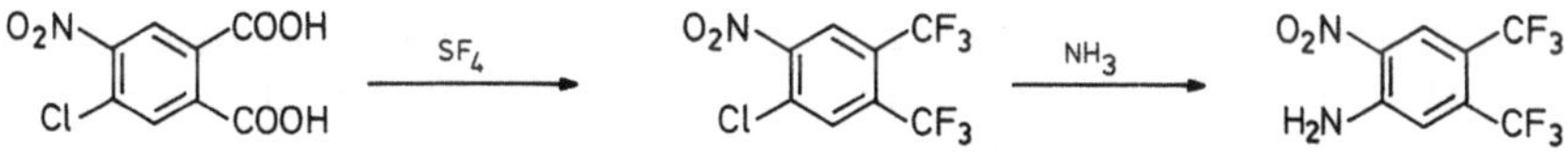

5-Nitro-4-chloro-1,2-bis(trifluoromethyl)benzene. In a 100-ml stainless steel autoclave provided with a needle valve, are placed 9.8 g (0.04 mol) of 5-nitro-4-chlorophthalic acid and 32 g (0.25 mol) of 85% SF_4 (see p. 2). The mixture is heated at 100 °C for 1 h, at 130 °C for 2 h, at 150 °C for 3 h, and at 160°–175 °C for 8 h. The autoclave is then cooled, the gaseous products and excess SF_4 are let out into an alkaline trap. The product is extracted with hexane and transferred to a flask fitted with a reflux condenser and a gas inlet tube. Ammonia is then passed into the solution which is kept cooled with ice-water, to convert acyl fluorides, formed in the reaction, to amides, which are then separated by filtration. Hexane is distilled from the filtrate and residual 5-nitro-4-chloro-1,2-bis(trifluoromethyl)benzene is purified by vacuum-distillation. A fraction boiling at 97 °C/12 mm Hg is collected. Yield is 9.8 g (83%) [43].

2-Nitro-4,5-bis(trifluoromethyl)aniline. In a 100 ml stainless steel autoclave are placed 8.8 g (0.003 mol) of 5-nitro-4-chloro-1,2-bis(trifluoromethyl)benzene and 30 ml of 30% aqueous ammonia. The mixture is heated at 150 °C for 8 h, the autoclave is cooled, its contents diluted with water, the precipitate filtered off, washed with water and dried. The yield of 2-nitro-4,5-bis(trifluoromethyl)aniline is 7 g (84%); mp 89°–90 °C (from aqueous ethanol) [43].

4-(Perfluoro-tert-butyl)aniline

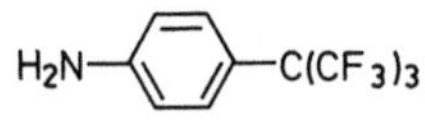

$C_{10}H_6F_9N$ MW 311

4-(Perfluoro-*tert*-butyl)aniline is a colourless crystalline substance, soluble in common organic solvents, insoluble in water. Mp 62.5°–63 °C [44].

Method of Preparation

4-(Perfluoro-*tert*-butyl)aniline is prepared by the reduction of 4-(perfluoro-*tert*-butyl)nitrobenzene with $SnCl_2$ [44].

Procedure

$$4\text{-}O_2NC_6H_4F \xrightarrow[\text{CsF}]{(CF_3)_2C{=}CF_2;} 4\text{-}O_2NC_6H_4C(CF_3)_3 \xrightarrow[\text{HCl}]{SnCl_2;}$$

$$\rightarrow 4\text{-}H_2NC_6H_4C(CF_3)_3$$

Caution! Perfluoroisobutylene is extremely toxic. It should be handled only under an efficient hood (see p. 2).

4-(Perfluoro-tert-butyl)nitrobenzene [45]. In a 100-ml stainless steel autoclave provided with a needle valve, are placed 7 g (0.05 mol) of 4-fluoronitrobenzene, 15.1 g (0.1 mol) of CsF (Note 1) and 30 ml of anhydrous sulpholane. The autoclave is closed, cooled with liquid nitrogen, evacuated to 1 mm Hg residual pressure, then 20 g (0.1 mol) of perfluoroisobutylene (Note 2) is condensed into it through the valve. The valve is shut off and the autoclave is heated at 80 °C for 8 h with intermittent shaking. Then it is cooled to 20 °C and unchanged perfluoroisobutylene is let out through the valve and a trap held at −70 °C. The autoclave is opened, the reaction mixture poured into water and extracted with ether (3 × 30 ml). The ether extract is washed with water and dried over $MgSO_4$. The ether is evaporated and the residue crystallized from hexane. The yield of 4-(perfluoro-tert-butyl)nitrobenzene is 12.7 g (75%); mp 77°–78 °C [45].

4-(Perfluoro-tert-butyl)aniline [44]. In a 250-ml three-necked flask provided with a stirrer, a reflux condenser, and a dropping funnel, are placed 17 g (0.05 mol) of 4-(perfluoro-*tert*-butyl)nitrobenzene and 70 ml of methanol; then a solution of 40 g (0.18 mol) of $SnCl_2 \cdot 2\,H_2O$ in 30 ml of 35% HCl is added dropwise with stirring, and the mixture is stirred at 95°–100 °C for 2 h. After the mixture has been cooled, it is poured into 100 g of ice placed in a 500 ml beaker immersed in an ice-bath. The crude amine is isolated by slowly adding 130 ml of 50% NaOH solution with stirring at 15°–20 °C or below. The product is extracted with ether (3 × 100 ml), the extract is washed with 40% NaOH solution (2 × 30 ml), with water, and dried over $MgSO_4$. The ether is distilled off, the amine crystallized from pentane. The yield of 4-(perfluoro-*tert*-butyl)aniline is 13 g (85%); mp 62°–63 °C [44].

Notes

1. CsF used in the reaction is previously dried in vacuum at 150 °C at 1 mm Hg during 8 h.
2. In order to cut off the required amount of perfluoroisobutylene and charge it into the autoclave, it is convenient to use the rubber bladder of a sports ball.

2,4,6-Tris(trifluoromethyl)Aniline

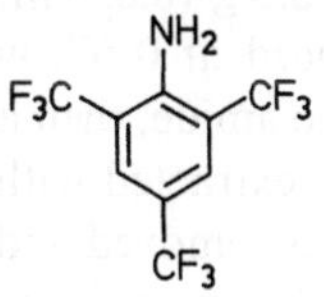

$C_9H_4F_9N$ MW 297

2,4,6-Tris(trifluoromethyl)aniline is a white crystalline substance, well soluble in common organic solvents, insoluble in water. Mp 58°–59 °C; bp 88 °C/60 mm Hg [46].

Method of Preparation

2,4,6-Tris(trifluoromethyl)aniline is prepared by partial fluorination of 1,2,3,5-benzenetetracarboxylic acid with SF_4, followed by conversion of 2,4,6-tris(trifluoromethyl)benzoyl fluoride to amide with subsequent Hoffmann conversion to the corresponding amine (total yield 56%) [46].

Procedure

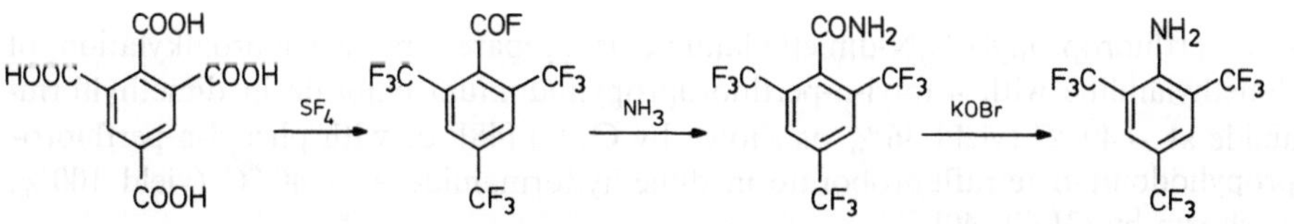

2,4,6-Tris(trifluoromethyl)benzoyl fluoride. In a 180-ml stainless steel autoclave fitted with a needle valve, are placed 10.2 g (0.04 mol) of 1,2,3,5-benzenetetracarboxylic acid [47] and 60 g of 85% (0.48 mol of 100%) SF_4 (see p. 2). The autoclave is heated at 100 °C for 1 h, at 150 °C for 3 h, at 170 °C for 5 h, and at 200 °C for 5 h. Then it is cooled, the gaseous products are led off through the valve into an alkaline trap, the autoclave is opened and its contents are boiled during 15 min with 50 ml of 10% aqueous KOH solution. The product is extracted with ether, the extract dried over Na_2SO_4, the ether distilled off and the residue vacuum-distilled. The yield of 2,4,6-tris(trifluoromethyl)benzoyl fluoride is 10.2 g (78%); bp 95 °C/80 mm Hg.

2,4,6-Tris(trifluoromethyl)benzoyl fluoride is a white crystalline substance; mp 35°–37 °C (from petroleum ether) [46].

2,4,6-Tris(trifluoromethyl)benzamide. 2,4,6-Tris(trifluoromethyl)benzoyl fluoride (9.85 g, 0.03 mol) and 60 ml of 25% aqueous ammonia are placed in a 50-ml stainless steel autoclave and heated at 120°–130 °C for 4 h. The contents of the autoclave is cooled, filtered, the precipitate is washed with water, then with hexane to remove the unconverted fluoride. The yield of 2,4,6-tris(trifluoromethyl)benzamide is 9.5 g (96%); mp 223°–224 °C (from aqueous methanol) in a sealed capillary [46].

2,4,6-Tris(trifluoromethyl)aniline. A solution of KOBr is prepared in a 150-ml three-necked flask provided with a stirrer and a thermometer, from 9 g (0.16 mol) of KOH, 80 ml of water, and 1.7 ml of bromine. Then 8.5 g (0.026 mol) of finely powdered 2,4,6-tris(trifluoromethyl)benzamide is introduced at 0 °C, with stirring. The mixture is stirred at 0 °C to complete dissolution of the amide, then at 75°–80 °C for several minutes. The amine is distilled off with steam, extracted with ether from the distillate and the extract dried over $MgSO_4$. The ether is removed and the residue vacuum-distilled. The yield of 2,4,6-tris(trifluoromethyl)aniline is 5.8 g (75%); bp 88 °C/60 mm Hg; mp 58°–59 °C [46].

4-(n-Perfluoropropyl)-N,N-dimethylaniline

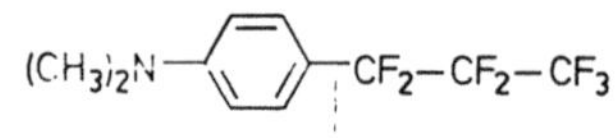

$C_{11}H_{10}F_7N$ MW 289

4-(n-Perfluoropropyl)-N,N-dimethylaniline is a colourless crystalline substance with a specific odour, soluble in common organic solvents. Mp 52 °C; bp 73 °C/0.6 mm Hg [48].

Method of Preparation

4-(n-Perfluoropropyl)-N,N-dimethylaniline is prepared by perfluoroalkylation of dimethylaniline with 4-tolyl-n-perfluoropropyliodonium chloride in dimethylformamide at —40 °C (yield 36%, as shown by GLC) [48], or with phenyl-n-perfluoropropyliodonium tetrafluoroborate in dimethylformamide at —40 °C (yield 100%, as shown by GLC) [49].

Procedure [49]

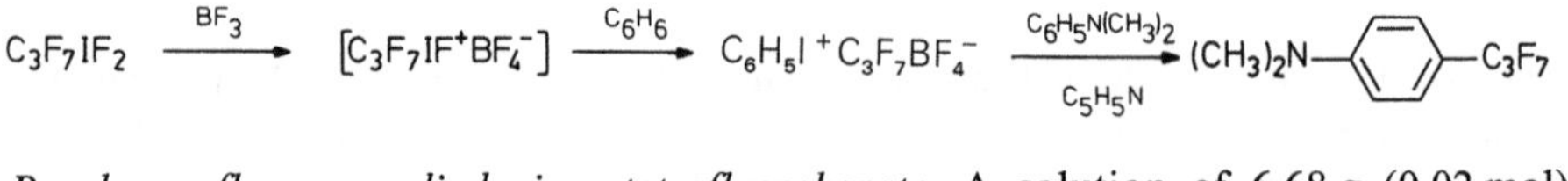

Penyl-n-perfluoropropyliodonium tetrafluoroborate. A solution of 6.68 g (0.02 mol) of difluoroiodo-n-perfluoropropane [50] in 50 ml of anhydrous methylene chloride is placed in a 100-ml three-necked dry quartz flask provided with a gas inlet tube, a dropping funnel and a bubble counter filled with concentrated H_2SO_4. The mixture is cooled to —60 °C and a fast flow of BF_3 is passed into the mixture stirred with a magnetic stirrer. After a precipitate has been formed, a solution of 2.2 ml (0.025 mol) of anhydröus benzene in 10 ml of anhydrous methylene chloride is dropped in with continuous passing of BF_3. Then the temperature of the reaction mixture is slowly raised to 25 °C and the mixture is allowed to stand for 1 h. The solution is filtered to remove impurities and the precipitate is washed on the filter with acetonitrile. The combined filtrates are evaporated in vacuum to dryness and the product is crystallized from acetonitrile with an addition of anhydrous ether (Note 1). The yield of phenyl-n-perfluoropropyliodonium tetrafluoroborate is 8.02 g (87%); decompn. p. 146° to 147 °C [49].

4-(n-Perfluoropropyl)-N,N-dimethylaniline. A solution of 1.21 g (0.01 mol) of dimethylaniline and 0.8 g (0.01 mol) of pyridine in 15 ml of anhydrous dimethylformamide is placed into a 100-ml three-necked flask fitted with a thermometer, a dropping funnel, and a bubble counter filled with concentrated H_2SO_4. The mixture is cooled to —40 °C, while it is stirred with a magnetic stirrer, and a solution of 4.6 g (0.01 mol) of phenyl-n-perfluoropropyliodonium tetrafluoroborate in 35 ml of anhydrous dimethylformamide is dropped in at that temperature. The temperature of the mixture is slowly raised to 25 °C and the mixture is allowed to stand for 5 h, whereupon it is diluted with 150 ml of water. The organic layer is extracted with ether (3 × 70 ml), the combined ether extracts are washed with water and dried over $MgSO_4$. The ether is distilled off, the residue is dissolved in 50 ml of anhydrous pentane, cooled to 0 °C and a fast flow of dry HCl is passed through it during 30 min. The oily layer is extracted with concentrated HCl (3 × 30 ml). The combined hydrochloric acid extracts are washed with pentane and neutralized with a 5 N NaOH solution, while the mixture is cooled with ice-water. The precipitate is extracted with ether (3 × 70 ml), the combined ether extracts are washed with water and dried over $MgSO_4$. The ether is distilled off and the residue is distilled in vacuum, with a small dephlegmator; a fraction boiling at 72°–73 °C/0.6 mm Hg is collected, which gradually crystallizes. The yield of 4-(n-perfluoropropyl)-N,N-dimethylaniline is 1.65 g (57%); mp 50°–51 °C [48] (Note 2).

Notes

1. For the sake of preventing decomposition of iodonium salt, the ether should be free from peroxides.
2. The product may be further purified by vacuum-sublimation at 80°–100 °C/0.5 mm Hg; mp 52 °C.

1-(4-Aminophenyl)Perfluoropropene

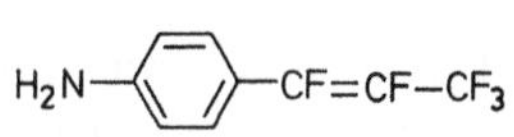

$C_9H_6F_5N$ MW 223

1-(4-Aminophenyl)perfluoropropene is a light cream-coloured crystalline substance, well soluble in ethanol, ether, acetone, benzene, chloroform, less soluble in petroleum ether, insoluble in water. Mp 35.5°–36.5 °C [51].

Method of Preparation

1-(4-Aminophenyl)perfluoropropene is prepared by the reduction of 1-(4-nitrophenyl)perfluoropropene, formed in nitration of 1-phenylperfluoropropene-1 [51,52], with $SnCl_2$ in HCl or with hydrogen on a platinum catalyst [51, 52]. 1-Phenylperfluoropropene-1 is obtained by the reaction of perfluoropropene with phenylmagnesium bromide or phenyllithium [53–55] (total yield 15%).

Procedure [51–53]

$$C_6H_5MgBr \xrightarrow{CF_2=CF-CF_3} C_6H_5CF{=}CF{-}CF_3 \xrightarrow{HNO_3}$$

$$\rightarrow 4\text{-}O_2NC_6H_4CF{=}CF{-}CF_3 \xrightarrow{H_2/Pt} 4\text{-}H_2NC_6H_4CF{=}CF{-}CF_3$$

1-Phenylperfluoropropene-1. A Grignard reagent is prepared from 190 g (1.2 mol) of bromobenzene and 30 g (1.23 mol) of magnesium turnings in 1.2 l of anhydrous ether, in a 2.5-l three-necked flask provided with a thermometer, a stirrer, a reflux condenser, and a bubbler. Gaseous perfluoropropene (~25 l) is passed through the column packed with freshly calcinated $CaCl_2$ and the column packed with P_2O_5, and bubbled through the solution, with vigorous stirring during 30–35 h. The reaction is conducted at 15°–17 °C for 25 h, then the temperature is raised to 25 °C. When all the gas has been absorbed, the contents of the flask are cooled with ice and 600 ml of 20% H_2SO_4 is gradually added with stirring. The ether layer is separated and the aqueous one extracted with ether (2 × 250 ml). The combined ether extracts are washed with 20% H_2SO_4, with water, with 5% Na_2CO_3 solution and dried over $MgSO_4$. The ether is distilled off and the residue distilled twice on a 20-cm Vigreux column (Note 1). The yield of 1-phenylperfluoropropene-1 is 130–140 g (52%–56%, based on bromobenzene); bp 86.5°–88 °C/94–95 mm Hg. Bp 77°–78 °C at 65 mm Hg; n_D^{20} 1.4482 [53].

1-(4-Nitrophenyl)perfluoropropene. In a 0.5-l three-necked flask fitted with a stirrer, a thermometer, and a dropping funnel, is placed 250 ml (5.9 mol) of 99% HNO_3 (d 1.5), which is then cooled to −15 °C. Phenylperfluoropropene (84 g, 0.4 mol) is added dropwise during 3 h; the temperature should not rise above −12 °C. Stirring is continued for 2 h, and the temperature is raised to 0 °C. The contents of the flask is poured onto ice, the oily mixture of the nitration products is extracted with ether, the ether extract is washed with water, with 5% Na_2CO_3 solution, again with water, and dried over $CaCl_2$. The ether is distilled off in vacuum. The resulting mixture of isomeric products of nitration (94 g, 93%) gradually solidifies upon standing. Crystallization from petroleum ether (bp 40°–60 °C) gives 30–32 g (30%–32%) of 1-(4-nitrophenyl)perfluoropropene (needle-shaped crystals); mp 51°–52 °C [51, 52].

1-(4-Aminophenyl)perfluoropropene. In a 0.5-l long-necked round-bottom flask are placed 10.12 g (0.04 mol) of 1-(4-nitrophenyl)perfluoropropene in 200 ml of methanol and 0.01–0.02 g of platinum catalyst (Note 2). The flask is evacuated to 10 mm Hg residual pressure and connected with a graduated gasometer filled with hydrogen under excess pressure of 60–70 mm Hg. The mixture is hydrated at 20 °C, while it is stirred with a magnetic stirrer, till all hydrogen has been absorbed; this takes 4 h. The crystalline precipitate is dissolved with heating to 40°–50 °C, the warm solution is filtered to remove the catalyst and evaporated in vacuum to the volume of 60–70 ml. The solution is cooled and the crystalline product is filtered off (Note 3). The filtrate is evaporated to dryness in vacuum, with gentle heating, then 75 ml of a (1 : 1) petroleum ether-benzene mixture is added, the solution is treated with activated charcoal at 40 °C, filtered and evaporated to remove the solvents. The product separates as a yellowish oil, which crystallizes upon standing. The yield of 1-(4-aminophenyl)perfluoropropene is 7.8 g (87.5%) [51] (Note 4).

Notes

1. After the first distillation of the product the still-residue contains diphenyl formed in 10% yield, based on bromobenzene.
2. The platinum hydration catalyst (PtO_2) is prepared as described in [56].
3. The product of partial hydration, 4,4′-bis(perfluoropropenyl)azobenzene, is formed in the amount of 0.5–0.6 g (6%–7%); mp 137.5°–138.5 °C (from methanol) [51].
4. The product is sufficiently pure to be used in further syntheses. To obtain an analytically pure compound, the product is crystallized from pentane. Mp 35.5° to 36.5 °C [51].

4-Amino-1-trifluoromethylnaphthalene

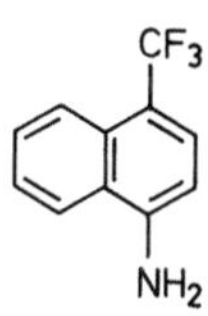

$C_{11}H_8F_3N$ MW 211

4-Amino-1-trifluoromethylnaphthalene is a colourless crystalline substance, soluble in common organic solvents, insoluble in water. Mp 85 °C [23], 86 °C [57].

Method of Preparation

4-Amino-1-trifluoromethylnaphthalene is prepared by fluorination of 4-nitro-1-naphthoic acid with SF_4 (53% yield) [57], or with SF_4 in anhydrous HF (83% yield) [42], with subsequent reduction of 4-nitro-1-trifluoromethylnaphthalene with hydrogen in the presence of platinum [57] (total yield 51%–79.5%) [42, 57, 58]; or by the Hofmann reaction, from 4-trifluoromethyl-1-naphthamide (69% yield) [23].

Procedure [42, 57]

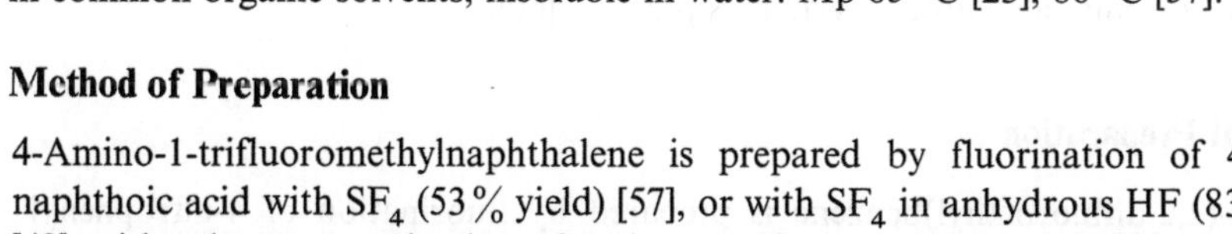

4-Nitro-1-trifluoromethylnaphthalene. 4-Nitro-1-naphthoic acid (11 g, 0.05 mol) is placed in a 100-ml stainless steel autoclave provided with a needle valve [59]. The autoclave is closed, cooled with liquid nitrogen, evacuated to 1 mm Hg residual pressure, whereupon 20 g of 85% (0.15 mol of 100%) SF_4 and 20 g of anhydrous HF is condensed into it through the valve (see pp. 1, 2). The autoclave is heated at 50 °C for 10 h, then cooled, and the gaseous products (SOF_2, SF_4, HF) are let out through the valve into an alkaline trap. The autoclave is then opened, its contents are poured onto 200–300 g of ice placed in a polyethylene bottle, extracted with ether (2 × 100 ml) and the ether is distilled off. The residue is boiled during 30 min with 100 ml of 10% KOH solution and steam-distilled. The distillate is extracted with ether (2 × 100 ml), dried over $MgSO_4$ and the ether distilled off. The product is crystallized from pentane. The yield of 4-nitro-1-trifluoromethylnaphthalene is 10 g (83%); mp 49 °C; bp 128° to 130 °C/5 mm Hg [57].

4-Amino-1-trifluoromethylnaphthalene. In a 100-ml round-bottom flask for hydration are placed 4.2 g (0.02 mol) of 4-nitro-1-trifluoromethylnaphthalene, 50 ml of methanol and 0.01 g of the platinum catalyst [56]. The flask is evacuated to 5–10 mm Hg residual pressure, filled with hydrogen and the mixture is hydrated, while it is stirred with a magnetic stirrer, at 20 °C under ~40 mm Hg hydrogen pressure till all the gas has been absorbed. The solution is filtered to remove platinum, methanol is distilled off in vacuum and the product crystallized from pentane. The yield of 4-amino-1-trifluoromethylnaphthalene is 3.6 g (96%); mp 86 °C [57].

1-Nitro-4-(2,2-difluoroethyl)Benzene

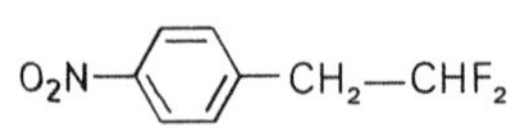

$C_8H_7F_2NO_2$ MW 187

1-Nitro-4-(2,2-difluoroethyl)benzene is a colourless crystalline substance, soluble in organic solvents, insoluble in water. Mp 55 °C [60].

Method of Preparation

1-Nitro-4-(2,2-difluoroethyl)benzene is prepared by fluorination of 4-nitrophenyloxyrane with SF_4 (47% yield) [60].

Procedure

$$4\text{-}O_2NC_6H_4\text{–}\underset{\diagdown O \diagup}{HC\text{–}CH_2} \xrightarrow{SF_4} 4\text{-}O_2NC_6H_4\text{—}CH_2CHF_2$$

4-Nitrophenyloxyrane (3.3 g, 0.02 mol) is placed in a 100-ml stainless steel autoclave equipped with a needle valve [61]. The autoclave is closed, evacuated to 1 mm Hg residual pressure, cooled with liquid nitrogen, whereupon 5 g of 85% (0.04 mol of 100%) SF_4 is condensed into it through the valve (see p. 2). The autoclave is heated at 85 °C for 3 h, then cooled, and the gaseous products are let out through the valve into an alkaline trap. The contents of the autoclave is extracted with anhydrous ether (3 × 20 ml) and anhydrous NaF is added to the coloured ethereal solution. In 3–4 h the solution is filtered, the ether distilled off and the product extracted from the dark viscous residue with 20–30 ml of hot hexane. One-third of the amount of hexane is distilled off, the residue is maintained in a refrigerator for 12 h, the crystalline product is filtered off and washed with a small amount of cool hexane. The yield of 1-nitro-4-(2,2-difluoroethyl)benzene is 1.8 g (47%); mp 55 °C [60].

(±)-1,2-Difluoro-1,2-bis(4-nitrophenyl)ethane and *meso*-1,2-Difluoro-1,2-bis(4-nitrophenyl)ethane

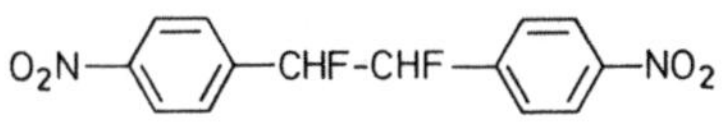

$C_{14}H_{10}F_2N_2O_4$ MW 308

(±)-1,2-Difluoro-1,2-bis(4-nitrophenyl)ethane is a crystalline substance, soluble in organic solvents, insoluble in water. Mp 153°–154 °C [62].

meso-1,2-Difluoro-1,2-bis(4-nitrophenyl)ethane is a yellowish crystalline substance, soluble in organic solvents, insoluble in water. Mp 153°–154 °C [62].

Method of Preparation

1,2-Bis(4-nitrophenyl)-1,2-difluoroethanes are prepared by fluorination of *cis*- and *trans*-2,3-bis(4-nitrophenyl)oxyranes, with SF_4 in the presence of NaF [62].

Procedure

$$O_2NC_6H_4{-}\underset{\diagdown O \diagup}{HC{-}CH}{-}C_6H_4NO_2 \xrightarrow{SF_4;\ NaF} (O_2NC_6H_4{-}CHF{-})_2$$

(±)-1,2-Difluoro-1,2-bis(4-nitrophenyl)ethane. In a 70-ml stainless steel autoclave fitted with a needle valve, are placed 2.9 g (0.01 mol) of *cis*-2,3-bis(4-nitrophenyl)-oxyrane [62–64], 2 g of calcinated NaF and 15 ml of anhydrous methylene chloride. The autoclave is closed, cooled with liquid nitrogen, evacuated to 1 mm Hg residual pressure, whereupon 9.5 g of 85% (0.07 mol of 100%) SF_4 is condensed into it through the valve (see p. 2). The autoclave is heated at 150 °C for 6 h, then cooled and the gaseous products are let out into an alkaline trap. The contents of the autoclave is extracted with 20 ml of anhydrous methylene chloride, then 2 g of calcinated NaF is added. After 4–5 h the solution is filtered, the solvent is distilled off and the product crystallized from glacial acetic acid. The yield of (±)-1,2-difluoro-1,2-bis(4-nitrophenyl)ethane is 2.9 g (98%); mp 153°–154 °C [62].

meso-1,2-Difluoro-1,2-bis(4-nitrophenyl)ethane. In a 50-ml stainless steel autoclave provided with a needle valve, are placed 0.58 g (0.002 mol) of *trans*-2,3-bis(4-nitrophenyl)oxyrane [62–64], 0.9 g of calcinated NaF, 15 ml of anhydrous methylene chloride, then 6 g of 85% (0.045 mol) SF_4 is introduced. The autoclave is heated at

150 °C for 8 h, then cooled and the gaseous products are let out into an alkaline trap. The contents of the autoclave are extracted with 10–15 ml of anhydrous methylene chloride, then 1.5 g of calcinated NaF is added and after 4–5 h the solution is filtered. Methylene chloride is distilled off and the residue is dissolved in a small amount of ethanol. The crystalline precipitate, unchanged *trans*-2,3-bis(4-nitrophenyl)oxyrane (0.12–0.15 g), is separated. The filtrate is evaporated to dryness and the residue crystallized from glacial acetic acid. The yield of *meso*-1,2-difluoro-1,2-bis(4-nitrophenyl)ethane is 0.23 g [50%, based on the converted amount of *trans*-2,3-bis(4-nitrophenyl)oxyrane]; mp 153°–154 °C [62].

cis- and *trans*-α,β-Difluorocinnamic Acids

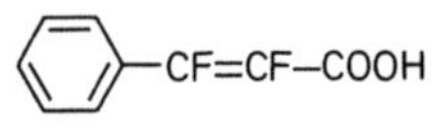

$C_9H_6F_2O_2$ MW 184

cis-α,β-Difluorocinnamic acid is a colourless crystalline substance, well soluble in methanol, ether, acetone, benzene, slightly soluble in water. Its potassium salt is soluble in water at 20 °C. Mp 140°–142 °C.

trans-α,β-Difluorocinnamic acid is a colourless crystalline substance, well soluble in methanol, ether, acetone, benzene. Its potassium salt is slightly soluble in water at 20 °C. Mp 125°–126 °C [73], 125.5°–127 °C [74].

Method of Preparation

A mixture of *cis*- and *trans*-α,β-difluorocinnamic acids is prepared by alkaline saponification of a mixture of the corresponding nitriles obtained by the reaction of α,β,β-trifluorostyrene with NaCN in acetonitrile, with subsequent separation of the acids as their potassium salts, which are then acidified with mineral acids to give *cis*- and *trans*-α,β-difluorocinnamic acids [74, 75].

trans-α,β-Difluorocinnamic acid is also prepared by carbonization, with solid CO_2, of α,β-difluorostyryllithium formed in metallation of α,β-difluoro-β-chlorostyrene with butyllithium [73].

Procedure

1. Conversion of α,β,β-Trifluorostyrene [74, 75]

$$C_6H_5CF{=}CF_2 \xrightarrow{\text{NaCN}} C_6H_5CF{=}CFCN \xrightarrow{\text{KOH}} C_6H_5CF{=}CFCOOK \xrightarrow{\text{H}^+}$$

$$\rightarrow C_6H_5CF{=}CFCOOH$$

cis- and trans-α,β-Difluorocinnamic nitriles. In a 250-ml three-necked flask provided with a stirrer, a thermometer, and a reflux condenser, are placed 82 g (0.52 mol) of α,β,β-trifluorostyrene [65–72], 63 g (1.3 mol) of powdered NaCN, 80 ml of acetonitrile and 15 drops of water. The mixture is stirred at 80°–85 °C for 3 h, then cooled, the precipitate is filtered off and washed with hexane. The filtrate is extracted several times with hexane. The extracts are combined, washed with water and dried over $MgSO_4$. The solvent is distilled off at atmospheric pressure, the residue is first distilled in vacuum, then distilled on a 10-cm column packed with glass helices. The following fractions are collected: fraction 1 (~40 g; bp 67°–70 °C/70 mm Hg) (unchanged α,β,β-trifluorostyrene); fraction 2 (~15 g); 75°–100 °C/14 mm Hg) (a mixture of *cis*- and *trans*-α,β-difluorocinnamic nitriles). Yield is 30%–33% (based on the converted α,β,β-trifluorostyrene) [75].

cis- and trans-α,β-Difluorocinnamic acids. In a 0.5-l two-necked flask provided with a stirrer, a thermometer, and a reflux condenser, are placed 24.9 g (0.15 mol) of a mixture of α,β-difluorocinnamic nitriles and 300 ml of 10% KOH solution. The mixture is stirred at 100 °C till the nitriles are converted to the salts of α,β-difluorocinnamic acids (Note 1), with the latter partially precipitated. The mixture is cooled to 10 °C, the potassium salt of *trans*-α,β-difluorocinnamic acid is filtered off and washed with a small amount of ice-water, then with ether. The filtrate is acidified and extracted with ether (3 × 50 ml). The ether extracts are combined, washed with water, dried over $MgSO_4$ and, after distilling off the solvent, the product is crystallized from benzene. The yield of *cis*-α,β-difluorocinnamic acid is 4.35 g (15%); mp 140°–142 °C [74].

The potassium salt of *trans*-α,β-difluorocinnamic acid is washed with ether and dissolved by boiling it with water. The solution is filtered and cautiously acidified with 5 ml of concentrated H_2SO_4. The mixture is cooled, the precipitate is filtered off, washed with water, dried and crystallized from benzene. The yield of *trans*-α,β-difluorocinnamic acid is 13.75 g (48%); mp 125°–127 °C [74].

2. Conversion of α,β-Difluoro-β-chlorostyrene [73]

$$C_6H_5CF{=}CFCl \xrightarrow{n\text{-}C_4H_9Li} C_6H_5CF{=}CFLi \xrightarrow{CO_2} C_6H_5CF{=}CFCOOH$$

In a 150-ml four-necked flask previously flushed with dry, oxygen-free argon and provided with a stirrer, a thermometer, a dropping funnel with dry-ice methylene chloride jacket, a gas inlet tube, and a reflux condenser with a calcium chloride tube, are placed 6.48 g (0.04 mol) of α,β-difluoro-β-chlorostyrene [14] and a mixture of anhydrous 30 ml of tetrahydrofuran and 30 ml of pentane (or hexane). Then a solution of n-butyllithium, prepared from 0.56 g (0.08 mol) of lithium turnings and 3.68 g (0.04 mol) of n-butyl chloride in 30 ml of anhydrous ether and cooled to —75 °C, is added dropwise at —90° ÷ —80 °C, with stirring during 20 min. The mixture is stirred at —90° ÷ —80 °C during 1 h and poured onto 40–50 g of solid CO_2. After the temperature of the mixture has risen to 20 °C, the lithium salt of *trans*-α,β-difluorocinnamic acid is extracted with water (4 × 10 ml), the combined extracts are washed with ether, filtered and acidified with HCl. The precipitated product is filtered off, dried and crystallized from benzene. The yield of *trans*-α,β-difluorocinnamic acid is 2.66 g (36%); mp 125°–126 °C [73] (Note 2).

Notes

1. The end of the reaction is indicated by the disappearance of the nitrile emulsion drops from the reaction mixture.
2. Similar conversions of the corresponding tetrasubstituted derivatives of α,β-difluoro-β-chlorostyrene afford 4-methoxy-*trans*-α,β-difluorocinnamic acid (25% yield; mp 185°–187 °C), 4-trifluoromethyl-trans-α,β-difluorocinnamic acid (49% yield: mp 204°–206 °C), 4-dimethylamino-*trans*-α,β-difluorocinnamic acid (47.5% yield; decompn. p. 168°–169 °C) [73].

trans,trans-Tetrafluoro-1,4-bis(4-carboxyphenyl)-butadiene

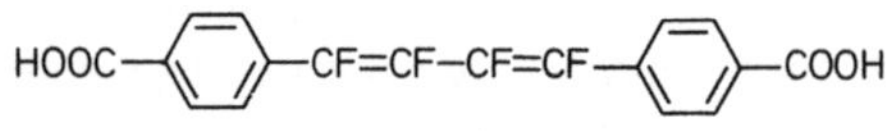

$C_{18}H_{10}F_4O_4$ MW 366

trans,trans-Tetrafluoro-1,4-bis(4-carboxyphenyl)butadiene is a colourless crystalline substance, slightly soluble in ethanol, ether, acetone, benzene, acetic acid; insoluble in water. Mp 345 °C (decompn.) [76].

Method of Preparation

trans,trans-Tetrafluoro-1,4-bis(4-carboxyphenyl)butadiene is prepared by oxidation of 1,2,3,4-tetrafluoro-1,2,3,4-tetrachloro-1,4-bis(4-tolyl)butane and 1,2,3,4-tetrafluoro-1,2,3,4-tetrachloro-1,4-bis(α-chloro-4-tolyl)butane formed in chlorination of 1,4-bis(4-tolyl)perfluorobutadiene, with subsequent dechlorination of 1,2,3,4-tetrafluoro-1,2,3,4-tetrachloro-1,4-bis(4-carboxyphenyl)butane. The product is purified by converting it to dichloride, which after crystallization is hydrolysed with water in dimethylformamide (17% yield) [76].

The starting *trans,trans*-1,4-bis(4-tolyl)perfluorobutadiene is obtained by the reaction of 4-tolyllithium with perfluorobutadiene (40% yield) [76].

Procedure

$$4\text{-}CH_3C_6H_4Li \xrightarrow{CF_2=CF-CF=CF_2} 4\text{-}CH_3C_6H_4CF{=}CF{-}CF{=}CFC_6H_4CH_3\text{-}4 \rightarrow$$

$$\xrightarrow{Cl_2} CH_3C_6H_4(CFCl)_4C_6H_4CH_3 + ClCH_2C_6H_4(CFCl)_4C_6H_4CH_2Cl$$

$$\xrightarrow[H_2SO_4]{CrO_3;} 4\text{-}HOOCC_6H_4(CFCl)_4C_6H_4COOH\text{-}4 \xrightarrow{Cu;\ C_5H_5N}$$

$$\rightarrow 4\text{-}HOOCC_6H_4CF{=}CF{-}CF{=}CFC_6H_4COOH\text{-}4$$

trans,trans-1,4-Bis(4-tolyl)perfluorobutadiene. The ethereal solution of 4-tolyllithium is prepared from 4.8 g (0.69 mol) of lithium and 58 g (0.34 mol) of 4-bromotoluene in 350 ml of anhydrous ether in a 1-l flask which has previously been flushed with dry, oxygen-free nitrogen. The solution is filtered and 27.5 g (0.17 mol) of perfluoro-

4-Trifluoroacetylbenzoyl chloride. In a 150-ml two-necked flask fitted with a stirrer and a reflux condenser with a bubble counter filled with concentrated H_2SO_4, are placed 8.72 g (0.04 mol) of 4-trifluoroacetylbenzoic acid and 24 g (0.2 mol) of thionyl chloride. The mixture is refluxed till HCl has ceased to evolve. Excess $SOCl_2$ is distilled off at atmospheric pressure and the residue distilled in vacuum. The yield of 4-trifluoroacetylbenzoyl chloride is 7.1 g (75%); bp 105°–107 °C/20 mm Hg; n_D^{20} 1.4909 [79].

Trifluoromethyl-4-aminophenylketone. 4-Trifluoroacetylbenzoyl chloride (11.82 g, 0.05 mol) and 50 ml of anhydrous toluene are placed in a 150-ml three-necked flask fitted with a stirrer and reflux condenser with a bubble counter filled with concentrated H_2SO_4. Then 3.32 g (0.051 mol) of activated NaN_3 is added (Note), the mixture is heated at 130 °C with vigorous stirring during 20 h and cooled. The precipitate is filtered off, 40 ml of 35% HCl is added to the filtrate and the mixture is heated at 100 °C for 3 h till the gas has ceased to evolve. The mixture is then cooled, the precipitate is filtered off and washed on a filter with hot water. The aqueous layer of the first filtrate is separated from the toluene one, which is then throughly extracted several times with warm diluted HCl. The combined aqueous-acid extracts and the washings are neutralized with 25% NH_3, while the mixture is being cooled and stirred. Precipitated product is filtered off, washed with water, dried and crystallized from hexane. The yield of trifluoromethyl-4-aminophenylketone is 4.06 g (43%); mp 94 °C [79].

Note. NaN_3 is activated as described on p. 78. Its activity is lost in a day and may be restored by treating NaN_3 with acetone to settle it from aqueous solution [80].

2-Difluoromethoxybenzaldehyde

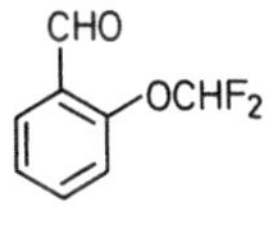

$C_8H_6F_2O_2$ MW 172

2-Difluoromethoxybenzaldehyde is a colourless liquid, soluble in common organic solvents, slightly soluble in water; easily oxidized by air to the corresponding acid. Bp 89 °C/5 mm Hg [81], 94°–96 °C/10 mm Hg [82]; n_D^{20} 1.4940 [81], n_D^{25} 1.4920 [82].

Method of Preparation

2-Difluoromethoxybenzaldehyde is prepared from 2-difluoromethoxybenzyl bromide (38%) [82], or by difluoromethylation of salicylic aldehyde (88–90% yield) [81].

Procedure [81]

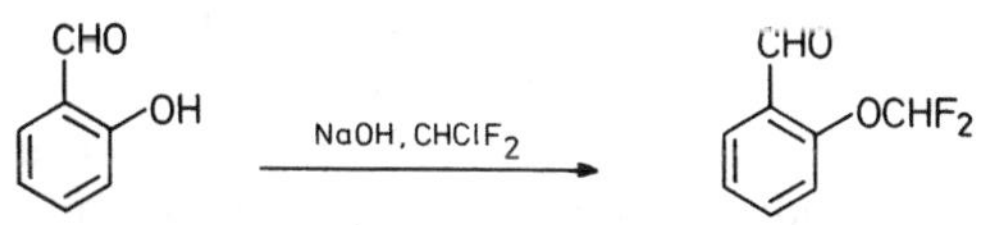

In a 1-l three-necked flask equipped with a stirrer, a thermometer, a bubbler connected with a gas-washing bottle, and a reflux condenser also connected with a gas-washing bottle, are placed 84 g (2.1 mol) of NaOH in 250 ml of water, 250 ml of dioxan (or propanol-2), 4 g of $Na_2S_2O_4 \cdot 2\,H_2O$, and 42.7 g (0.35 mol) of salicylic aldehyde. Difluorochloromethane (21–22 l) is bubbled through the mixture at 60°–65 °C with stirring during 5–6 h (Note 1). The mixture is cooled, poured into 500 ml of water, and the product extracted with ether (3 × 100 ml). The combined ether extracts are washed with water till neutral and dried over $MgSO_4$. The ether is distilled off and the product distilled in vacuum, in a stream of oxygen-free nitrogen. The yield of 2-difluoromethoxybenzaldehyde is 52.8–54.2 g (88%–90%); bp 89 °C/5 mm Hg; n_D^{20} 1.4940 [81] (Note 2).

Notes

1. Difluorochloromethane is bubbled in at such a rate that all the gas should be absorbed.
2. 2-Difluoromethoxybenzaldehyde should be stored in sealed tubes that have been flushed with oxygen-free nitrogen, with an addition of 0.1%–0.2% triethylamine as a stabilizer.

4-Chlorophenyltrifluoromethyl Ether

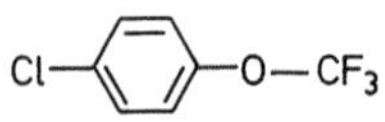

$C_7H_4ClF_3O$ MW 196.5

4-Chlorophenyltrifluoromethyl ether is a colourless liquid, well soluble in organic solvents, insoluble in water. Bp 141°–142 °C; n_D^{20} 1.4359; d_4^{20} 1.368 [83].

Method of Preparation

4-Chlorophenyltrifluoromethyl ether is prepared by fluorination of 4-chlorophenyltrichloromethyl ether with SbF_3 in the presence of $SbCl_5$ (80% yield) [83] or with anhydrous HF (83% yield) [84].

Procedure

$$4\text{-}ClC_6H_4OCH_3 \xrightarrow[PCl_5]{Cl_2} 4\text{-}ClC_6H_4OCCl_3 \xrightarrow[SbCl_5]{SbF_3;} 4\text{-}ClC_6H_4OCF_3$$

4-Chlorophenyltrichloromethyl ether. 4-Chloroanisole (142.5 g, 1 mol) is placed in a 0.5-l three-necked flask fitted with a thermometer, a reflux condenser, and a chlorine inlet tube, then 10.4 g (0.05 mol) of PCl_5 is added. The mixture is heated to 140°–150 °C in an oil-bath and dry chlorine is let in; the temperature in the flask should be maintained at 190°–200 °C. Chlorination is conducted to a 103.5 g gain in weight; this takes about 6 h. The yield of the crude product is 246 g. It is distilled in vacuum at 16 mm Hg; the following fractions are collected: fraction 1 — dichlorobenzene, bp below 110 °C, 15 g; fraction 2 — the products of partial chlorination, bp below 131 °C, 16 g;

fraction 3 — 4-chlorophenyltrichloromethyl ether, bp 132°–136 °C. Yield is 190 g (76%); n_D^{21} 1.552 [83].

4-Chlorophenyltrifluoromethyl ether. In a 0.5-l Würtz flask connected with a condenser and a receiver, are placed 123 g (0.5 mol) of 4-chlorophenyltrichloromethyl ether, 106.2 g (0.6 mol) of SbF_3, then 10.6 g of $SbCl_5$ is added. The mixture immediately heats and boils up. The flask is heated during several minutes so that the mixture continues to boil but does not distil. Then the fluorination product is distilled off together with $SbCl_3$ formed in the reaction. The distillate is dissolved in ether, washed several times with 15% HCl, then with water till neutral, and dried over Na_2SO_4. The ether is distilled off and the product distilled on a small column. The yield of 4-chlorophenyltrifluoromethyl ether is 79 g (80%); bp 141°–142 °C; n_D^{20} 1.4359 [83].

4-Aminophenyltrifluoromethyl Ether

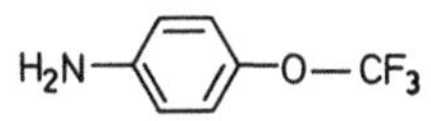

$C_7H_6F_3NO$ MW 177

4-Aminophenyltrifluoromethyl ether is a colourless liquid, well soluble in organic solvents, hydrochloric acid, insoluble in water. Bp 56°–58 °C/4 mm Hg [85], 73° to 75 °C/10 mm Hg [86].

Method of Preparation

4-Aminophenyltrifluoromethyl ether is prepared from 4-trifluoromethoxybenzamide (70% yield) [83, 86], or by the reduction of 4-nitrophenyltrifluoromethyl ether with $SnCl_2$ in HCl (90% yield) [85].

Procedure

$$\text{4-CH}_3\text{OC}_6\text{H}_4\text{COCl} \xrightarrow[\text{PCl}_5]{\text{Cl}_2;} \text{4-CCl}_3\text{OC}_6\text{H}_4\text{COCl} \xrightarrow{\text{SbF}_3} \text{4-CF}_3\text{OC}_6\text{H}_4\text{COF} \rightarrow$$

$$\xrightarrow{\text{NH}_3} \text{4-CF}_3\text{OC}_6\text{H}_4\text{CONH}_2 \xrightarrow{\text{KOBr}} \text{4-CF}_3\text{OC}_6\text{H}_4\text{NH}_2$$

4-(Chlorformyl)phenyltrichloromethyl ether. In a 0.25-l flask of an apparatus used for chlorination of 4-chloroanisole (see p. 264) is placed 85 g (0.5 mol) of anisoyl chloride 86,87, then 5.2 g (0.03 mol) of PCl_5 is added. The mixture is heated in an oil-bath to 140°–150 °C, whereupon dry chlorine is let in. Chlorination is conducted at 190°–200 °C for 4 h to a 50–51 g gain in weight. The reaction mixture is then distilled in vacuum and a fraction boiling at 138°–140 °C/4 mm Hg is collected (Note). The yield of 4-(chloroformyl)phenyltrichloromethyl ether is ~120 g (~87%).

4-(Fluoroformyl)phenyltrifluoromethyl ether. In a 0.5-l Claisen flask are placed 137 g (0.5 mol) of 4-(chloroformyl)phenyltrichloromethyl ether and 135 g (0.75 mol) of SbF_3. The mixture heats and SbF_3 dissolves. Then the mixture is heated for several minutes so that it boils but does not distil, is cooled, and the fluorination product is distilled off in vacuum of a water-jet pump, while the mixture is heated with a burner. Redistillation on a Vigreux column (20 cm long) gives 4-(fluoroformyl)-phenyltrifluoromethyl ether (74 g, 71%); bp 161°–162 °C.

4-(Carbamoyl)phenyltrifluoromethyl ether. In a 250-ml three-necked flask provided with a stirrer and cooled in an ice-bath, is placed 41.6 g (0.2 mol) of 4-(fluoroformyl)-phenyltrifluoromethyl ether, then 52 ml of 25% NH_3 solution is slowly poured in with stirring. After 30 min the precipitate is filtered off, washed with water and crystallized from aqueous ethanol. The yield of 4-(carbamoyl)phenyltrifluoromethyl ether is 34.8 g (85%); mp 152°–154 °C [86].

4-Aminophenyltrifluoromethyl ether. A solution of KOBr is prepared from 33.6 g (0.6 mol) of KOH in 280 ml of water and 5 ml (0.1 mol) of bromine, in a 1-l three-necked flask provided with a thermometer, a reflux condenser, and a stirrer, then 15.4 g (0.075 mol) of finely ground 4-(carbamoyl)phenyltrifluoromethyl ether is immediately introduced at 0 °C with vigorous stirring. After the amide has dissolved, the reaction mixture is heated at 75°–80 °C during 20 min, then steam-distilled. The distillate is extracted with ether (3 × 100 ml) and the ethereal solution dried over Na_2SO_4. The ether is distilled and the residue distilled in vacuum. The yield of 4-amino-phenyltrifluoromethyl ether is 9.3 g (70%) bp 73°–75 °C/10 mm Hg [86].

Note. The first fraction, bp below 130 °C/4 mm Hg, involves the products of partial chlorination.

2,4,6-Tris(pentafluoroethoxy)Aniline

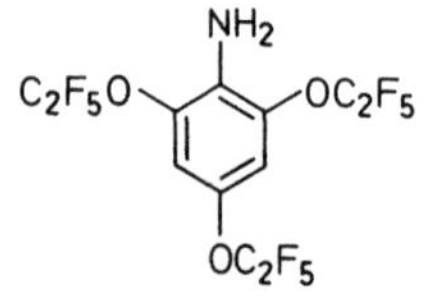

$C_{12}H_4F_{15}NO_3$ MW 495

2,4,6-Tris(pentafluoroethoxy)aniline is a colourless crystalline substance, well soluble in organic solvents, insoluble in water and mineral acids. Mp 31°–31.5 °C; bp 92° to 93 °C/20 mm Hg [88].

Method of Preparation

2,4,6-Tris(pentafluoroethoxy)aniline is prepared by the reduction of 2,4,6-tris(pentafluoroethoxy)nitrobenzene (82% yield) [88].

Procedure [88]

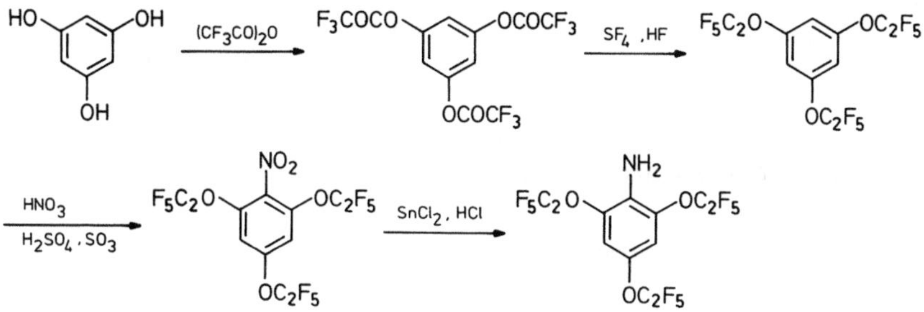

Phloroglucinol tris(trifluoroacetate). In a glass tube (or steel autoclave) are placed 12.6 g (0.1 mol) of phloroglucinol and 94.5 g (0.45 mol) of trifluoroacetic anhydride, the tube is sealed and heated at 150 °C for 20 h. Then it is cooled and its contents are transferred to a 100-ml Würtz flask. Excess trifluoroacetic anhydride and trifluoroacetic acid formed in the reaction are distilled off and the residue is distilled in vacuum; a fraction boiling at 100–102 °C/2 mm Hg is collected, which crystallizes upon cooling it with ice. The yield of phloroglucinol tris(trifluoroacetate) is 27 g (64.7%); mp 17° to 18 °C (from pentane); n_D^{20} 1.3973; d_{20}^{20} 1.6044 [88].

1,3,5-Tris(pentafluoroethoxy)benzene. In a 100-ml stainless steel autoclave fitted with a needle valve, is placed 20.7 g (0.05 mol) of phloroglucinol tris(trifluoroacetate), then 32 g of 85% (0.25 mol of 100%) SF_4 and 5 g of HF are introduced (see pp. 1, 2). The autoclave is heated in an oil-bath at 150 °C for 3 h, and at 175 °C for 6 h. Then it is cooled and the gaseous products are let out through the valve into an alkaline trap. The contents of the autoclave are steam-distilled and the fluorination product extracted from the distillate with ether (3 × 50 ml). The ether extract is washed with 10% NaOH solution, with water, and dried over Na_2SO_4. The ether is distilled off and the product vacuum-distilled. The yield of 1,3,5-tris(pentafluoroethoxy)benzene is 17.9 g (74.8%); bp 78°–80 °C/20 mm Hg; n_D^{17} 1.3355; d_{17}^{17} 1.6129 [88].

2,4,6-Tris(pentafluoroethoxy)nitrobenzene. In a 50-ml three-necked flask provided with a stirrer, a dropping funnel, and a reflux condenser with a calcium chloride tube, are mixed 5 ml of HNO_3 (d 1.52) and 15 ml of 20% oleum. Then 14.4 g (0.03 mol) of 1,3,5-tris(pentafluoroethoxy)benzene is added dropwise at 20 °C, with stirring. The flask is heated at 100 °C and stirred at that temperature for 2 h, and at 120 °C for 6 h. The reaction mixture is cooled and poured onto ice. The organic layer is extracted with ether (3 × 50 ml), the ether extract washed with water, with 10% Na_2CO_3 solution, with water, and dried over Na_2SO_4. The ether is distilled off and the product vacuum-distilled. The yield of 2,4,6-tris(pentafluoroethoxy)nitrobenzene is 9.7 g (62%); bp 103°–104 °C; n_D^{20} 1.3538; d_{20}^{20} 1.6602 [88].

2,4,6-Tris(pentafluoroethoxy)aniline. In a 250-ml flask fitted with a reflux condenser are placed 5.25 g (0.01 mol) of 2,4,6-tris(pentafluoroethoxy)nitrobenzene, 200 ml of methanol, and a solution of 12.3 g (0.05 mol) of $SnCl_2 \cdot 2\,H_2O$ in 15 ml of 35% HCl. The mixture is refluxed during 1 h, then 150–160 ml of methanol is distilled off. The amine is distilled off together with the solvent. The distillate is diluted with water and the amine extracted with ether (3 × 50 ml). The ether extract is washed with water and dried over Na_2SO_4. The solvent is distilled off and the residue vacuum-distilled; a fraction boiling at 91°–94 °C at 20 mm Hg is collected, which gradually crystallizes. The product is purified by recrystallization from hexane. The yield of 2,4,6-tris(pentafluoroethoxy)aniline is 4.1 g (82%); mp 31°–31.5 °C [88].

4-(Perfluoro-n-heptoxy)Benzoic Acid

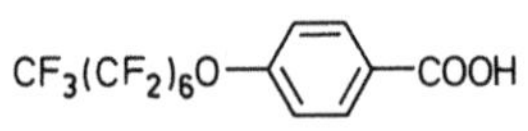

$C_{14}H_5F_{15}O_3$ MW 506

4-(Perfluoro-n-heptoxy)benzoic acid is a colourless crystalline substance, partially soluble in common organic solvents, insoluble in water, slightly soluble in aqueous alkalis, better soluble in diethanolamine. Mp 185°–186 °C [89].

Method of Preparation

4-(Perfluoro-n-heptoxy)benzoic acid is prepared by fluorination of methyl 4-(perfluoroenanthoyloxy)benzoate formed in acylation of methyl 4-hydroxybenzoate with perfluoroenanthoyl chloride, with subsequent hydrolysis of trifluoromethyl group of 4-(perfluoro-n-heptoxy)benzotrifluoride with a mixture of H_2SO_4 and HSO_3Cl in the presence of silica gel (50%) [89].

Procedure

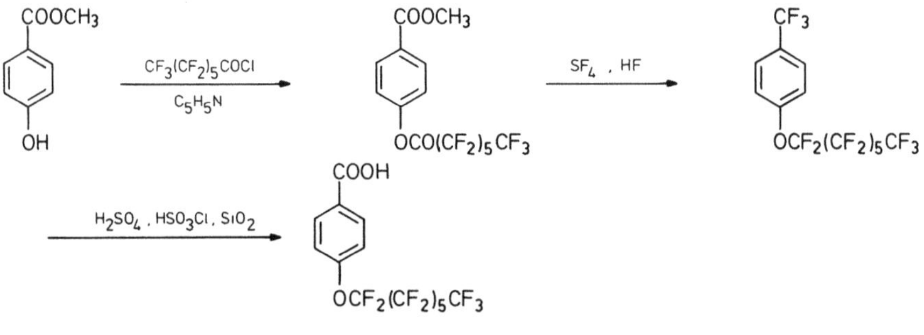

Methyl 4-(perfluoroenanthoyloxy)benzoate. In a 250-ml three-necked flask provided with a stirrer, a reflux condenser with a bubble counter filled with a concentrated H_2SO_4, and a dropping funnel, are placed 15.2 g (0.1 mol) of methyl 4-hydroxybenzoate, 7.9 g (0.1 mol) of pyridine and 150 ml of anhydrous ether. Perfluoroenanthoyl chloride (37 g, 0.1 mol) [90] is dropped in the solution with vigorous stirring, and the mixture is refluxed during 1 h. Then the mixture is cooled and filtered; the crystalline residue on the filter is washed with 20 ml of ether. The solvent is distilled off from the combined ether filtrates and the residue crystallized from pentane. The yield of methyl 4-(perfluoroenanthoyloxy)benzoate is 41.8 g (86%); mp 40°–42 °C [89].

Procedure

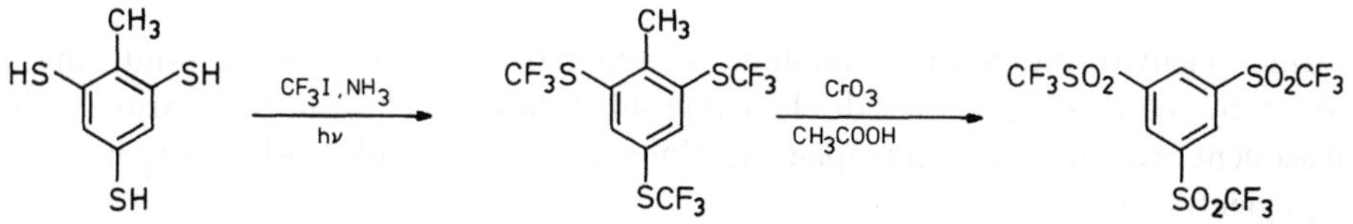

2,4,6-Tris(trifluoromethylthio)toluene. 2,4,6-Trimercaptotoluene (15.6 g, 0.083 mol) 93 is placed in a 0.5-l four-necked flask previously flushed with argon and equipped with a stirrer, gas inlet tubes and a dry-ice isopropanol reflux condenser. Ammonia (250 ml), then trifluoroiodomethane (62 g, 0.316 mol), are condensed into the flask kept cooled with dry-ice chloroform mixture. The reaction mixture is irradiated during 1.5 h, with vigorous stirring; the temperature should be regulated so that the ammonia boiled. After several minutes, a voluminous precipitate is formed. After the lamp is switched off and the cooling bath removed, the ammonia gradually evaporates and the residue is dissolved in ether, washed with 3% NaOH solution, with 5% Na_2SO_3 solution, with water till neutral, and dried over $MgSO_4$. The ether is distilled off and the product vacuum-distilled; a fraction boiling at 98°–100 °C/10 mm Hg is collected. The yield of 2,4,6-tri(trifluoromethylthio)toluene is 29.1 g (90%); n_D^{24} 1.4660 [93].

1,3,5-Tris(trifluoromethylsulphonyl)benzene. In a 1-l three-necked flask fitted with a stirrer and a reflux condenser, are placed 25.8 g (0.066 mol) of 2,4,6-tris(trifluoromethylthio)toluene and 200 ml of glacial acetic acid (Note). To the resulting solution, 100 g (1 mol) of CrO_3 is gradually added with vigorous stirring at 20°–25 °C, the reaction mixture is cautiously (to avoid its ejection from the flask) heated to reflux, and refluxed during 6 h. In 1 h after the beginning of the reaction, a precipitate is formed. The mixture is cooled, 300 ml of water is added, the precipitate is filtered off, washed with several portions of water and dried at 100°–110 °C. The yield of 1,3,5-tris(trifluoromethylsulphonyl)benzene is 25 g (80%); mp 199°–200 °C [93].

Note. Oxidation of 2,4,6-tris(trifluoromethylthio)toluene with CrO_3 in H_2SO_4 at 25 °C gives 2,4,6-tris(trifluoromethylsulphonyl)-toluene (66% yield; mp 135 °C from a 1:2 benzene-hexane mixture) [94].

4-Aminophenyltrifluoromethyl Sulphide

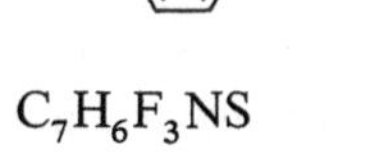

$C_7H_6F_3NS$ MW 193

4-Aminophenyltrifluoromethyl sulphide is a colourless liquid, well soluble in organic solvents, hydrochloric acid, insoluble in water. Bp 102°–103 °C/8 mm Hg [95], 102°–103 °C/11 mm Hg [96]; n_D^{20} 1.5253 [95], n_D^{21} 1.5250 [96].

Method of Preparation

4-Aminophenyltrifluoromethyl sulphide is prepared by methylation of 4-nitrothiophenol, by chlorination of the side chain and substitution of chlorine by fluorine, with subsequent reduction of 4-nitrophenyltrifluoromethyl sulphide with $SnCl_2$ in HCl (34% yield) [95].

The following procedure involves trifluoromethylation of 4-aminothiophenol with trifluoromethyl iodide in liquid ammonia at —30° ÷ —70 °C, under UV-irradiation (87% yield) [96].

Procedure

$$4\text{-}H_2NC_6H_4SH + CF_3I \xrightarrow[NH_3]{h\nu} 4\text{-}H_2NC_6H_4SCF_3$$

4-Aminothiophenol (8 g, 0.06 mol) is placed in a three-necked flask previously flushed with dry, oxygen-free nitrogen (Note 1), and provided with a stirrer, a gas inlet tube, and a dry-ice isopropanol reflux condenser. The flask is cooled with dry-ice chloroform mixture to —60° ÷ —70 °C, then 40 ml of liquid ammonia and 16 g (0.08 mol) of trifluoromethyl iodide are condensed into it. The mixture is irradiated during 1 h with a mercury lamp, with stirring. In 5–10 min after irradiation has started, the temperature of the reaction mixture is gradually raised till ammonia starts to reflux (Note 2), and the mixture refluxed under irradiation during 1 h. Ammonia is then evaporated and 25–30 ml of 5% NaOH solution is added to the mixture. The product is extracted with ether (2 × 100 ml), the combined ether extracts are washed with water, with 5% solution of $Na_2S_2O_3$, with water, and dried over $MgSO_4$. The ether is distilled off and the product vacuum-distilled. The yield of 4-aminophenyltrifluoromethyl sulphide is 10.8 g (87%); bp 102°–103 °C/11 mm Hg; n_D^{21} 1.5250 [96] (Note 3).

Notes.

1. During the whole process the flask is flushed with dry, oxygen-free nitrogen or argon.
2. Quick boiling-up of ammonia and ejection of the mixture from the flask may be prevented by immersing the flask deeper into the cooling bath. The ice should be intermittently wiped off the walls of the flask with ethanol or acetone.
3. Similar conversions of the corresponding derivatives of thio- and selenophenol afford phenyltrifluoromethyl sulphide (76% yield, bp 141.5 °C; n_D^{19} 1.4653 [97]); 4-chlorophenyltrifluoromethyl sulphide (72% yield; bp 173°–174 °C; n_D^{18} 1.4915 [96]); 2-aminophenyltrifluoromethyl sulphide (71% yield; mp 33 °C; bp 89° to 89.5 °C at 17 mm Hg [97]); 4-hydroxyphenyltrifluoromethyl sulphide (69.5% yield; mp 56 °C; bp 97°–100 °C/16 mm Hg [97]); 2-trifluoromethylthiobenzothiazole (87.5% yield; mp 35°–36 °C; bp 68°–69 °C/0.05 mm Hg [97]); phenyltrifluoromethyl selenide (69% yield; bp 152 °C; n_D^{23} 1.4872 [98]); 4-bromophenyltrifluoromethylselenide (34% yield, bp 95 °C/16 mm Hg, n_D^{17} 1.5380 [98]).

4-Nitrophenyltrifluoromethyl Sulphoxide

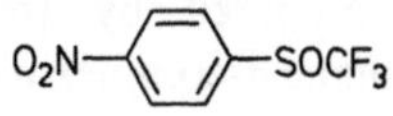

$C_7H_4F_3NO_3S$ MW 239

4-Nitrophenyltrifluoromethyl sulphoxide is a light yellow crystalline substance, soluble in organic solvents, insoluble in water. Bp 129°–130 °C/3 mm Hg; mp 60 °C [99].

Method of Preparation

4-Nitrophenyltrifluoromethyl sulphoxide is prepared by oxidation of 4-nitrophenyltrifluoromethyl sulphide with HNO_3 [99].

Procedure

$$4\text{-}O_2NC_6H_4SCF_3 \xrightarrow{HNO_3} 4\text{-}O_2NC_6H_4SOCF_3$$

In a 1.5-l two-necked flask fitted with a stirrer and a reflux condenser, are placed 200 g (0.9 mol) of 4-nitrophenyltrifluoromethyl sulphide (Note) and 1 l of HNO_3 (d 1.46). The mixture is refluxed with stirring during 10 h, then cooled, poured, with stirring, onto 3 kg of ice and extracted with ether. The extract is washed with water and dried over Na_2SO_4. The ether is distilled off and the residue vacuum-distilled; a fraction boiling at 126°–130 °C/3 mm Hg is collected. This gives 170–180 g of the crude product, mp 44°–49 °C, which is purified by crystallization from methanol. The yield of 4-nitrophenyltrifluoromethyl sulphoxide is 147 g (68 %); mp 60 °C [99].

Note. 4-Nitrophenyltrifluoromethyl sulphide is prepared by fluorination, with SbF_3, of 4-nitrophenyltrichloromethyl sulphide (70 % yield; bp 115 °C/20 mm Hg), formed in chlorination of 4-nitrophenylmethyl sulphide [95], or by trifluoromethylation of 4-nitrothiophenol with CF_3I [97].

2,4,6-Tris(trifluoromethylsulphonyl)Chlorobenzene

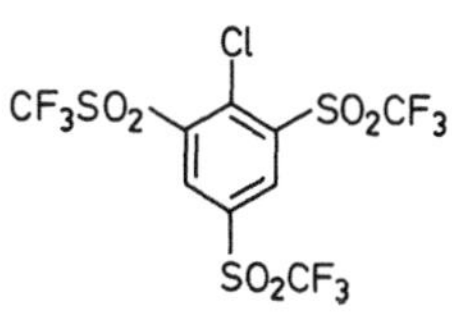

$C_9H_2ClF_9O_6S_3$ MW 509

2,4,6-Tris(trifluoromethylsulphonyl)chlorobenzene is a colourless crystalline substance, soluble in acetone, acetonitrile, insoluble in water. Mp 174°–175 °C [100].

Method of Preparation

2,4,6-Tris(trifluoromethylsulphonyl)chlorobenzene is prepared by treatment of 1,3,5-tris(trifluoromethylsulphonyl)benzene — nitromethane σ-complex with chlorine in aqueous dioxan [100].

Procedure

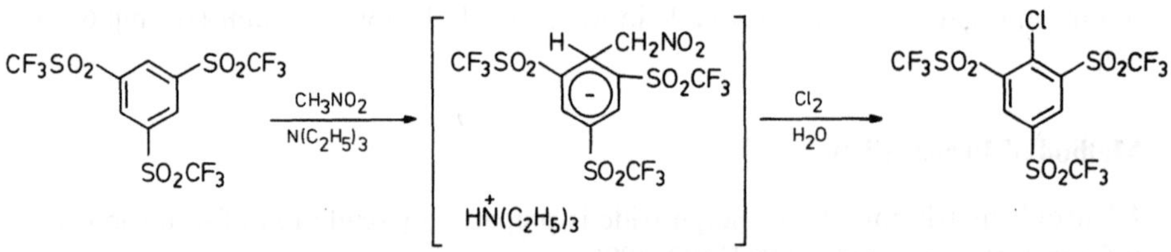

In a four-necked flask provided with a stirrer, gas inlet and outlet tubes, and a dropping funnel, the end of which is immersed into the liquid, are placed 30 ml of dioxan and 15 ml of distilled water. The aqueous dioxan is saturated with chlorine, with vigorous stirring and cooling of the flask in an ice-water bath (Note 1), till cloudiness appears. Then the dioxan solution of σ-complex (Note 2) is gradually added, while the strong flow of chlorine is continued to be let in. In several minutes, after all the σ-complex has been added, a precipitate is formed and the mixture undergoes almost complete decolorization. Stirring is continued for another 3 min, whereupon the flow of chlorine is cut off, the mixture is diluted with 100 ml of distilled water, the precipitate is filtered off, washed with distilled water, then with ether-hexane (1:1) mixture, and dried. The yield of 2,4,6-tris(trifluoromethylsulphonyl)chlorobenzene is 2.32 g (72%); mp 173°–174 °C [100].

Notes

1. Absorption of chlorine and oxidation proceed with considerable evolution of heat. At a higher temperature, the dioxan vapours are inflammable.
2. The σ-complex solution is prepared by adding 1.38 ml (0.0255 mol) of anhydrous nitromethane, then 1.2 ml (0.0086 mol) of anhydrous triethylamine, to 3 g (0.0063 mol) of 1,3,5-tris(trifluoromethylsulphonyl)benzene in 30 ml of anhydrous dioxan, with stirring at 25 °C. After the addition is complete, the mixture is stirred at 25 °C for 4 h.

Phenylsulphonyldifluoroacetic Acid

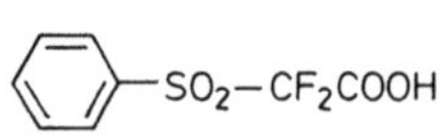

$C_8H_6F_2O_4S$ MW 236

Phenylsulphonyldifluoroacetic acid is a white crystalline substance, well soluble in organic solvents, slightly soluble in water. Mp 119°–120 °C [101].

Method of Preparation

Phenylsulphonyldifluoroacetic acid is prepared by the reaction of sodium thiophenoxide with $CF_2ClCOONa$, with subsequent conversion of α,α-difluoro-S-phenylthioglycolic acid to α,α-difluoro-S-phenylthiogylcolyl chloride [102], then to ester, oxidation of the ester to sulphone and its saponification (11% yield) [101].

Procedure

$$C_6H_5SNa \xrightarrow{ClCF_2COONa} C_6H_5SCF_2COONa \xrightarrow{PCl_5}$$

$$\rightarrow C_6H_5SCF_2COCl \xrightarrow[N(C_2H_5)_3]{CH_3OH;} C_6H_5SCF_2COOCH_3 \xrightarrow[CH_3COOH]{CrO_3;}$$

$$\rightarrow C_6H_5SO_2CF_2COOCH_3 \xrightarrow{HCl} C_6H_5SO_2CF_2COOH$$

α,α-Difluoro-S-phenylthioglycolyl chloride. Thiophenol (5.5 g, 0.05 mol) and chlorodifluoroacetic acid (6.45 g, 0.05 mol) (Note) are dissolved in 40 ml of anhydrous dioxan, in a 100-ml flask previously flushed with dry, oxygen-free nitrogen and provided with a reflux condenser with an Ascarite tube. Sodium lumps (2.3 g, 0.1 mol) are then gradually added to the solution and the mixture refluxed during 3 h. Dioxan is evaporated in vacuum of a water-jet pump, with gentle heating, and 11.6 g (0.055 mol) of PCl_5 is added in one portion to the solid residue. The mixture heats and liquefies. After 30 min the mixture is filtered through a porous glass filter protected from air moisture, into a 30-ml Claisen flask. The flask is fitted with a thermometer and a capillary connected with a P_2O_5 column for drying air. The product is distilled in vacuum. The yield of α,α-difluoro-S-phenylthioglycolyl chloride is 3.5 g (31%); bp 79°–81 °C/10 mm Hg; n_D^{29} 1.5110; d_4^{29} 1.3442 [102].

Methyl α,α-difluoro-S-phenylthioglycolate. A solution of 6.7 g (0.03 mol) of α,α-difluoro-S-phenylthioglycolyl chloride in 20 ml of anhydrous ether is placed in a 50-ml four-necked flask provided with an air condenser with a P_2O_5 column, a thermometer, a stirrer, and a dropping funnel. Then 1.12 g (0.035 mol) of anhydrous methanol is added at 20 °C, with stirring, whereupon 3.53 g (0.035 mol) of triethylamine is gradually introduced. After 30 min $N(C_2H_5)_3 \cdot HCl$ is filtered off. The filtrate is evaporated in vacuum of a water-jet pump, with gentle heating, and the residue vacuum distilled. The yield of methyl α,α-difluoro-S-phenylthioglycolate is 5.25 g (80.5%); bp 92°–93 °C/2 mm Hg; n_D^{20} 1.5032; d_4^{20} 1.2796 [101].

Methyl phenylsulphonyldifluoroacetate. Methyl α,α-difluoro-S-phenylthioglycolate (6.6 g, 0.03 mol) is dissolved in 30 ml of glacial acetic acid, in a 50-ml three-necked flask fitted with a thermometer, a stirrer, and a reflux condenser. Then 9 g (0.09 mol) of CrO_3 is added in small portions, while stirring, so that the temperature should not rise above 70 °C. The mixture is stirred at 70 °C for 1.5 h, then poured into water. The precipitate is filtered off, washed with water, dried and crystallized from a benzene petroleum-ether mixture. The yield of methyl phenylsulphonyldifluoroacetate is 4.2 g (56%); mp 55°–56 °C [101].

Phenylsulphonyldifluoroacetic acid. In a 25-ml round-bottom flask provided with a reflux condenser are placed 2 g (0.008 mol) of methyl phenylsulphonyldifluoroacetate and 10 ml of concentrated HCl. The mixture is refluxed for 1.5 h, then cooled. The product is poured into water, extracted with ether and dried over Na_2SO_4. The ether is distilled off and the residue crystallized from benzene. The yield of phenylsulphonyldifluoroacetic acid is 1.45 g (77%); mp 119°–120 °C [101].

Note. Chlorodifluoroacetic acid is prepared by saponification of methyl chlorodifluoroacetate [103] with 10% aqueous solution of NaOH. The product is purified by distillation; a fraction boiling at 120°–125 °C is collected.

4-Aminophenyltrifluoromethyl Selenide

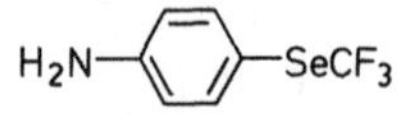

$C_7H_6F_3NSe$ MW 240

4-Aminophenyltrifluoromethyl selenide is a colourless liquid, well soluble in organic solvents and hydrochloric acid, insoluble in water. Bp 108 °C/8 mm Hg; mp 20 °C; n_D^{20} 1.5503; d_4^{20} 1.638 [104].

Method of Preparation

4-Aminophenyltrifluoromethyl selenide is prepared by the reaction of aniline with trifluoromethylselenyl chloride (76.7% yield) [104]. Trifluoromethylselenyl chloride is obtained from selenium and silver trifluoroacetate with subsequent splitting of bis(trifluoromethyl)diselenide with chlorine (10% yield) [105].

Procedure

$$2\,CF_3COOAg \xrightarrow[-2\,CO_2;\,-2\,Ag]{2\,Se} CF_3Se{-}SeCF_3 \xrightarrow{Cl_2} CF_3SeCl \xrightarrow{C_6H_5NH_2}$$
$$\rightarrow 4\text{-}H_2N{-}C_6H_4SeCF_3$$

Bis(trifluoromethyl)diselenide. In a test tube 300 mm long and 30 mm in diameter connected to a U-tube with a receiver held at —70 °C, is placed a mixture of colloidal selenium (39.5 g, 0.5 mol) (Note 1) and silver trifluoroacetate (40 g, 0.18 mol) (Note 2), thoroughly dried (over P_2O_5) and powdered. The tube is gradually heated on a metallic bath to 280 °C and maintained at 280°–300 °C for 3 h. The distillable products are condensed in the receiver. After the process is complete (Note 3), the contents of the receiver are heated to 20 °C and washed with water. A heavy yellow liquid is separated from the aqueous layer, dried over Na_2SO_4 and distilled; a fraction boiling at 66°–74 °C is collected (Notes 4, 5). The yield of bis(trifluoromethyl)diselenide is 6.4 g (24%, based on silver trifluoroacetate) [105].

Trifluoromethylselenyl chloride. In a 20-ml three-necked round-bottom flask provided with a stirrer, a reflux condenser, and a bubbler, is placed 3.5 g (0.012 mol) of bis(trifluoromethyl)diselenide, which is then cooled, with stirring, to —30 °C. Dry chlorine is then bubbled through the resulting thick mass to complete solidification of the reaction product. To remove excess chlorine, the flask is flushed with dry nitrogen. The bubbler is removed, 7.1 g (0.024 mol) of bis(trifluoromethyl)diselenide is added to the resulting yellow crystals of trifluoromethylselenium trichloride and the mixture is allowed to stand at 20 °C for 12 h, with intermittent shaking. During that time the mixture turns into a dark cherry-red liquid. The product is distilled on a column; a fraction boiling at 28°–32 °C is collected. The yield of trifluoromethylselenyl chloride is 11.9 g (90%); n_D^{14} 1.3930; d_4^{14} 1.9899 [105].

4-Aminophenyltrifluoromethyl selenide. Trifluoromethylselenyl chloride is dissolved in 50 ml of anhydrous ether in a 100-ml four-necked flask equipped with an efficient stirrer, a dropping funnel, a thermometer, and a reflux condenser protected from air moisture by a P_2O_5 tube, and the solution is cooled to —30 °C. Then a solution of 5.6 g (0.06 mol) of aniline in 25 ml of anhydrous ether is added dropwise at that temperature with stirring. The temperature is gradually brought to 20 °C and the mixture is stirred during 1 h. The anilinium hydrochloride is filtered off, the ether is distilled off and the residue vacuum-distilled. A fraction boiling at 108°–109 °C/8 mm Hg is collected. The yield of 4-aminophenyltrifluoromethyl selenide is 5.5 g (76%); mp 20 °C [104].

Notes

1. Colloidal selenium is prepared by passing SO_2 through the saturated aqueous solution of selenous acid or its soluble salt to complete precipitation of selenium. The original selenium precipitate is of reddish-brown colour; upon being dried over P_2O_5 it slightly darkens.
2. Silver trifluoroacetate is prepared by dissolution of an aqueous suspension of freshly prepared, washed (with water) silver oxide in trifluoroacetic acid, while the solution is being cooled with water. The solution is evaporated to dryness in vacuum of a water-jet pump at 60°–70 °C, while the residue is intermittently ground in a mortar, then dried over P_2O_5 in a vacuum-desiccator.
3. After the reaction is complete, a solid cake remains in the test tube, which consists of silver selenides. It is best withdrawn by heating the test tube till the residue has melted.
4. The product is sufficiently pure to be used in further preparations. Pure bis(trifluoromethyl)diselenide boils at 71 °C; n_D^{18} 1.4035; d_4^{18} 2.203 [105].
5. The crude product contains a small amount of bis(trifluoromethyl)selenide which boils at 0°–2 °C and can be easily separated by redistillation.

N,N-Bis(trifluoromethyl)-4-phenylenediamine

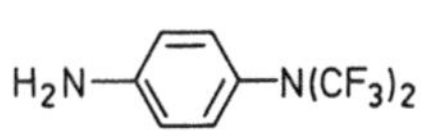

$C_8H_6F_6N_2$ MW 244

N,N-Bis(trifluoromethyl)-4-phenylenediamine is a colourless liquid with a pleasant odour, soluble in common organic solvents, insoluble in water. Bp 85 °C/15 mm Hg; n_D^{21} 1.4320; d_4^{21} 1.4562 [106].

Method of Preparation

N,N-Bis(trifluoromethyl-4-phenylenediamine is prepared by fluorination of bis(N-methyl-N-4-chlorophenyl)thiuram disulphide [107] with SF_4, with subsequent chlorination of the product, fluorination of the resulting 4-chloro-N-trifluoromethyl-N-trichloromethylaniline with SbF_3 [108], and final replacement of chlorine with the amino group [106] (26.2% yield).

Procedure

$$\text{4-ClC}_6\text{H}_4\text{NHCH}_3 \xrightarrow[\text{I}_2;\,\text{C}_5\text{H}_5\text{N}]{\text{CS}_2;} [\text{ClC}_6\text{H}_4\underset{\substack{|\\ \text{CH}_3}}{\text{N}}\text{CSS}-]_2 \xrightarrow{\text{SF}_4} \text{ClC}_6\text{H}_4\underset{\substack{|\\ \text{CH}_3}}{\text{N}}\text{CF}_3 \rightarrow$$

$$\xrightarrow{\text{Cl}_2} \text{ClC}_6\text{H}_4\overset{\substack{\text{CCl}_3\\ |}}{\text{N}}\text{CF}_3 \xrightarrow{\text{SbF}_3} \text{ClC}_6\text{H}_4\text{N(CF}_3)_2 \xrightarrow[\text{Cu}]{\text{NH}_3} \text{H}_2\text{NC}_6\text{H}_4\text{N(CF}_3)_2$$

Bis(N-methyl-N-4-chlorophenyl)thiuram disulphide. In a 2-l three-necked flask fitted with a stirrer, a dropping funnel, and a reflux condenser with a calcium chloride tube, are placed 141.5 g (1 mol) of 4-chloro-N-methylaniline, 90 ml of anhydrous CS_2, and 80 g of anhydrous pyridine (1 mol). The mixture is heated to reflux, then 127 g (0.5 mol) of iodine in 400 ml of anhydrous CS_2 is dropped in with vigorous stirring. The mixture is stirred at boiling point till the iodine colouring disappears, then cooled, filtered to remove the pyridinium salt, and the filtrate evaporated to dryness. The residue is dissolved in 700 ml of benzene, washed with 5% HCl, with water, with 5% Na_2SO_3 solution to decolorization, then with water again, and dried over Na_2SO_4. The benzene solution is passed through an Al_2O_3 column and evaporated to dryness. The yield of bis(N-methyl-N-4-chlorophenyl)-thiuram disulphide is 155.5 g (72%); mp 171 °C [107].

4-Chloro-N-methyl-N-trifluoromethylaniline. Bis(N-methyl-N-4-chlorophenyl)-thiuram disulphide (7.4 g, 0.02 mol) is placed in a 50-ml stainless steel autoclave fitted with a needle valve. The autoclave is cooled with liquid nitrogen and evacuated to 0.1 mm Hg residual pressure, then 52 g of 85% (0.04 mol of 100%) SF_4 are condensed into it through the valve see p. 2). The autoclave is heated at 120 °C for 8 h, then cooled, the valve is opened and the gaseous products are let out into an alkaline trap. The product is extracted with ether, the ethereal solution is washed with water, with 5% solution of $NaHCO_3$, with water again, and dried over $MgSO_4$. The solvent is distilled off and the product vacuum-distilled. The yield of 4-chloro-N-methyl-N-trifluoromethylaniline is 5.9 g (70%); bp 76°–77 °C/17 mm Hg; n_D^{20} 1.4678; d_4^{20} 1.3183 [109] (Note 1).

4-Chloro-N-trifluoromethyl-N-trichloromethylaniline. 4-Chloro-N-methyl-N-trifluoromethylaniline (75 g, 0.35 mol), azadiisobutyronitrile (0.5 g), and urotropine (0.1 g) are placed in a 200-ml quartz flask provided with a stirrer, a thermometer, a reflux condenser connected with an HCl absorbing system, and a chlorine inlet tube 5 mm in diameter. The mixture is chlorinated on a bath of quartz glass filled with H_2SO_4, under UV-irradiation, at 60°–70 °C for 2 h, at 90 °C for 5 h, at 100°–105 °C for 13 h and at 110 °C for 4 h. The contents of the flask are then vacuum-distilled at 0.4 mm Hg, with a short dephlegmator; the following fractions are collected: fraction 1 — mainly monochloroderivative, bp 80°–82 °C, n_D^{20} 1.5039, 8.5 g; fraction 2 — partial chlorination products, bp 82–85 °C, n_D^{20} 1.5054, 42 g; fraction 3 — 4-chloro-N-trifluoromethyl-N-trichloromethylaniline, bp 85°–88 °C, n_D^{20} 1.5074, 40 g. Fractions 1 and 2 are combined and further chlorinated at 100 °C for 2 h with an addition of 0.2 g of azodiisobutyronitrile and 0.05 g of urotropine, then distilled in vacuum at 0.4 mm Hg; a fraction boiling at 86°–88 °C is collected, which is combined with the main product. The combined fractions are vacuum-distilled. The yield of 4-chloro-N-trifluoromethyl-N-trichloromethylaniline is 80.6 g (71 %); bp 86°–88 °C/ 0.4 mm Hg; n_D^{20} 1.5071 [108].

4-Chloro-N,N-bis(trifluoromethyl)aniline. 4-Chloro-N-trifluoromethyl-N-trichloromethylaniline (47 g, 0.15 mol), calcinated SbF_3 (30 g, 0.168 mol), and SbF_5 (0.3 g) are placed in a 200-ml round-bottom quartz flask provided with a reflux condenser with a P_2O_5 tube. The mixture is cautiously heated with a burner, then refluxed for 4–5 min. After having been cooled, the mixture is treated with 200 ml of 20 % HCl and the organic layer is extracted with ether (3 × 100 ml). The ether extract is washed with 15 % HCl to remove the antimony salts, then with water, and dried over Na_2SO_4. The ether is distilled off and the product vacuumdistilled. The yield of 4-chloro-N,N-bis(trifluoromethyl)aniline is 33 g (85 %); bp 75°–76 °C/50 mm Hg; n_D^{20} 1.4122; d_4^{20} 1.4670 [110] (Note 2).

N,N-Bis(trifluoromethyl)-4-phenylenediamine. In a 150-ml stainless steel autoclave equipped with a needle valve, are placed 20 g (0.076 mol) of 4-chloro-N,N-bis(trifluoromethyl)aniline and 50 ml of H_2O, and the mixture is cooled to —70 °C. Then 50 ml of liquid NH_3, 10 g of CuCl, and 4 g of copper powder are added, the autoclave is closed and heated at 240 °C for 50 h. After the autoclave has been cooled, excess NH_3 is let out through the valve, the autoclave is opened and its contents treated with ether. The ethereal solution is washed with water and dried over Na_2SO_4. After the ether has been removed, the product is distilled. The yield of N,N-bis(trifluoromethyl)-4-phenylenediamine is 16.8 g (86.5 %); bp 85 °C/15 mm Hg; n_D^{21} 1.4320; d_4^{21} 1.4562 [106].

Notes

1. 4-Chloro-N-methyl-N-trifluoromethylaniline can also be obtained by fluorination of N-methyl-N-4-chlorophenylthiocarbamoyl chloride with SF_4 [107].
2. 4-Chloro-N-trifluoromethyl-N-trichloromethylaniline may be fluorinated with HF in an autoclave [108].

N,N-Bis(trifluoromethylsulphonyl)-4-anisidine

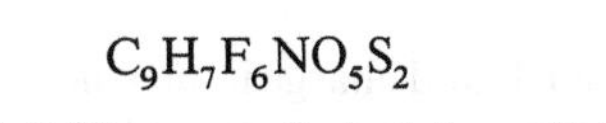

$C_9H_7F_6NO_5S_2$ MW 387

N,N-Bis(trifluoromethylsulphonyl)-4-anisidine is a colourless crystalline substance, well soluble in common organic solvents, insoluble in water. Mp 66°–67 °C [111].

Method of Preparation

N,N-Bis(trifluoromethylsulphonyl)-4-anisidine is prepared by the stepwise acylation of 4-anisidine with trifluoromethanesulphonyl fluoride (37% yield) [111].

Procedure [111]

$$4\text{-}CH_3OC_6H_4NH_2 \xrightarrow[N(C_2H_5)_3]{CF_3SO_2F;} 4\text{-}CH_3OC_6H_4NHSO_2CF_3 \xrightarrow[THF]{Li;}$$

$$\rightarrow 4\text{-}CH_3OC_6H_4N(Li)SO_2CF_3 \xrightarrow{CF_3SO_2F} 4\text{-}CH_3OC_6H_4N(SO_2CF_3)_2$$

N-Trifluoromethylsulphonyl-4-anisidine. In a 50-ml stainless steel autoclave fitted with a needle valve, are placed 7.4 g (0.06 mol) of 4-anisidine and 15 ml of triethylamine. The autoclave is cooled with liquid nitrogen, evacuated to 1–2 mm Hg residual pressure, then 10 g (0.07 mol) of trifluoromethanesulphonyl fluoride is condensed into it through the valve (Note 1). The autoclave is heated in an oil-bath at 100 °C for 20 h, then cooled, excess gas is let out through the valve and the contents of the autoclave are treated with 50 ml of 20% HCl. The product is extracted with ether (3 × 50 ml). The ether extracts are washed with water and dried over Na_2SO_4. The ether is distilled off and the product vacuum-distilled. The yield of N-trifluoromethylsulphonyl-4-anisidine is 8.5 g (55.5%); bp 125 °C/1.5 mm Hg; mp 38°–40 °C [112].

N,N-Bis(trifluoromethylsulphonyl)-4-anisidine. N-Trifluoromethylsulphonyl-4-anisidine (6 g, 0.0235 mol) and anhydrous tetrahydrofuran (30 ml) are placed in a 100-ml three-necked flask, previously flushed with dry, oxygen-free nitrogen and provided with a stirrer and a reflux condenser with a calcium chloride tube. The solution is heated to reflux, then 0.3 g (0.044 mol) of thin metallic lithium bands 10–12 mm long and 1.5–2 mm wide, are added with stirring. The mixture is refluxed for 1 h, with stirring. The hot solution is separated from the precipitate by decanting and evaporated in vacuum to dryness to remove the solvent.

Dry lithium salt and 15 ml of anhydrous sulpholane are placed in a 50-ml stainless steel autoclave provided with a needle valve. The autoclave is cooled with liquid nitrogen, evacuated to 1–2 mm Hg residual pressure, then 7 g (0.046 mol) of trifluoromethanesulphonyl fluoride (Note 1) is condensed in it through the valve. The autoclave is heated at 100 °C for 20 h, then cooled and its contents are poured into water. The product is extracted with ether (3 × 30 ml), the ether extracts are washed with 5% NaOH solution (Note 2), with water, and dried over Na_2SO_4. The ether is distilled off and the residue vacuum-distilled (Note 3). The yield of N,N-bis(trifluoromethylsulphonyl)-4-anisidine is 3.35 g (67%, based on the converted N-trifluoromethylsulphonyl-4-anisidine); bp 97 °C/0.15 mm Hg; mp 66°–67 °C (from pentane) [111].

Notes

1. In order to cut off the required amount of CF_3SO_2F and charge it into the autoclave, a rubber bladder of a sports ball may be used. The weighted dry bladder is filled with CF_3SO_2F from a cylinder.

2. The alkaline washings may be acidified with HCl to give 2.7 g of the starting N-trifluoromethylsulphonyl-4-anisidine.
3. The receiver is intermittently heated with caution during distillation to prevent crystallization of the product.

Tris(trifluoromethyl)Orthothiobenzoate

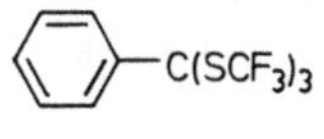

$C_{10}H_5F_9S_3$ MW 392

Tris(trifluoromethyl)orthothiobenzoate is a colourless liquid, readily soluble in organic solvents, insoluble in water. Bp 104°–105 °C/10 mm Hg; n_D^{25} 1.4682; d_{25}^{25} 1.559 [113].

Method of Preparation

Tris(trifluoromethyl)orthothiobenzoate is prepared by the reaction of benzotrichloride or benzotribromide with $AgSCF_3$ (75–80% yield) [113].

Procedure [113]

$$3\,AgF + CS_2 \rightarrow AgSCF_3 + Ag_2S$$

$$C_6H_5CCl_3 \xrightarrow{AgSCF_3} C_6H_5C(SCF_3)_3$$

Silver trifluoromethylmercaptide [114]. In a 1-l rocking steel autoclave fitted with a manometer, a thermocouple in a well, and a valve, are placed 270 g (2.1 mol) of anhydrous AgF (Note 1) and 265 ml (4.4 mol) of CS_2 previously distilled over P_2O_5 (Note 2). The autoclave is heated at 140 °C for 12 h, then cooled, the valve is cautiously opened and the gaseous products are let out (Note 3). Excess CS_2 is decanted and silver trifluoromethylmercaptide is extracted from the precipitate in a Soxhlet apparatus with 1 l of anhydrous acetone during 5–6 h. Acetone is distilled off in vacuum of a water-jet pump at 50 °C, to 80–100 ml volume. To the resulting solution is added 300 ml of acetone at 20 °C and the mixture is thoroughly stirred. Silver trifluoromethylmercaptide precipitate (cream-coloured shining tablets) is filtered off, thoroughly pressed and dried in vacuum of a water-jet pump at 50 °C for 1 h. The product is stored in a dark bottle. The yield of $AgSCF_3$ is 95 g (65%) [114].

Tris(trifluoromethyl)orthothiobenzoate. In a 25-ml flask provided with a reflux condenser with a calcium chloride tube, are placed 3 g (0.015 mol) of benzotrichloride, 12.5 g (0.06 mol) of $AgSCF_3$, and 20 ml of anhydrous acetonitrile. The mixture is heated at 50 °C for 10 h, the AgCl precipitate is filtered off and washed with 10 ml of acetonitrile; the washings are combined with the main product. The filtrate is poured into a 200-ml beaker, then 20 ml of ether, 20 ml of heptane (Note 4), and 150 ml of water are added. Unconverted $AgSCF_3$ precipitates. Tris(trifluoromethyl)-

orththiobenzoate is extracted with an ether-heptane mixture (3:1; 3 × 50 ml). The extract is washed with water (4 × 50 ml) and dried over Na_2SO_4. The ether is distilled off with a dephlegmator and the product vacuum-distilled. The yield of tris(trifluoromethyl)orthothiobenzoate is 4.4 g (75%); bp 104–105 °C/10 mm Hg [113].

Notes.

1. Silver fluoride is triturated in a platinum dish with 10 ml of anhydrous methanol, the solvent is decanted and the residue treated with 150 ml of anhydrous ether; the product is dried over P_2O_5 in a vacuum desiccator during 1 week.
2. *Caution*! Carbon disulphide is toxic and flammable.
3. This operation should be carried out under a hood, in the absence of any sources of heat or open flame.
4. Heptane is added to cause precipitation of $AgSCF_3$, partially dissolved in ether.

Phenyltetrafluorophosphorane

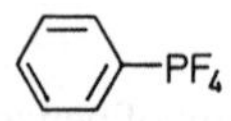

$C_6H_5F_4P$ MW 187

Phenyltetrafluorophosphorane is a colourless liquid that fumes in air; well soluble in common organic solvents; vigorously reacts with water; etches glass. It should be stored in hermetically closed platinum or Teflon bottles. Bp 58 °C/60 mm Hg [115], 133°–134 °C [116], 134.5°–136 °C [115].

Method of Preparation

Phenyltetrafluorophosphorane is prepared by fluorination of phenyltetrachlorophosphorane with SbF_3 (66% yield) or ZnF_2 (35%–40% yield) [116]; by oxidative fluorination of phenyldichlorophosphine with SbF_3 [116], AsF_3 [117], SbF_5 (63% yield) or a mixture of SbF_3 and $SbCl_5$ [118]; by the reaction of phenyldifluorophosphine with SbF_3 (60% yield) [119]; or by fluorination of phenylphosphinic acid with SF_4 (58% yield) [120].

Procedure [116]

$$C_6H_5PCl_2 \xrightarrow{SbF_3} C_6H_5PF_4$$

Phenyldichlorophosphine (179 g, 1 mol) is placed in a 1-l three-necked flask equipped with a stirrer, a Vigreux column (15 cm) and a downward condenser connected to a Würtz flask with a P_2O_5 tube. Then 250 g (1.4 mol) of SbF_3 is gradually added in small portions with stirring, at such a rate that the mixture should slightly boil but not distil. The mixture is then boiled during 15 min, phenyltetrafluorophosphorane is distilled off to the Würtz flask and a slow stream of chlorine is passed into that flask during several minutes (Note). The product is redistilled into the receiver with a P_2O_5 tube; a fraction boiling at 133°–134 °C is collected. Yield is 122 g (66%).

Note. Chlorine is passed to convert traces of phenyldichlorophosphine to nonvolatile phenyltetrachlorophosphorane.

Phenyltetrafluoroantimony

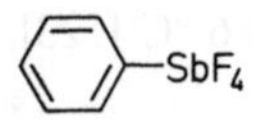

$C_6H_5F_4Sb$ MW 275

Phenyltetrafluoroantimony is a colourless crystalline substance, insoluble in organic solvents; easily hydrolysed by air moisture. Decompn. p. 125°–126 °C (in a sealed quartz capillary) [121].

Method of Preparation

Phenyltetrafluoroantimony is prepared by fluorination of phenylstibonic acid with SF_4 (94% yield) [121].

Procedure

$$C_6H_5SbO(OH)_2 \xrightarrow{SF_4} C_6H_5SbF_4 \cdot SF_4 \xrightarrow{Heat} C_6H_5SbF_4$$

In a dried quartz test tube (Note 1) is placed a solution of 0.75 g (0.003 mol) of phenylstibonic acid in 20 ml of anhydrous methlyene chloride. The tube is cooled to —70 °C, evacuated to 1 mm Hg residual pressure, then 1.1 g (0.01 mol) of SF_4 is introduced (see p. 2). The cooling bath is removed and the mixture is allowed to warm to 0 °C. After several minutes, a precipitate is formed. The gaseous products are let out in vacuum at 20 mm Hg; together with them, part of the solvents evaporates off. Then air is let into the apparatus through a P_2O_5 tube. The precipitate is quickly filtered off on a porous filter, in a dry box, washed with 8 ml of anhydrous methylene chloride and transferred to a 10-ml quartz flask. The flask is evacuated to 1 mm Hg residual pressure, then heated in vacuum at 60°–65 °C during 4 h to decompose the complex and remove SF_4. The yield of phenyltetrafluoroantimony is 0.77 g (94%); decompn. p. 125°–126 °C (in a sealed quartz capillary) (Note 2) [121].

Notes

1. The quartz test tube is 180 mm long and 35 mm in diameter; to it is sealed a vertical tube 60 mm long, 10–12 mm in diameter, which is connected, via a thick-walled rubber tube, with a T-bore stopcock.
2. The product is sufficiently pure. It is stored in thoroughly dried and sealed quartz tubes.

(Tetrafluoroiodo)Benzene

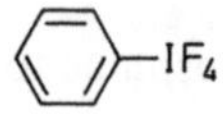

$C_6H_5IF_4$ MW 280

(Tetrafluoroiodo)benzene is a white crystalline substance, well soluble in ether, methylene chloride, chloroform, dichloroethane, less soluble in benzene, CCl_4, insoluble in hexane; easily hydrolysed by air moisture. mp 86 °C [123].

Method of Preparation

(Tetrafluoroiodo)benzene is prepared by fluorination of iodylbenzene with SF_4, in a quantitative yield [123].

The following method involves fluorination of difluoroiodosylbenzene with SF_4, since fluorination of iodylbenzene with SF_4 occasionally proceeds with an explosion.

Procedure

$$C_6H_5IO_2 \xrightarrow{HF} C_6H_5IOF_2 \xrightarrow{SF_4} C_6H_5IF_4$$

Difluoroiodosylbenzene. In a platinum dish is placed 250 ml of 40% HF, which is heated to 60°–70 °C in a water-bath; then 11.8 g (0.05 mol) of powdered iodylbenzene [124] (Note 1) is dissolved in it in small portions, with continuous stirring. The hot solution is filtered through cotton wool on a platinum funnel into a polyethylene bottle, then cooled. The product crystallizes in the form of needles, which are filtered off on a platinum Büchner funnel and thoroughly pressed. Then it is dried over P_2O_5 and solid KOH in a vacuum-desiccator to give 7–8 g of difluoroiodosylbenzene. Evaporating the filtrate in a platinum dish in a water-bath affords further 1.5–2 g of the product. The total yield is 8–10 g (62%–77%) (Note 2); decompn. p. 213 °C (in a capillary closed with a polyethylene cap).

(Tetrafluoroiodo)benzene. In a 50-ml stainless steel autoclave fitted with a needle valve, are placed 7.8 g (0.03 mol) of difluoroiodosylbenzene and 4.3 g (0.04 mol) of SF_4 (see p. 2). The autoclave is heated to 150 °C in an oil-bath during 3–4 h, and maintained at that temperature for 3 h, then cooled and the gaseous products are cautiously let out. The autoclave is discharged in a dry box and the product is placed in a Teflon or platinum bottle. The yield of (tetrafluoroiodo)benzene is quantitative; mp 86 °C [123].

Notes

1. Iodylbenzene should be ground cautiously in small portions, in a porcelain mortar; the pestle should be pressed weakly, since iodylbenzene can decompose with an explosion.
2. Difluoroiodosylbenzene should be stored over P_2O_5, since it is hydrolysed by air moisture to form iodylbenzene and HF.

Tetrakis(trifluoroacetoxy)Iodobenzene

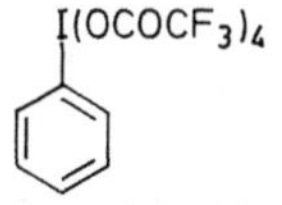

$C_{14}H_5F_{12}IO_8$ MW 656

Tetrakis(trifluoroacetoxy)iodobenzene is a colourless crystalline substance, soluble in methylene chloride, chloroform, slightly soluble in ether; hydrolysed by air moisture. Decompn. p. 83 °C [125]; IR and [^{19}F] NMR spectra are found in [125].

Method of Preparation

Tetrakis(trifluoroacetoxy)iodobenzene is prepared by the reaction of iodylbenzene or (tetrafluoroiodo)benzene with excess trifluoroacetic anhydride in the presence of trifluoroacetic acid. Yield is quantitative [125].

Procedure

$$C_6H_5IO_2 + 2\,(CF_3CO)_2O \xrightarrow{CF_3COOH} C_6H_5I(OCOCF_3)_4$$

$$C_6H_5IF_4 + 2\,(CF_3CO)_2O \xrightarrow{CF_3COOH} C_6H_5I(OCOCF_3)_4 + 4\,CF_3COF$$

In a 50-ml round-bottom flask provided with a reflux condenser with a P_2O_5 tube, are placed 0.24 g (0.001 mol) of iodylbenzene [124], or 0.28 g (0.001 mol) of (tetrafluoroiodo)benzene (see p. 284) (Note 1), 30 ml of trifluoroacetic anhydride, and 1 ml of trifluoroacetic acid. The mixture is stirred at 20 °C with a magnetic stirrer during 2 h, and the solution evaporated in vacuum at 15–20 mm Hg; the distillable trifluoroacetic anhydride is collected in a trap cooled to —70 °C (Note 2). The residue is tetrakis(trifluoroacetoxy)iodobenzene, yield is 100%; decompn. p. 8 3 °C [125].

Notes

1. For (tetrafluoroiodo)benzene as a substrate, a quartz flask is used.
2. Trifluoroacetic anhydride collected in the trap may be regenerated by distilling it over P_2O_5.

2-Trifluorovinylbenzothiazol

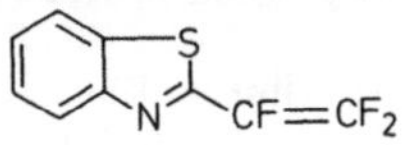

$C_9H_4F_3NS$ MW 215

2-Trifluorovinylbenzothiazol is a white crystalline substance, well soluble in benzene, ether, less soluble in petroleum ether, insoluble in water; soblimes in vacuum upon heating; easily polymerized; upon standing in air for 15–20 min, it turns red, then becomes a dark viscous mass. Mp 57°–58 °C [126].

Method of Preparation

2-Trifluorovinylbenzothiazol is prepared by the pyrolysis of the sodium salt of β-(benzothiazolyl-2)perfluoropropionic acid (62.2% yield) [126].

Procedure

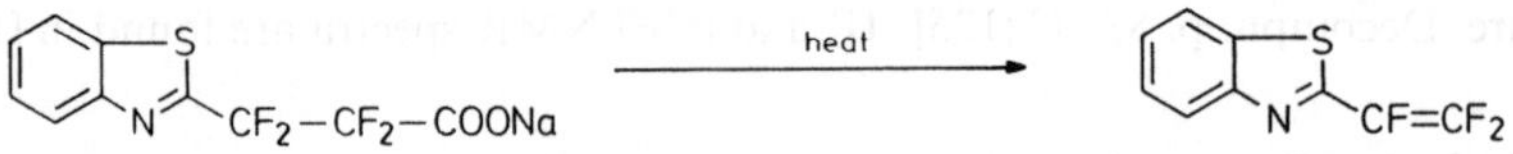

In a 15-ml pear-shaped flask fitted with a Würtz column and a wide air-condenser connected with a receiver, is placed 4.5 g (0.015 mol) of dry sodium salt of β-(benzothiazolyl-2)perfluoropropionic acid (Note 1), upon which is placed a pad of glass wool (Note 2). The set is flushed with dry, oxygen-free nitrogen and evacuated to 3 mm Hg residual pressure. The flask is immersed in a metallic bath heated to 260 °C. The salt metls at 280 °C (in the bath), the reaction product distils off, collecting on the walls of the condenser and the receiver as light-yellow crystals. The yield of 2-trifluorovinylbenzothiazol is 1.7 g (52%); mp 54°–55 °C (Note 3) [126].

Notes

1. The sodium salt of β-(benzothiazolyl-2)perfluoropropionic acid is prepared by dissolving the acid [127] in equal amount of 10% aqueous NaOH solution, with heating. The filtered solution is evaporated to dryness in vacuum in a water-bath, the residue is ground and dried at 100 °C in vacuum at 1–2 mm Hg. Mp 256 °C (decompn.) [126].
2. Glass wool is to prevent the loss of the salt in vacuum.
3. 2-Trifluorovinylbenzothiazol is sufficiently pure to be used in further preparations. It may be stored without being polymerized over P_2O_5 in a desiccator filled with nitrogen, at +5 °C or below, during 1 day.

References

1. Lukmanov, V. G. et al.: Zh. Org. Khim. *9*, 1019 (1973)
2. Lukmanov, V. G., Alekseeva, L. A., Yagupolskii, L. M.: Zh. Org. Khim. *10*, 2000 (1974)
3. USSR Pat. 401,127 (1974)
4. Markarian, M. J.: J. Am. Chem. Soc. *74*, 1858 (1952)
5. Blakitny, A. N., Fialkov, Yu. A., Yagupolskii, L. M.: Zh. Org. Khim. *6*, 2055 (1970)
6. Blakitny, A. N. et al.: Zh. Org. Khim. *10*, 503 (1974)
7. Yagupolskii, L. M., Kunshenko, B. V., Alekseeva, L. A.: Zh. Obshch. Khim. *38*, 2592 (1968)
8. Kunshenko, B. V., Alekseeva, L. A., Yagupolskii, L. M.: Zh. Org. Khim. *6*, 1286 (1970)
9. Yagupolskii, L. M., Matuyshecheva, G. I.: Zh. Obshch. Khim. *35*, 1422 (1965)
10. Hart, H., Fish, R. W.: J. Am. Chem. Soc. *83*, 4460 (1961)
11. Yagupolskii, L. M. et al.: Zh. Org. Khim. *13*, 1561 (1977)
12. Yurchenko, V. M., et al.: Teor. Eksperim. Khim. *18*, 745 (1982)
13. Yurchenko, V. M., Antipin, M. Yu., Yagupolskii, L. M.: Cryst. Struct. Commun. *7*, 77 (1977)

14. Sauvetre, R., Norman, J. F.: Bull. Soc. Chim. Fr. *1972*, 3202
15. Henne, A. L., Postelneck, W.: J. Am. Chem. Soc. *77*, 2334 (1955)
16. Yagupolskii, L. M.: Zh. Obshch. Khim. *30*, 3444 (1960)
17. Yagupolskii, L. M., Fialkov, Yu. A., Yufa, P. A.: Zh. Obshch. Khim. *31*, 3962 (1961)
18. Ott, E.: Ber. dtsch. Chem. Ges. *55*, 2108 (1922)
19. Brewester, J. H. et al.: J. Org. Chem. *28*, 498 (1963)
20. Yagupolskii, L. M., Klyushnik, G. I., Troitskaya, V. I.: Zh. Obshch. Khim. *34*, 307 (1964)
21. Barger, G. J.: J. Chem. Soc. (Lond.) *93*, 563 (1908)
22. Baker, W.: J. Chem. Soc. (Lond.) *1931*, 1765
23. Kunshenko, B. V., Alekseeva, L. A., Yagupolskii, L. M.: Zh. Org. Khim. *10*, 1698 (1974)
24. Yagupolskii, L. M., Mospan, V. S.: Ukr. Khim. Zh. *21*, 81 (1955)
25. US Pat. 2,257,093; C.A. *36*, 664 (1942)
26. Bunnet, J. F., Moe, H., Knutson, D.: J. Am. Chem. Soc. *76*, 3936 (1954)
27. Burmakov, A. I., Alekseeva, L. A., Yagupolskii, L. M.: Zh. Org. Khim. *5*, 1892 (1969)
28. Yagupolskii, L. M., Burmakov, A. I., Alekseeva, L. A.: Zh. Obshch. Khim. *39*, 2053 (1969)
29. Töhl, A.: Ber. dtsch. Chem. Ges. *25*, 1521 (1892)
30. Wirth, H., Herrman, F., Kern, W.: Macromol. Chem. *80*, 120 (1964)
31. Lukmanov, V. G.: Ph. D. Dissertation, Odessa, 1976
32. Jones, R. G.; J. Am. Chem. Soc. *69*, 2346 (1947)
33. Pouterman, E., Girardet, A.: Helv. Chim. Acta *30*, 107 (1947)
34. Germ. Pat. 637,318; C.A. *31*, 706 (1937)
35. Raasch, M. S.: J. Org. Chem. *27*, 1406 (1962)
36. Forbes, E. J., Stacey, M., Tatlow, J. C., Wragg, R. T.: Tetrahedron *8*, 67 (1960)
37. Fukui, K. et al.: Nippon Kagaku Zasshi *79*, 889 (1958); C.A. 54, 4430 (1960)
38. Freifelder, M., Stone, G. R.: J. Org. Chem. *26*, 1477 (1961)
39. Drake, N. L. et al.: J. Am. Chem. Soc. *68*, 1602 (1946)
40. Roberts, J. D. et al.: J. Am. Chem. Soc. *78*, 611 (1956)
41. Yagupolskii, L. M., Pavlenko, N. T., Solodushenkov, S. N., Fialkov, Yu. A.: Ukr. Khim. Zh. *32*, 852 (1966)
42. Kunshenko, B. V. et al.: Zh. Org. Khim. *10*, 886 (1974)
43. Troitskaya, V. I., Burmakov, A. I., Kudryavtseva, L. S., Yagupolskii, L. M.: Ukr. Khim. Zh. *37*, 808 (1971)
44. Yagupolskii, L. M. et al.: Zh. Org. Khim. *9*, 649 (1973)
45. Delyagina, N. I. et al.: Zh. Org. Khim. *8*, 851 (1972)
46. Burmakov, A. I., Alekseeva, L. A., Yagupolskii, L. M.: Zh. Org. Khim. *6*, 144 (1970)
47. Jacobsen, O.: Ber. dtsch. Chem. Ges. *17*, 2516 (1884)
48. Yagupolskii, L. M., Maletina, I. I., Kondratenko, N. V., Orda, V. V.: Synthesis *1978*, 835
49. Yagupolskii, L. M., Mironova, A. A., Maletina, I. I., Orda, V. V.: Zh. Org. Khim. *16*, 232 (1982)
50. Lyalin, V. V., Orda, V. V., Alekseeva, L. A., Yagupolskii, L. M.: Zh. Org. Khim. *6*, 329 (1970)
51. Fialkov, Yu. A., Yagupolskii, L. M.: Zh. Obshch. Khim. *36*, 1968 (1966)
52. Fialkov, Yu. A.: Ph. D. Dissertation, Kiev, 1965
53. Knunyants, I. L., German, L. S., Dyatkin, B. L.: Izv. Akad. Nauk SSSR Otd. Khim. Nauk *1960*, 221
54. Dixon, S.: J. Org. Chem. *21*, 400 (1956)
55. Andreades, S.: J. Am. Chem. Soc. *84*, 864 (1962)
56. Organic Synthesis, Vol. 1. G. Gilman, (ed). New York: 1947
57. Kunshenko, B. V., Alekseeva, L. A., Yagupolskii, L. M.: Zh. Org. Khim. *8*, 830 (1972)
58. Kunshenko, B. V.: Ph. D. Dissertation, Odessa, 1973
59. Leuck, G. J., Perkins, R. P., Whitmore, F. C.: J. Am. Chem. Soc. *51* 1831 (1929)
60. Yagupolskii, L. M. et al.: Zh. Org. Khim. *7*, 726 (1971)
61. Rubtsov, M. B., Baichikov, V. G.: Sinteticheskiye Khimiko-farmatsevticheskiye Preparaty. Meditsina, Moskwa 1971
62. Golikov, B. I. et al.: Zh. Org. Khim. *8*, 1078 (1972)
63. Golikov, B. I.: Ph. D. Dissertation, Kiev, 1973
64. Bergmann, E., Hervey, J.: Ber. dtsch. Chem. Ges. *62*, 893 (1929)
65. Cohen, S., Wolosinski, H., Sheuer, P.: J. Am. Chem. Soc. *71*, 3439 (1949)
66. Cohen, S., Wolosinski, H., Sheuer, P.: J. Am. Chem. Soc. *72*, 3952 (1950)

67. Prober, M.: J. Am. Chem. Soc. *75*, 968 (1953)
68. Dixon, S.: J. Org. Chem. *21*, 400 (1956)
69. US Pat. 2,874,197 (1955)
70. Meier, R., Bohler, F.: Chem. Ber. *90*, 2344 (1957)
71. US Pat. 2,752,400 (1956)
72. US Pat. 3,489,807 (1970)
73. Yagupolskii, L. M. et al.: Zh. Org. Khim. *12*, 1372 (1976)
74. Sevastyan, A. P. et al.: Zh. Org. Khim. *10*, 413 (1974)
75. Sevastyan, A. P.: Ph. D. Dissertation, Kiev, 1978
76. Yagupolskii, L. M. et al.: Zh. Obshch. Khim. *38*, 1500 (1968)
77. Haszeldine, R. N.: J. Chem. Soc. (Lond.) *1952*, 4423
78. Klabunde, K. J., Burton, D. J.: J. Org. Chem. *35*, 1711 (1970)
79. Yagupolskii, L. M. et al.: Zh. Org. Khim. *12*, 166 (1976)
80. Nelles, J.: Ber. dtsch. Chem. Ges. *65*, 1345 (1932)
81. US Pat. 4,316,989 (1982)
82. Fialkov, Yu. A. et al.: Ukr. Khim. Zh. *49*, 187 (1983)
83. Yagupolskii, L. M.: Dokl. Akad. Nauk SSSR *105*, 100 (1955)
84. Brit. Pat. 765,527; C.A. *51*, 14803 (1957)
85. Yagupolskii, L. M., Troitskaya, V. I.: Zh. Obshch. Khim. *31*, 915 (1961)
86. Yagupolskii, L. M., Troitskaya, V. I.: Zh. Obshch. Khim. *27*, 518 (1957)
87. Wilds, A. L., Meader, A. L.: J. Org. Chem. *13*, 763 (1948)
88. Alekseeva, L. A., Belous, V. M., Yagupolskii, L. M.: Zh. Org. Khim. *10*, 1053 (1974)
89. Fialkov, Yu. A., Moklyarchuk, L. I., Kremlev, M. M., Yagupolskii, L. M.: Zh. Org. Khim. *16*, 1476 (1980)
90. US Pat. 2,559,630; C.A. *46*, 3064 (1952)
91. Sedova, L. N. et al.: Zh. Org. Khim. *6*, 568 (1970)
92. Zincke, T., Krüger, G.: Ber. dtsch. Chem. Ges. *45*, 3472 (1912)
93. Boiko, V. N., Shuchupak, G. M., Ignatiev, N. V., Yagupolskii, L. M.: Zh. Org. Khim. *15*, 1245 (1979)
94. Yagupolskii, L. M. et al.: Zh. Org. Khim. *18*, 2390 (1982)
95. Yagupolskii, L. M., Kiprianov, A. V.: Zh. Obshch. Khim. *22*, 2216 (1952)
96. USSR Pat. 687,067 (1979)
97. Boiko, V. N., Shchupak, G. M., Yagupolskii, L. M.: Zh. Org. Khim. *13*, 1057 (1977)
98. Voloshchuk, V. G., Boiko, V. N., Yagupolskii, L. M.: Zh. Org. Khim. *13*, 2008 (1977)
99. Yagupolskii, L. M., Marenets, M. S., Kondratenko, N. V.: Zh. Obshch. Khim. *35*, 377 (1965)
100. Ignatiev, N. V., Boiko, V. N., Yagupolskii, L. M.: Zh. Org. Khim. *16*, 1501 (1980)
101. Yagupolskii, L. M., Korinko, V. A.: Zh. Obshch. Khim. *39*, 202 (1969)
102. Yagupolskii, L. M., Korinko, V. A.: Zh. Obshch. Khim. *37*, 1717 (1967)
103. Yarovenko, N. N., Motorny, S. P., Kirenskaya, L. I.: Zh. Obshch. Khim. *27*, 2796 (1957)
104. Yagupolskii, L. M., Voloshchuk, V. G.: Zh. Obshch. Khim. *36*, 165 (1966)
105. Yarovenko, N. N., Shemanina, V. N., Gazieva, G. B.: Zh. Obshch. Khim. *29*, 942 (1959)
106. Dronkina, M. I., Yagupolskii, L. M.: Zh. Org. Khim. *9*, 2167 (1973)
107. Dronkina, M. I., Yagupolskii, L. M.: Zh. Org. Khim. *7*, 2000 (1971)
108. Dronkina, M. I.: Ph. D. Dissertation, Kiev, 1973
109. Yagupolskii, L. M., Dronkina, M. I.: Zh. Obshch. Khim. *36*, 1309 (1966)
110. Yagupolskii, L. M., Dronkina, M. I.: Zh. Obshch. Khim. *36*, 1343 (1966)
111. Yagupolskii, L. M., Gandelsman, L. Z., Nazaretyan, V. P.: Zh. Org. Khim. *10*, 889 (1974)
112. Belg. Pat. 718,129 (1972)
113. Yagupolskii, L. M., Kondratenko, N. V.: Zh. Obshch. Khim. *39*, 1755 (1969)
114.. Emeleus, H. J., MacDuffie, D. E.: J. Chem. Soc. (Lond.) *1961*, 2597
115. Schmutzler, R.: Inorg. Chem. *3*, 410 (1964)
116. Yagupolskii, L. M., Ivanova, Zh. M.: Zh. Obshch. Khim. *29*, 3766 (1959)
117. Schmutzler, R.: Chem. Ind. (Lond.) *1962*, 1868
118. US Pat. 2,904,588; C.A. *54*, 2254 (1960)
119. Schmutzler, R.: Chem. Ber. *98*, 552 (1965)
120. Smith, W. C.: J. Am. Chem. Soc. *82*, 6176 (1960)
121. Yagupolskii, L. M., Kondratenko, N. V., Popov, V. I.: Zh. Obshch. Khim. *46*, 620 (1976)

N-Sulphinylpentafluoroaniline 211
Sulphur tetrafluoride 18

1,1,3,3-Tetrachloro-1,3-dihydroisobenzofuran 240
1,2,3,4-Tetrachlorohexafluorobutane 13
1,1,1,2-Tetrachloro-2-(pentafluorophenyl)ethane 135
Tetrachlorophthalyl dichloride 156
Tetradecafluoro-1,2,7,8-tetrachlorooctane 16
1,1,2,2-Tetrafluoroacenaphthene 236
2,3,5,6-Tetrafluoro-4-aminobenzoic acid 165
2,3,5,6-Tetrafluoro-4-aminophenol 201
2,4,5,6-Tetrafluoro-3-aminophenol 200
3,4,5,6-Tetrafluoro-2-aminophenol 177
2,3,4,5-Tetrafluoroaniline 193
2,3,5,6-Tetrafluoroaniline 194
3,4,5,6-Tetrafluoro-2-anilinobenzoic acid 164
3,4,5,6-Tetrafluoroanthranylic acid 164
Tetrafluorobenzobarrelene 221
5,6,7,8-Tetrafluorobenzobicyclo[2,2,2]octatriene 221
5,6,7,8-Tetrafluoro-1,4-benzodioxan 175
2,3,4,5-Tetrafluorobenzoic acid 160
2,3,5,6-Tetrafluorobenzoquinone-1,4 149
3,4,5,6-Tetrafluorobenzoquinone-1,2 148
trans,trans-Tetrafluoro-1,4-bis(4-carboxyphenyl)butadiene 260
trans,trans-Tetrafluoro-1,4-bis(4-chloroformylphenyl)butadiene 261
1,3,3,3-Tetrafluoro-1-bromo-2-trifluoromethylpropylene 11
Tetrafluoro-1-chloro-1,2-epoxy-2-(difluorochloromethyl)butene-3 36
Tetrafluoro-3-chloropropenol-2 27
1,3,3,3-Tetrafluoro-1-chloro-2-trifluoromethylpropylene-1 10
Tetrafluoro-1,1-dichloro-2-(difluorochloromethyl)buten-3-ol-2 36
1,1,3,3-Tetrafluoro-1,3-dihydroisobenzofuran 240
Tetrafluoro-1,2-diiodoethane 15
Tetrafluoroethane-β-sultone 95
1,1,1,2-Tetrafluorohexanedione-3,5 44
2,3,5,6-Tetrafluorohomoterephthalic acid 159
2,3,5,6-Tetrafluorohydroquinone 176
2,3,5,6-Tetrafluoro-4-hydroxybenzaldehyde 143
2,3,5,6-Tetrafluoro-4-hydroxybenzoic acid 163
4,5,6,7-Tetrafluoroindole 222
4,5,6,7-Tetrafluoroindoline 223
(Tetrafluoroiodo)benzene 284
5,6,7,8-Tetrafluoro-2-methyl-3-carbethoxychromone 162
4,5,6,7-Tetrafluoro-2-methyl-3-carbethoxycoumarone 224
2,2,9,9-Tetrafluoro-5-nitronaphtho[1,8-c,d]-pyrane 242
2,3,5,6-Tetrafluoro-4-nitrophenol 202
3,4,5,6-Tetrafluoro-2-nitrophenol 177
Tetrafluoro-2-nitrosoethyl ether of hexafluoroacetone oxime 86
3,4,5,6-Tetrafluorophenylenediamine-1,2 198
2,4,5,6-Tetrafluorophenylenediamine-1,3 199
1,1,3,3-Tetrafluorophthalane 240
Tetrafluorophthalic acid 155
Tetrafluorophthalic anhydride 151
Tetrafluorophthalyl difluoride 156
1,1,1,2-Tetrafluoropropiophenone 44
3,4,5,6-Tetrafluoropyrocatechol 174
2,4,5,6-Tetrafluororesorcine 175
3,4,5,6-Tetrafluorosalicylic acid 161
Tetrafluoroterephthalic acid 157
Tetrafluoroterephthalodinitrile 158
Tetrafluoroterephthalyl difluoride 158
1,2,3,4-Tetrafluoro-1,2,3,4-tetrachloro-1,4-bis-(4-carboxyphenyl)butane 261
2,3,5,6-Tetrafluoro-4-trifluoromethylphenol 171
Tetrakis(trifluoroacetoxy)iodobenzene 285
Tetrakis(trifluoromethyl)allene 17
Tetrakis(trifluorovinyl)silane 98
2,3,5,6-Tetramethylbenzotrifluoride 246
4,4′-Thiobis(tetrafluorobenzenesulphonyl) difluoride 187
N-(p-Toluenesulphonyl)-β,β-bis(trifluoromethyl)-β-propiolactam 57
1,1,2-Trichloro-2-(pentafluorophenyl)ethylene 135
1,1,3-Trichloroperfluoro(2-phenylpropanol-2) 36
Trifluoroacetaldehyde 37
Trifluoroacetic oxide 209
3,3,3-Trifluoroacetoacetic ester 69
1,1,1-Trifluoroacetone 37, 69
4-Trifluoroacetylbenzoic acid 262
4-Trifluoroacetylbenzoyl chloride 263
Trifluoroacetylnitrite 84
Trifluoroacrylonitrile 71
α,α,α-Trifluoro-4-aminoacetophenone 262
4,4,4-Trifluoro-3,3-bis(trifluoromethyl)butene-1 11
1,1,1-Trifluoro-2,2-bis(trifluoromethyl)hexane 5
5,5,5-Trifluoro-4,4-bis(trifluoromethyl)pentene-1 5
4,4,4-Trifluoro-3,3-bis(trifluoromethyl)-1-phenylbutyne-1 21
1,1,1-Trifluoro-2,2-bis(trifluoromethyl)-3-phenylpropane 5
1,1,1-Trifluoro-2,2-bis(trifluoromethyl)propane 5
1,1,2-Trifluoro-2-chlorodiethyl sulphide 90
Trifluoro-1-chloro-1-nitroso-2-nitroethane 85
Trifluoro-1-chloropropyne 21

1,3,3-Trifluoro-1,3-dichloro-1,2-epoxy-2-methylpropane 35
Trifluoro-1,3-dichloro-1,2-epoxy-2-phenylpropane 36
Trifluoro-1,2-dichloroethylfluorosulphate 96
3,3,3-Trifluoro-1,1-dichloro-2-trifluoromethylpropylene 11
Trifluoroethanol 23
2,2,2-Trifluoroethylamine hydrochloride 79
Trifluoro-2-hydroxy-2-phenylpropionic acid 58
Trifluoromethyl-4-aminophenylketone 262
1-Trifluoromethyl-2,3,5,6-benzenetetracarboxylic acid 246
Trifluoromethyl-β-diethylaminovinylketone 44
p-Trifluoromethylphenylbis(trifluoromethyl)carbinol 235
3-Trifluoromethylphthalic acid 245
3-Trifluoromethylphthalic anhydride 244
3-Trifluoromethylphthalyl difluoride 245
2-Trifluoromethyl-3-propylhexen-2-oic acid 60
Trifluoromethylselenyl chloride 276
N-Trifluoromethylsulphonyl-4-anisidine 280
2-Trifluoromethylthiobenzothiazole 272
Trifluoromethyl-4-tolylketone 262
2,2,2-Trifluoronitroethane 79
Trifluoronitrosomethane 84
1,1,1-Trifluoropropiophenone 44
3,3,3-Trifluoropropyne 18
1,1,3-Trifluoro-1,3,3-trichloro-2-methylpropanol-2 36
1,1,3-Trifluoro-1,3,3-trichloro-2-phenylpropanol-2 36
1,1,1-Trifluoro-3,3,3-trichloropropane 21
2-Trifluorovinylbenzothiazol 285
2,4,6-Trimethylbenzotrichloride 237
2,4,6-Trimethylbenzotrifluoride 237
2,4,6-Tris(pentafluoroethoxy)aniline 266
1,3,5-Tris(pentafluoroethoxy)benzene 267
2,4,6-Tris(pentafluoroethoxy)nitrobenzene 267
Tris(pentafluorophenyl)phosphate 218
Tris(pentafluorophenyl)phosphine 215
Tris(pentafluorophenyl)phosphine oxide 215
2,4,6-Tris(trifluoromethyl)aniline 251
2,4,6-Tris(trifluoromethyl)benzamide 252
1,2,3-Tris(trifluoromethyl)benzene 233
2,4,6-Tris(trifluoromethyl)benzoyl fluoride 251
Tris(trifluoromethyl) orthothiobenzoate 281
1,3,5-Tris(trifluoromethylsulphonyl)benzene 270
2,4,6-Tris(trifluoromethylsulphonyl)chlorobenzene 273
2,4,6-Tris(trifluoromethylthio)toluene 271

MIX
Papier aus verantwortungsvollen Quellen
Paper from responsible sources
FSC® C105338

If you have any concerns about our products,
you can contact us on
ProductSafety@springernature.com

In case Publisher is established outside the EU,
the EU authorized representative is:
Springer Nature Customer Service Center GmbH
Europaplatz 3, 69115 Heidelberg, Germany

Printed by Libri Plureos GmbH
in Hamburg, Germany